Martin J. Forrest
Rubber Analysis

Also of interest

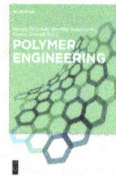

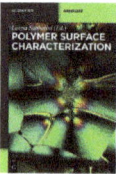

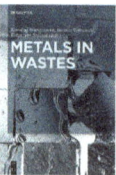

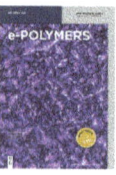

Martin J. Forrest

Rubber Analysis

Characterisation, Failure Diagnosis and Reverse
Engineering

DE GRUYTER

Author
Dr. Martin J. Forrest
Shrewsbury Laboratory
Shrewsbury
SY4 4NR
Great Britian

ISBN 978-3-11-064027-4
e-ISBN (PDF) 978-3-11-064028-1
e-ISBN (EPUB) 978-3-11-064044-1

Library of Congress Control Number: 2018966615

Bibliographic information published by the Deutsche Nationalbibliothek
The Deutsche Nationalbibliothek lists this publication in the Deutsche Nationalbibliografie;
detailed bibliographic data are available on the Internet at http://dnb.dnb.de.

© 2019 Walter de Gruyter GmbH, Berlin/Boston
Typesetting: Integra Software Services Pvt. Ltd.
Printing and binding: CPI books GmbH, Leck
Cover image: https://www.science-photo.de - 11550499 Differential scanning calorimetry - Science
Photo Library / Brookes, Andrew / National Physical Laboratory

www.degruyter.com

Preface

This book provides a comprehensive review of rubber analysis and its many applications and uses in the rubber industry and in other industrial sectors. These applications include fundamental characterisation work that can be carried out on rubber polymers, the various levels of investigative work that can be undertaken to obtain quality control and compositional information on rubber products, and the contributions that rubber analysis makes to occupational hygiene studies in the rubber industry, regulatory studies in the medical and pharmaceutical industries, and regulatory compliance work in the food industry. An overview of the wide range of analytical techniques (e.g., spectroscopic, chromatographic, thermal and elemental) that can be used in the pursuit of the different objectives of these industrially orientated applications of rubber analysis is included. The analytical techniques section also provides extensive referencing for readers who wish to find out more information on the fundamental principles and attributes of the individual techniques.

A practical, application-based approach has been used throughout the book, with the emphasis on how the analytical techniques can be used, both individually and in combination, to provide analysts and researchers in industry, consultancy organisations and test houses with important and commercially useful information. For example, in addition to the use of these techniques for routine quality control and quality assurance purposes, their application to the identification and quantification of specific additives is described, as is their important contribution to the industrially important areas of reverse engineering, failure diagnosis and the identification of contaminants.

The majority of rubber products are cured, crosslinked materials, and a specific section is devoted to the study of the curing process and characterisation of cure state due to their vital role in assuring that a rubber product will achieve its optimum properties and provide satisfactory long-term performance in service.

In order to fully reflect the important role that rubber analysis and the analytical techniques associated with it have today, a number of important specific applications have been detailed. These include the studies that have been undertaken to characterise rubber process fumes and to identify relationships between fumes and the composition of the rubber compounds that generate them. Also included is risk assessment work such as the profiling and determination of extractables and leachables from rubber products used in medical devices and pharmaceutical packaging, as well as the identification and quantification of potential and actual migrants from food-contact rubber products.

Finally, the inclusion of a section on the analysis and characterisation of rubber latex, both the natural and synthetic versions, and the products that are manufactured from them, ensures that this important class of rubber products is not overlooked.

https://doi.org/10.1515/9783110640281-201

In writing this book, the up-to-date, extensive information contained in the Smithers Rapra Polymer Library has been used. This has been complemented by many other sources of information, including the published findings of relevant research projects, trade literature references, and the considerable experience that the author has acquired during 40 years in the rubber industry.

Contents

1 Introduction

This book follows on from a review of rubber analysis published by the author as a *Rapra Review Report* in 2001 [1]. In the intervening 17 years, there have been number of important developments in the field. Some of these were associated with the development and commercialisation of new analytical techniques and others led to the appearance of improved versions of established techniques, both of which have increased the speed, efficiency and capability available to the rubber analyst. Other developments were associated with new testing methodologies created to help protect the environment and to safeguard human health. Alongside these advances and developments, there has also been a significant amount of new research and investigative work carried out and published. Taking everything into account, it was, therefore, considered to be timely and important to reflect and represent all of the these changes in a new, and very much enlarged edition of the book.

The analysis of rubber products to demonstrate compliance with regulations is one area that has grown significantly since 2001. In addition to the many regulations that existed prior to 2001, new regulations have been published, such as those concerning the presence of polyaromatic hydrocarbons in extender oils [2] and a new Chinese standard for the use of rubber products for food-contact applications [3]. As well as these specific, rubber-related regulations, major new European Union regulations have been published which have had a significant and wider impact on industry, such as the Registration, Evaluation, Authorisation and restriction of Chemicals [4] and the Restriction of Hazardous Substances [5]. Regulatory bodies, such as the US Food and Drug Administration, have put more emphasis since 2001 on identifying and quantifying the substances that could migrate from the rubber products used in medical devices, and in drug storage, dispensing and manufacturing systems.

It is for these reasons that dedicated sections that deal specifically with the analysis work that has been undertaken in these subject areas have been included in this book, as well as other sections that address additional specific areas of interest, such as the characterisation of rubber process fume, studying the curing behaviour of rubbers, and the identification of blooms and contaminants in rubber products.

To ensure that the major 'themes' of rubber analysis work are fully represented, there are also extensive sections in the book that deal with the identification, quantification and characterisation of the principal constituents in a rubber compound (i.e., polymer, plasticiser/oil and filler), and of the wide range of additives that can be used in rubber compounds (e.g., cure system and antidegradant system species). To provide a thorough background to all the investigative and themed sections, there is also a major section which describes the extensive range of analytical techniques available to rubber analysts when they are in pursuit of their various and diverse goals.

https://doi.org/10.1515/9783110640281-001

The overall objective of this book, therefore, is to present an introduction and overview of rubber analysis and the analytical methods and techniques used to achieve a wide range of objectives, including:
- Quality-control work
- Failure investigations
- Deformulation studies
- Characterisation work (e.g., curing behaviour)
- Work on regulatory compliance (e.g., for food-contact applications)
- Work to address health and safety or environmental concerns (e.g., characterisation of rubber fume)

To set the scene for a book on rubber analysis, it is useful to provide a brief introduction to rubber technology and, hence, provide the background to the complexity of the challenges that the rubber analyst often encounters. This is the intended function of the next six paragraphs.

To obtain useful products that perform under various demanding conditions, a rubber compound can be 'tailor-made' by selecting from a broad range of polymers and additives. With rubbers, the possible compositional permutations are made more numerous by an extensive array of plasticisers, fillers, process aids, antidegradants and cure systems. The technology of rubbers is, therefore, a mature one, and allows 'fine tuning' of a compound to fit a number of seemingly conflicting design criteria and product requirements. Hence, rubber compounds and products present analysts with one of their most difficult, but satisfying, challenges.

Natural and synthetic rubber compounds are used in a highly diverse range of rubber products for many applications and are manufactured throughout the world for various sectors of industry and end users. The examples of the products that can be manufactured from rubber compounds vary from bridge bearings to rubber seals, gaskets, adhesives and elastic cord (Table 1.1). The majority of these products are manufactured using traditional thermoset rubbers (i.e., vulcanised rubbers), but there are also a number of thermoplastic rubbers available in which the crosslinks are physical as a result of crystalline domains, or those that have a glass transition temperature above ambient. Of the sectors where rubber is used, the automotive industry is of particular importance in that tyre-related products, account for ~60% of the synthetic rubber and ~75% of the natural rubber used worldwide [6].

Table 1.1 provides an indication of the diverse range of rubber products that are manufactured throughout the world today. Both the composition of a rubber compound and the manufacturing processes that are used to produce it will depend upon the individual product that is being manufactured.

As mentioned above, rubber products have complex compositions and the company or individual who designs a rubber formulation for a specific product has a large number of ingredients to choose from [7]. Although there are general guidelines for designing rubber compounds, the final formulation depends upon the knowledge

Table 1.1: Examples of commercial rubber products.

Generic class of product	Examples of commercial products
Tyres	Passenger car, truck tyres, racing, cycle, off-road tyres, inner tubes, curing bladders
Conveyor/transmission belting	Steel cord conveyor belting, repair material for conveyor belting, scrapers, mining conveyors, V-belts, flat belts, synchronous belts
Industrial hose	Water hose, high-pressure hose, welding hose, hydraulic hose, spiral hose, offshore hose, oil hose, chemical hoses
Automotive products	Hoses and gaskets, anti-vibration mounts, timing belts, window and door seals, transmission and engine components, wiper blades
General mouldings/sheeting	Seals and gaskets, floor and roller coverings, sheeting, protection linings, microcellular products, rubberised fabric, wire and cable jackets, pump impellors, rail mounts, bridge bearings
Medical/pharma products	Surgeons' gloves, medical tubing, gaskets, catheters, dialysis products, implants, contraceptives, baby soothers, tubing and valves, masks and respirators
Clothing	Footwear, protective suits, household gloves, industrial gloves, footwear/boot heels and soling, cellular rubber soles, diving suits, coated fabrics, sports footwear and clothing
Food contact products	Transportation belts and hoses, gloves, pipe and machinery seals and gaskets, plate heat exchanger gaskets, seals/gaskets for cans and bottles, tank linings
Miscellaneous products	Adhesives, rubberised asphalt, high-vacuum and radiation components, carpet backing, latex thread, sealants and caulking, toys

and expertise of each rubber compounder, i.e., there are no 'standard' rubber compounds.

One of the reasons that the composition of a rubber compound can vary enormously is because there are many types of base polymer (>20), each having a unique chemical structure which has a direct influence on its inherent chemical and physical properties, processing behaviour and the applications and products in which it can be used [8]. It is also the case that for each generic type of rubbery polymer (e.g., ethylene propylene diene monomer rubber), there are a significant number of different grades available, each of which may differ in a number of aspects (e.g., molecular weight or comonomer type). For the ethylene propylene class of rubbers, for example, over 100 different grades are available.

In addition to the base polymer, rubbers are formulated with a range of additives (typically between 5 and 20) to achieve a compound that has the desired processing and service properties [9]. There are also many cases where two or more base polymers are mixed and blended together in order to achieve the desired properties (e.g., some of the compounds used in the manufacture of tyres). Also, within each class of ingredient (e.g., antioxidant) there is a wide range of different chemical compounds available on the market, each having different structures and properties, from a number of manufacturers. Within another additive group, the carbon blacks, there are in excess of 30 possible products. The composition of rubber compounds can also alter during manufacturing because some of the ingredients are volatile at typical processing temperatures and others (e.g., cure accelerators and antioxidants) generate breakdown and reaction products [10–12], all of which makes accurate quantification of the original concentration extremely difficult to derive. It is also common industry practice to use more than one ingredient from a particular class in a formulation to achieve the desired properties (e.g., two or more accelerators). Certain rubber compounds may also contain reactive ingredients (e.g., pre-vulcanisation inhibitors) at a relatively low (<0.5%) level, which can make detection difficult.

The rubber analyst, therefore, has to bear in mind all these complexities to ensure an effective analytical strategy is devised and, also consider the possible interferences and difficulties that may arise and influence the data. The usual result is that when, for example, reverse engineering work is carried out on a rubber compound, a number of wet chemistry, elemental, spectroscopic, chromatographic and thermal techniques have to be used in an integrated and structured approach in order to ensure a successful result. It is still the case, though, that even with the greatest care, and in the most advantageous situations, >90% of a complex rubber formulation can be completely elucidated by analytical means alone. Some compound development work will still be required in order to fill knowledge gaps and find a satisfactory match for the processing, curing, physical properties and in-service performance of the target sample.

To assist the practicing analyst, the emphasis throughout the book is of an empirical nature, with the author drawing on his 40 years of technical experience within the polymer industry, to provide practical illustrations of what can be achieved using certain analytical techniques and methodologies. For example, to assist analysts with certain specific areas of investigation (e.g., pyrolysis gas chromatography–mass spectrometry analysis), examples of the experimental conditions that have been used to carry out particular analyses have been provided and the results obtained by the use of these conditions displayed to illustrate the capability and diagnostic power of a technique.

In order for the analyst to have a full understanding of the capability and efficacy of the analytical methods and techniques that are available today, it is extremely useful for them to have an understanding of rubber technology itself, beyond the very brief introduction that is provided in the paragraphs above. Also, with respect to characterisation and deformulation studies, the analysts' job when analysing 'unknowns' is always made easier if at least some information on the sample is available at the outset (e.g., the generic rubber type and the intended application). Armed with this information, and guided by their background knowledge of rubber technology, analysts can work more efficiently and often obtain more comprehensive results.

To assist the reader in obtaining a greater understanding of rubber analysis and rubber technology, relevant books, academic articles, guidance documents, testing standards and other information references are cited throughout each of the sections in this book. There are also Appendices at the end of the book which summarise the standard nomenclature system for rubbers (Appendix 1), International Standards (International Organization for Standardization) for rubber analysis (Appendix 2), and the specific gravities of rubbers and compounding ingredients (Appendix 3).

References

1. M.J Forrest in *Rubber Analysis – Polymers, Compounds and Products*, Rapra Review Report No.139, Smithers Rapra, Shawbury, UK, 2001.
2. EU Directive 2005/69/EC: PAH's in Tread Rubber.
3. GB 4806.11-2016: Chinese National Food Safety Standard – Food Contact Rubber Materials and Articles.
4. EU Regulation (EC) 1907/2006: REACH.
5. EU Directive 2002/95/EC: Restriction of Hazardous Substances.
6. *Rubber Statistical Bulletin*, 2008, **60**, 4–5, 2.
7. *Rubber Compounding Ingredients Sourcebook*, Rapra Technology Ltd, Shawbury, UK, 2000.
8. J.R. White and S.K. De in *Rubber Technologists Handbook*, Smithers Rapra, Shawbury, UK, 2001.
9. H. Long in *Basic Compounding and Processing of Rubber*, ACS Rubber Division, Akron, OH, USA, 1985.
10. N. Grassie and G. Scott in *Polymer Degradation and Stabilisation*, Cambridge University Press, Cambridge, UK, 1985.
11. A.G. Davies in *Organic Peroxides*, Butterworths, Oxford, UK, 1961.
12. A.Y. Coran in *Vulcanisation, Science and Technology of Rubber*, Ed., F.R. Eirich, Academic Press, New York, NY, USA, 1978.

2 Analytical techniques

2.1 Introduction

The objective of this section is to provide an introduction and overview to the extensive range of analytical techniques that are available to the rubber analyst. Although there are a very large number available, an attempt has been made to include as many types and classes, and variants within specific classes, as possible. If the reader would like more information on a particular analytical technique and its capabilities and relevant applications, there are a number of specialist books available, and examples have been provided throughout this section. For a broader coverage, there are books that cover a wide range of techniques and methodologies within general topics, such as the characterisation of polymers [1–3], or address particular themes, including:
- Practical techniques [4, 5].
- Determination of additives [6, 7].
- Application of thermal techniques [8, 9].
- Application of specific analytical techniques, for example, infrared (IR) and Raman spectroscopy [10].
- Study of particular fundamental properties, for example, polymer morphology [11].

2.2 Wet chemistry techniques

Before the advent of modern instrumental techniques, rubber analysis relied on traditional wet chemistry (e.g., colorimetric) techniques to determine information such as the generic polymer type of a compound and the nature of the cure system present; classical organic chemistry methods were used constantly. The amount of wet chemistry carried out in the modern rubber analysis laboratory is minimal, but there are some tests which can still be grouped under this heading, although even some of these employ an instrumental technique to provide the final answer.

The determination of total sulfur and combined sulfur in rubber can be carried out using essentially the same method; the only difference is that the combined sulfur is the sulfur that remains after a solvent extraction has been carried out on the sample. There are a number of methods that can be used and workers have been developing approaches to the determination of sulfur in organic compounds since the 19th century. Two standardised methods that are still in use today in some laboratories are the tube furnace combustion method and oxygen flask combustion method.

The tube furnace method, which is described in International Organization for Standardizations' (ISO) standard, ISO 6528-3, can be used on rubber samples that are fully compounded and contain chlorine and/or nitrogen. The sample is burnt in a stream of oxygen at very high temperatures (e.g., 1,350 °C) to ensure that all the

https://doi.org/10.1515/9783110640281-002

sulfur in the compound (even that in inorganic fillers such as barium sulfate) has reacted and is included in the final value. Adding catalysts into the method can reduce this temperature to $\approx$1,000 °C. If the sulfur contribution from inorganic additives is not desirable then catalysts can be left out and the test carried out at 1,000 °C. The volatile products are trapped in a solution of hydrogen peroxide (H_2O_2) and then a titration carried out to a permanent pink colour using 0.01 M barium perchlorate in a mixture of 80:20 (v/v) propan-2-ol and water.

The oxygen flask method, described in ISO 6528-1 and the British Standard (BS), BS 7164-23.1, also uses an initial burn in oxygen, although the sample is trapped between two platinum gauge 'flags', and the trapping of the volatile products in an absorbent solution of H_2O_2. Once the volatiles have been trapped, however, there is a choice of analytical method. A titration can be undertaken in the same way as for the tube furnace method, or ion chromatography (IC) can be used to quantify the sulfate ion. In the case of the titration approach, care has to be taken because a number of interferences can take place. For example, zinc sulfate [produced during the combustion of zinc oxide (ZnO) or due to the oxidation of zinc sulfide] will dissolve in the H_2O_2 solution, and zinc ions, in general, will interfere with the end point; these have to be removed by sending the H_2O_2 solution down through an ion-exchange column. Even small amounts (e.g., 0.5%) of chlorine in a rubber will give poor results due to the volatility of zinc chloride and hence its take-up by the absorption solution. If the sample contains calcium carbonate ($CaCO_3$), a change in the absorption solution is required to a mixture of 0.25 ml of concentrated hydrochloric acid, 2 ml of distilled water and 1 ml of 6% H_2O_2. This will dissolve the calcium sulfate formed and decompose any $CaCO_3$. An ion-exchange column is then used prior to the titration. These problems, which are also common to the tube furnace titration step, are completely removed by using IC to quantify the sulfur *via* the sulfate ion (Section 2.5.7).

The oxygen flask combustion technique can also be used for the determination of total halogens (titration method) or quantification of specific halogens (IC method) in a rubber. The absorbing solution will typically comprise 1.5 ml of 0.05 M potassium hydroxide, 0.2 ml of 30% H_2O_2 and 10 ml of distilled water.

It is sometimes necessary to determine the amount of elemental sulfur remaining in a rubber. One method, the copper spiral method, is the subject of a standard [12]; other methods include the sulfite method, polarography, and reverse-phase high-performance liquid chromatography (HPLC) using acetonitrile/water gradient elution and ultraviolet (UV) detection at 270 nm.

2.3 Elemental analysers and elemental analysis

The quantification of certain elements is required so often in so many branches of science that for a number of years now commercial instruments have been available for their specific determination.

Several configurations are available depending on user requirements. Examples which are useful for the analysis of rubbers and other polymeric samples include:
- Carbon:hydrogen:nitrogen
- Carbon:nitrogen:sulfur
- Carbon:hydrogen:nitrogen:sulfur
- Carbon:hydrogen:nitrogen:oxygen:sulfur

These dedicated analysers are able to provide quick, accurate and easy quantification of these elements on samples which do not require much preparation work and, in some cases, at sample masses as low as 1 mg.

An international standard has been published [13] which deals with the determination of the total nitrogen content of a sample using an automated analyser. No specific manufacturer's instrument is specified, and the standard focuses on the experimental procedure to be used. A standard has also been published [14] which deals with the determination of the total sulfur content of a sample using an automated analyser. Again, no specific instrument is specified and the standard focuses on the experimental procedure.

These instruments, therefore, offer an efficient and cost-effective way to obtain important quantitative elemental data that can be used for a variety of purposes, for example:
- Quantifying the amount of acrylonitrile in a nitrile rubber (NBR).
- Calculating the ratio of NBR and polyvinyl chloride in blends of those polymers.
- Determinations of total sulfur and bound sulfur.
- Differentiating between a sulfur-cured rubber and peroxide-cured rubber.

The determination of the concentration of other elements can also be very useful in rubber analysis work. In particular, it is often necessary to determine the amount of a halogen, such as chlorine or bromine, in the calculation of polymer content or polymer-blend proportions. Also, the targeting of specific elements can enable the quantity of certain additives (e.g., phosphate flame retardant or titanium dioxide pigment) to be calculated. The other techniques and instruments that can be used to identify and quantify elements within rubber samples are covered in Section 2.2 and in all the sections from 2.4.7–2.4.11.

2.4 Spectroscopic techniques

2.4.1 Infrared spectroscopy

IR spectroscopy or, Fourier–Transform infrared (FTIR) spectroscopy as it is often called due to the ubiquitous nature of the Fourier–Transform version, is a relatively cost-effective, quick analytical technique that is extensively used in the analysis of rubbers to identify the base polymer as well as the main (i.e., bulk) additives (e.g., plasticisers, oils and fillers) [15–18].

Although the IR spectra of polymers and other organic substances can be recorded using lasers emitting IR light in the near-, far- and mid-ranges, the majority are recorded in the mid-range (i.e., 400–4,000 $^{cm^{-1}}$) and commercial instruments and IR libraries reflect this fact. To record transmission IR data it is necessary to place a sample on an IR transparent 'window' and, typically, sodium chloride plates are used in standard IR accessories, whereas diamond 'windows' are commonly used in 'single bounce'-type attenuated total reflectance (ATR) accessories due to their greater durability. A number of other sampling accessories are available, and one that can be useful for direct rubber analysis work is a 'multi-reflectance' ATR accessory equipped with a germanium crystal due to its relatively small degree of penetration into the sample. This property can help reduce the interference that results due to the presence of carbon black (CB), which is also referred to below.

To assist in the identification of polymers, IR spectral libraries are available, such as those published by Forrest and co-workers [19] and the Smithsonian Institute [20]. In addition to these IR spectral libraries, there is a major multi-volume library of spectra published by Hummel and Scholl which, in addition to polymers, also includes a wide range of additives used in rubbers, plastics and other polymer-based products [21].

The majority of rubber compounds contain additives such as CB and plasticisers that can interfere with the IR data obtained if an attempt is made to analyse the compound directly. It is also common for a rubber product to have been subjected to a vulcanisation procedure, which renders it insoluble. For both reasons it is, therefore, necessary to undertake some simple preparation work prior to analysis. For example, it is possible to remove plasticisers from the sample matrix using solvent extraction (e.g., using acetone or methanol) and then to pyrolyse the extracted portion and record an IR spectrum from the collected pyrolysis condensates. This operation eliminates any interference from the CB, inorganic additives (e.g., ZnO) or fillers. For this approach to be successful, a database of pyrolysate IR spectra must be available. It is a useful feature of the database published by Forrest and co-workers [19] that a large number of pyrolysis spectra of pre-extracted rubbers are included. An example of a pyrolysis IR spectrum, in this case of a natural rubber (NR) sample, is shown in Figure 2.1.

Analysis of the solvent extract that was obtained in preparing the rubber sample for the pyrolysis work by transmission FTIR spectroscopy could, with the assistance of the IR databases mentioned above, identify the plasticiser or oil present in the compound. IR spectroscopy can also be used in this way to identify specialist organic additives, such as fire retardants; the inorganic types of fire retardants can be determined by IR using a similar approach to that described below for inorganic fillers.

With regard to the identification of inorganic additives, although FTIR can be used directly, usually employing an ATR accessory, to identify the inorganic fillers that are

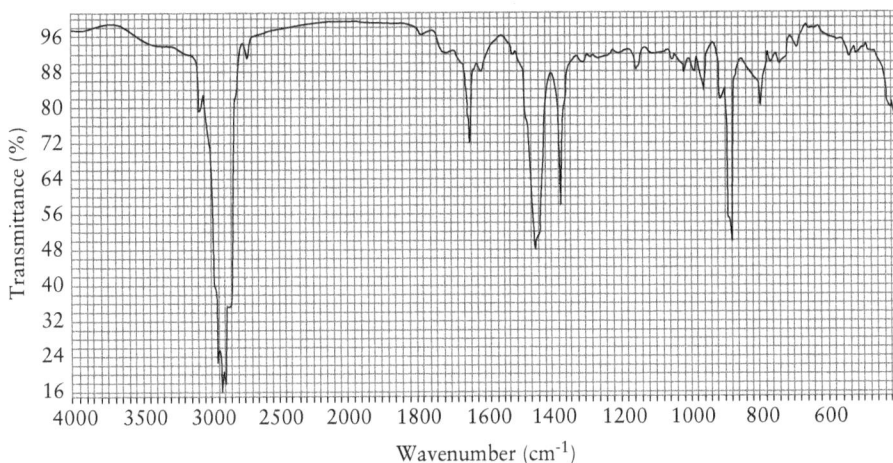

Figure 2.1: Pyrolysis IR spectrum of a NR sample. Reproduced with permission from M.J. Forrest in *Rubber Analysis – Polymers, Compounds and Products*, Rapra Review Report No.139, Smithers Rapra, Shawbury, UK, 2001, p.7. ©2001, Smithers Rapra [22].

present in a polymer [23], the CB in rubbers can cause problems by absorbing the IR radiation. Hence, a better approach is to isolate the inorganic filler by ashing. An FTIR analysis of the ash of a rubber sample obtained by heating to 550 °C for ≈60 min can identify the inorganic filler that is present. If it is suspected that a mixture of inorganic fillers has been used, or that other inorganic additives are also present, it is advisable to complement the FTIR data with elemental data on the ash, for example, by using X-ray fluorescence spectroscopy (XRF) or inductively coupled plasma (ICP) analysis.

If the raw, uncompounded polymer is to be investigated, it is possible to record an IR spectrum from either a film cast from solution, by direct analysis using ATR, or the pyrolysate approach. The transmission IR spectrum obtained by casting a film of NR from a chloroform solution is shown in Figure 2.2.

The examples mentioned above refer to the common form of IR spectroscopy which relies on absorption of the IR radiation. However, it is also possible to use IR emission spectroscopy in the analysis of rubbers. This is essentially a qualitative technique in which an IR beam causes vibrations in the polymer molecules and the radiation emitted is analysed by a highly sensitive spectrometer. The advantage of the technique is that it can be used on CB-filled vulcanisates, and it has been employed in degradation studies. It is possible to qualitatively determine chemical changes, such as carbonyl formation (e.g., in acid groups), weight loss and volatilisation (*via* loss of emission intensity due to the 'unzipping' of the polymer), as well as the formation of conjugated double bonds and specific polymer reactions. The technique has been used to study surface oxidation after natural and accelerated heat ageing and the results obtained provided a rate of degradation by studying the vibration of chemical groups associated with the degradation. Although the data obtained by emission IR

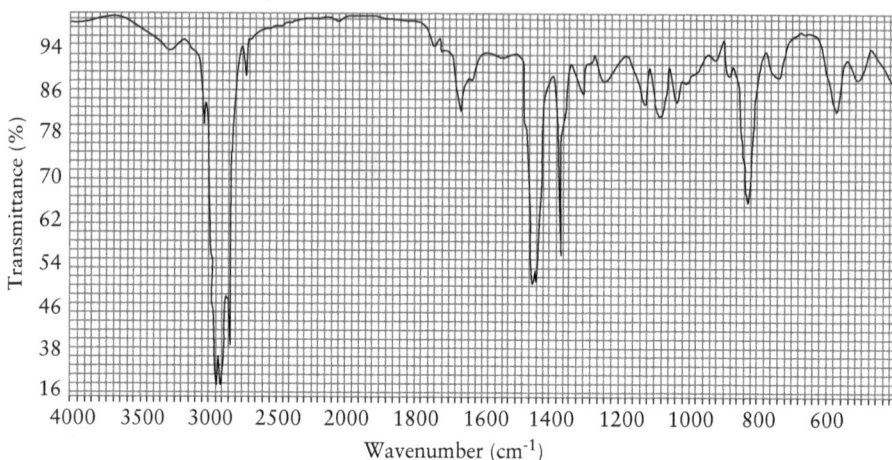

Figure 2.2: Transmission IR spectrum of a NR sample. Reproduced with permission from M.J. Forrest in *Rubber Analysis – Polymers, Compounds and Products*, Rapra Review Report No.139, Smithers Rapra, Shawbury, UK, 2001, p.7. ©2001, Smithers Rapra [22].

spectroscopy is not quantitative due to interferences such as sample reflectivity and re-absorbance in thicker samples, Pell and co-workers [24] have suggested an approximate relationship between absorbance and emittance.

Another use for FTIR, this time in the near-infrared (NIR) region, has been reported by researchers drawn from Chiang Mai University and Maejo University [25]. Those researchers have assessed the capability of Fourier–Transform (FT)–NIR to determine the mechanical properties of rubbers as an alternative to using conventional physical tests (e.g., hardness, tensile tests). A range of commercially important rubbers [e.g., NR, styrene-butadiene rubber (SBR), NBR and ethylene propylene diene monomer (EPDM) rubber], and their blends, were analysed by an FT–NIR instrument that was fitted with an integration sphere working in diffused-reflectance mode. The spectra that resulted were found to correlate with hardness and tensile properties, and the workers concluded that FT–NIR could be used to determine these properties of rubbers and rubber blends.

2.4.2 Infrared microspectrometry

For over 30 years it has been possible to obtain systems in which an FTIR instrument has been interfaced with a microscope. Initially optical microscopes, such as those marketed by Olympus, were used to examine and target the sample to be analysed and then an IR objective lens and a sensitive IR detector brought into line to record an IR spectrum. Modern instruments are now integrated packages that use digital imaging and have additional capabilities (e.g., the mapping of surfaces).

IR microspectrometry is an extremely useful tool for the following types of investigative work:

- Identification of fibres.
- Characterisation of surface blooms and discolourations.
- Identification of contaminants and inclusions in samples.
- Analysis of any sample that is too small to enable a standard IR accessory/bench combination to be used [e.g., extracts from microextractions, small drops of fluid, particles of dust, debris and other small solid particles, fractions separated by chromatographic analysis, and small amounts of inorganic residue from thermogravimetric analysis (TGA)].

2.4.3 Raman spectroscopy

Raman spectroscopy, available in both dispersive and FT forms, has a number of uses and is often used in conjunction with the other vibrational spectroscopy technique, IR spectroscopy, due to their complementary nature [26, 27]. The Raman effect occurs when a sample is irradiated by monochromatic light from a laser, causing a small fraction of radiation to exhibit shifted frequencies that correspond to the vibrational transitions of the sample.

Some absorption bands that are relatively weak in an IR spectrum (e.g., those due to sulfur–sulfur, carbon–carbon, and halogen–carbon bonds) are more intense in Raman spectroscopy and this is one of its advantages. The reverse is the case for substances such as silicate fillers, and this means that they are much less likely to obscure important diagnostic bands as they are prone to in an IR spectrum. This capability of Raman spectroscopy can aid the determination of additives in a rubber, but its sensitivity to particular absorption bands can also make it a useful tool for structural analysis [28–30].

A library of dispersive Raman spectra of polymers and plasticisers has been published by the Smithsonian Institute [20] and a handbook published by Kuptsov and Zhizhin [31] includes examples of FT Raman spectra of polymers.

2.4.4 Ultraviolet light spectroscopy

These instruments can usually scan over both the UV and ultraviolet visible (UV-Vis) portions of the electromagnetic spectrum and so are referred to as 'UV-Vis spectrometers' [32]. Although not as universally used as IR spectroscopy, or as versatile, UV-Vis spectroscopy still has a part to play in the analysis of polymers because it is well suited for particular tasks. For example, it is often used in the qualitative and quantitative analysis of additives that possess strong chromophores (e.g., antioxidants with aromatic rings) and in analytical methods that employ a derivatisation step to produce such species with this characteristic (e.g., determination of unreacted peroxides in silicone rubber).

One of the principal limitations of the technique is that the analyte needs to be in the form of a solution and the UV absorbance of the solvent must not interfere with that of the analyte.

2.4.5 Nuclear magnetic resonance spectroscopy

Nuclear magnetic resonance (NMR) spectroscopy is a powerful tool for the determination of polymer structure [33–36]. It is at its most sensitive when used on analytes in solution and so is very useful for the analysis of the raw polymers. The vast majority of rubber compounds are, however, vulcanised. Hence, either an NMR analysis has to be carried out on the fraction of the rubber matrix which can be extracted using a strong solvent (e.g., chloroform) and, because rubbers are lightly crosslinked structures, a sufficient amount of sample can usually be obtained to make this possible, or a solid-state NMR (e.g., magic angle) instrument has to be used. The disadvantage of the latter is its lack of sensitivity, which can make the analysis of blends in which one polymer is present at a low level difficult, as well as the detection of cure site monomers in rubbers such as halobutyl rubbers and EPDM.

Of the conventional NMR instruments, both proton and carbon-13 variants are commonly used in rubber analysis.

2.4.6 Mass spectroscopy techniques

Mass spectrometry (MS) is an extremely useful and powerful tool for identifying chemical compounds and for elucidating their chemical structure [36]. It has a number of features and advantages, including:
- Only small amounts of sample are required.
- The molar mass of a substance can be obtained directly by measuring the mass of the molecular ion.
- Molecular structures can be derived by means of the molar masses and ion-fragmentation patterns.
- Atomic compositions can be derived using accurate mass instruments.
- Mixtures of substances can be characterised using 'soft' ionisation methods in conjunction with chromatography techniques.

With regard to the last example in the list above, the use of mass spectrometers [e.g., quadrupole or time-of-flight (ToF)] to enhance the information that can be provided by techniques such gas chromatography (GC), liquid chromatography (LC) and ICP is mentioned in Sections 2.5.2, 2.5.3, 2.5.6 and 2.4.9.

Due to the advances made in the field in recent years (e.g., development of powerful FT instruments) and the increase in the use of ionisation methods for the desorption of ions direct from the surface of a sample [e.g., matrix-assisted laser desorption/ionisation (MALDI)], the technique has gained in importance, with its contribution to structural analysis complementing the data provided by IR and NMR. Various MS techniques have been applied to the analysis of rubbers and polymer additives, either as extracts or on the sample surface by laser techniques, and these have been reviewed by Lattimer and Harris [37].

As well as those examples where a mass spectrometer is used as an integral part of an analytical system (e.g., ICP–MS) There are now a large number of MS-based analytical systems available to the analyst:
- Electrospray ionisation MS
- Chemical ionisation MS
- Electron-impact MS
- Direct probe MS
- Field desorption–MS
- Fast atom bombardment–MS
- Tandem MS–MS
- MALDI–MS
- Laser desorption/laser photoionisation ToF MS
- Secondary-ion mass spectrometry (SIMS)

2.4.7 Atomic absorption spectrometry

Atomic absorption spectrometry (AAS) has been around since 1955 and for most of this time it has been the standard analytical tool employed by analysts for the determination of trace amounts of metals [38]. The elements in an aqueous solution are converted to atomic vapour in a flame (e.g., oxygen-acetylene flame) and radiation at the appropriate wavelength is passed through the vapour to excite the ground-state atoms to the first excited electronic level. The amount of radiation that is absorbed is directly related to the atom concentration.

AAS cannot 'survey' or 'screen' for a range of elements in a single operation, unlike ICP-based techniques (Section 2.4.8), but it can determine a particular element with little interference from other elements. Another limitation of AAS is that it is not as sensitive as ICP, and it is these two main limitations which have restricted its use in modern laboratories.

Variants of the AAS technique that are available commercially include:
- Graphite furnace AAS
- Vapour generation AAS
- Zeeman AAS

These variants provide advantages in important areas such as improved sensitivity and enhancement of the elemental range.

2.4.8 Inductively coupled plasma–atomic emission spectroscopy

This technique can also be referred to as ICP–optical emission spectroscopy and has a number of advantages over AAS, including [39]:
– Greater sensitivity.
– It can survey a sample for a large number (e.g., >50) of elements in a single operation.

As in the case of the AAS technique, initially, the rubber sample is digested in an strong acid, or mixture of acids, to create a solution, which is then filtered to remove particulates and insoluble matter. This operation can be labour-intensive and involve a considerable amount of heating if a rubber has a good inherent resistance to acids, but fortunately there are automated digestion systems available on the market.

Different ICP–atomic emission spectroscopy (AES) instruments available include those that operate in a sequential manner or which can scan a number of elements simultaneously. As well as providing advantages relating to efficiency and low limits of detection, these instruments can be used to obtain large amounts of elemental data on unknown samples, which is particularly useful for reverse engineering and failure diagnosis. In this mode, it is usual to operate the instrument semi-quantitatively and then to repeat the analysis using a full calibration if specific elements need to be quantified.

Applications of semi-quantitative work include the determination of a flame-retardant system in a rubber where additives such as antimony trioxide, zinc borate, hydrated alumina, or triphenyl phosphate may have been used, or the search for metals that may have catalysed the degradation of a rubber (e.g., copper or manganese). Regarding quantitative work, as well as the above examples, ICP–AES systems are used regularly for the accurate quantification of substances to check for compliance with regulations (e.g., food-contact regulations).

2.4.9 Inductively coupled plasma–mass spectrometry

ICP–MS uses the established ICP technique to break the sample into a stream of positively charged ions that are subsequently analysed on the basis of their mass. The ICP–MS systems use a compact quadruple mass analyser of the type employed by bench-top GC–MS and LC–MS systems.

The technique has a number of attractive attributes, including:
- Large dynamic range (e.g., solutions containing 2% solids can be analysed).
- High sensitivity.
- Unambiguous spectra.
- Ability to directly measure different isotopes of the same element.

2.4.10 X-Ray fluorescence spectroscopy and X-ray diffraction

Characteristic fluorescence X-rays are emitted from a sample when it has been excited by bombardment with high-energy X-rays or gamma rays, and these can be used to obtain elemental data of a sample [40]. The standard technique usually requires a relatively large amount of sample ($\approx$0.5 g). In principle, the lightest element that can be detected is beryllium (atomic number = 4) but, due to instrumental limitations and the low X-ray yields for the light elements, it has not yet been possible to achieve this. However, over the years the working range of XRF instruments has increased and it is now possible for routine elemental scans to be carried out from at least sodium (atomic number = 11) upwards.

The XRF instrument is usually operated in a semi-quantitative way and care must be taken in the interpretation of the results because they can be influenced by the nature of the matrix. In order to use the technique quantitatively, a range of standards of the element of interest in the same matrix as is present in the sample has to be prepared.

XRF spectroscopy, in conjunction with IR spectroscopy, is used routinely to identify inorganic fillers and inorganic pigments in rubber samples and their ashes. The semi-quantitative elemental data from XRF spectroscopy indicates the types of elements present in the sample, and the IR spectrum helps identify the type of compound(s). X-ray diffraction is also mentioned in this section because it can be used in conjunction with XRF and IR spectroscopy on these types of samples if it is important to identify the particular crystalline form (and hence type) of certain inorganic fillers (e.g., silica and silicates).

2.4.11 Energy-dispersive X-ray spectroscopy

Energy-dispersive X-ray spectroscopy (EDX) has a number of other names (e.g., energy dispersive spectroscopy and X-ray energy dispersive spectroscopy) and can also be referred to as energy-dispersive X-ray analysis or energy-dispersive X-ray microanalysis [41].

As in the case of XRF, characteristic fluorescence X-rays are emitted from a sample when it is impacted using a high-energy beam of charged particles (e.g., electrons, protons or X-rays). However, in EDX, the fluorescence radiation is analysed by sorting the energies of the photons as opposed to separating the wavelengths of the radiation as with XRF.

In the instruments that are commonly used to analyse rubber samples and other polymeric samples, a scanning electron microscope is often employed as the source of electrons. Using this source provides the added benefit of obtaining scanning electron microscopy (SEM) images of the sample or feature within the sample (e.g., a contaminant) as well as elemental data.

The EDX technique is extremely useful for the characterisation of very small samples (e.g, contaminants and inclusions removed from the surface and bulk of rubber samples), the analysis of surface defects, such as stains and blemishes, and areas of discolouration (e.g., blooms). In addition to obtaining data on these features, it is common to obtain data on a freshly cut or undiscoloured portion of the sample in order to obtain background or baseline values. In this way, it is possible to determine if the defect feature is chemically unrelated to the rubber sample, or if it may have originated from it in some way.

2.4.12 Chemiluminescence spectroscopy

Chemiluminescence analysis is suitable for studying the early stages of the thermal oxidation of rubbers. A weak emission of light formed by chemical reactions appears during the oxidative degradation of hydrocarbons. The weak light may originate from the deactivation of an excited carbonyl group formed during the oxidation, or from the formation of peroxide radicals. The use of a sensitive photon detector enables the chemiluminescence to be detected accurately and with good reproducibility. The intensity of the light emitted by the polymer is proportional to the concentration of the originating source. Thus, the degree of oxidation of the polymer can be assessed and the effectiveness of antioxidants evaluated. From the decrease in luminescence intensity it is possible to calculate the rate constant of the decay of the radicals in the polymer undergoing degradation and the effective activation energy for the initiation of oxidation.

The imaging chemiluminescence technique has proved to be a useful method to depth-profile the oxidation of rubbers.

The advantages of chemiluminescence are the relative simplicity of the equipment required, its high sensitivity and the short duration of the experiments. Tests can be carried out either dynamically (at heating rates from 1 to 15 °C/min) or isothermally, with an upper temperature limit of 250 °C.

2.5 Chromatographic techniques

2.5.1 Gas chromatography

GC can be used for the identification and quantification of a large range of low-molecular weight (MW) substances and additives [42, 43]. Two important criteria that species need to have in order to be amenable to its application are thermal stability and relatively volatility. Standards of high purity also need to be available to enable the identification of retention time (RT) and plotting of calibration curves. The mobile phase is often helium and the modern form of the technique employs capillary columns to achieve higher resolution than that possible with traditional packed columns. The standard detector is the flame ionisation type and other detectors are available which improve the specificity and detection limit of the technique (e.g., nitrogen phosphorus detector).

GC can be used as a relatively quick and simple technique to identify and quantify a number of analytes, including:
- Residual monomers and oligomers
- Solvents
- Monomeric antioxidants
- Antiozonants
- Monomeric plasticisers
- Organic flame retardants

GC can also be used to obtain polymer-type information by use of a pyrolysis approach, although it is less powerful than GC–MS because of the absence of a mass spectrometer. A pyrogram 'fingerprint' must be used and a comparison made against a library of known pyrolysis standards recorded under the same conditions (e.g., pyrolysis temperature and column temperature programme) as the 'unknown'.

2.5.2 Gas chromatography–mass spectrometry

Although it is bound by the same sample requirements as GC (e.g., thermally stability), this powerful, versatile technique can be used to obtain a significant amount of compositional information on rubber samples [44]. It is particularly useful in identifying minor components of the sample, such as the breakdown products of the cure system and antidegradants. GC–MS also plays a part in failure diagnosis by being able to provide information on chemicals which may cause odours, and organic contaminants that may have degraded the rubber matrix.

There are a variety of ways that the sample or fractions of the sample (e.g., an extract or volatiles) can be analysed by GC–MS:

- Static headspace
- Dynamic headspace
- Solution injection
- Pyrolysis

With regard to the last example listed above, this approach can be used to investigate the polymer fraction within a sample [45]. The GC–MS instrument is fitted with a pyrolysis unit set at ≈600 °C and the pyrolysis products are 'swept' into the instrument using an inert gas (e.g., helium). This method has the advantage over the pyrolysis FTIR approach in that it has higher resolution and sensitivity and so has greater capability in the identification of complex polymer blends.

One of the limitations of the standard mass spectrometer detectors (e.g., quadrupole or ion trap) used on common GC–MS instruments is that they have a limited accuracy when it comes to determining the actual mass of the ions in the mass spectrum of a compound (e.g., ±0.5 mass units). Hence, compounds are not identified by the mass of their molecular ions, but by obtaining matches on commercially available 70-eV electron-impact ionisation mass spectral libraries. If no match is found, it is usually only possible to comment on the possible generic class of compound from the overall appearance of the mass spectrum and from knowledge of how chemical structures within molecules fragment under 70 eV.

More sophisticated and expensive GC–MS systems that have high-resolution mass spectrometers are commercially available. These mass spectrometers can provide the value of the molecular ion to an accuracy that enables the empirical formula of the organic compound to be obtained. More accurate masses of the fragment ions in a mass spectrum are also obtained. In addition to GC–MS, high-resolution detectors are also available for other techniques that use MS, including LC–MS (Section 2.5.5). The range and capability of GC–MS instruments can also be enhanced by adding to electron-impact ionisation the option of chemical ionisation.

2.5.3 Gas chromatography–gas chromatography–time-of-flight mass spectrometry

This technique has been commercially available for ≈20 years and can be referred to as a three-dimensional (3D) chromatography technique. This comes about because of the combination of two GC columns in the instrument (a main, primary one and a shorter, secondary one), the mass spectrometer, and powerful software that can be used to reconstruct chromatograms and assist in the deconvolution of species peaks.

Despite its high resolution, GC–MS (Section 2.5.2) can encounter problems when trying to resolve all of the components present in complex samples, such as rubber extracts. Although improvements in resolution can be achieved (e.g., by using longer columns) the results obtained are limited because only one type of column (i.e., polar or non-polar) can be used at a time and so any improvements are mainly achieved by influencing the RT due to volatility or polarity. A GC×GC–ToF–MS instrument has the potential to overcome this limitation by interfacing two columns using a modulator at the end of the first, non-polar column (30 m × 0.25 mm), which concentrates the retained compounds over a short (1–5 s) period to focus them and send them down the second, polar (2 m × 0.10 mm) column. The latter is situated in its own compartment within the main oven.

This focusing operation is repeated throughout the length of the analytical run and, typically, 1,500 separate modulations can be undertaken. The mass spectrometer is located at the end of the second column and this has to scan the complete mass spectrum range selected (e.g., 35 to 600 Da) at a very fast rate (e.g., 50 scans/s) as the analysis of the second column is completed within the modulation period. A To F mass spectrometer is used in the instrument because it can scan faster than a quadrupole mass spectrometer. The software present in the instrument also aids data analysis and interpretation by possessing a powerful deconvolution programme.

The two GC columns allow separation by volatility on the first analytical column and by polarity on the second column so that peaks with a similar (or identical) RT on the first column can be separated by the second column. The practical benefit of this scenario to the rubber analyst is that complex mixtures can be separated more thoroughly. Hence, higher-quality mass spectra are obtained on the resolved components, assisting in the identification of the components *via* mass spectral library searches or by manual interpretation. Mass spectra are produced by these instruments using electron-impact ionisation at 70 eV and so matches can be found using the same commercial libraries that are used with conventional GC–MS systems.

Because of its ability to separate complex mixtures of substances, GC×GC–ToF–MS is a particularly useful tool for the analysis of rubber extracts in reverse engineering and food simulant samples for migration studies (Chapters 5 and 8).

To provide a simple illustration of the data obtained using GC×GC–ToF–MS, a chromatogram of a test mixture, in this case a Grob Mix, is shown in Figure 2.3. The composition of the Grob Mix, together with the relevant RTs of the first and second column, is shown in Table 2.1.

The Grob Mix is obviously a relatively simple mixture but the chromatogram in Figure 2.3 clearly illustrates the separation that is obtained by virtue of volatility on the first, main column and by polarity on the second column.

2.5.4 High-performance liquid chromatography

HPLC and LC–MS (Section 2.5.5) instruments are usually employed to identify and quantify additives in rubber and plastic compounds that are either too involatile or insufficiently thermally stable to be analysed by either GC or GC–MS. HPLC is, therefore, the first-choice instrument for antioxidants that have high MW (e.g.,

Table 2.1: First and second column retention times of substances in a Grob mix.

	Retention (s)	Retention time (s)
Decane	713.0	1.97
Octanal	796.9	2.35
Undecane	828.8	1.94
Nonanol	836.9	2.34
Dimethylphenol	840.9	2.83
Ethylhexanoic acid	848.9	2.34
Dimethylbenzamine	904.9	3.02
Methyldecanoate	1052.9	2.20
Methyldodecanoate	1220.9	2.18
Cyclohexylcyclohexamine	1144.9	2.34

Reproduced with permission from M. Forrest, S. Holding and D. Howells, *Polymer Testing*, 2006, **25**, 63. ©2006, Elsevier [46]

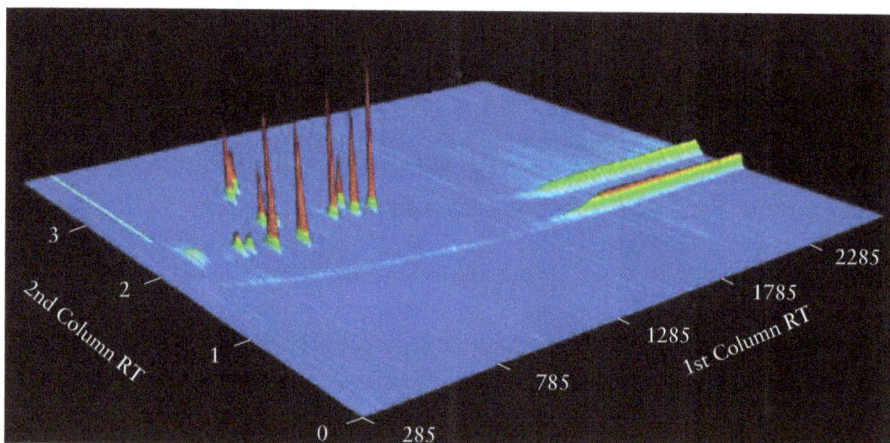

Figure 2.3: GC×GC–ToF–MS chromatogram of a 10-ppm solution of the Grob Mix. Reproduced with permission from M.J. Forrest, S. Holding and D. Howells, *Polymer Testing*, 2006, **25**, 63. ©2006, Elsevier [46].

Irganox® 1010) or are oligomeric (e.g., poly-2,2,4-trimethyl-1,2-dihydroquinoline) [47]. Such antioxidants are often used in rubber products to minimise loses during curing or in long-term end uses at high temperatures.

HPLC is also very useful for the detection and quantification of other relatively high-MW additives, for example, those that fall into the plasticiser (e.g., diisononyl phthalate) or process aid category (e.g., polyethylene glycol). The main limitation of HPLC, however, is that the standard detector is UV and so the analyte needs to have a chromophore to be detected. Hence, standard HPLC cannot be used to determine so high-MW additives such as paraffin waxes.

As with the majority of analytical techniques, a number of modes of operation are available relating to variables such as the column mobile-phase combination (e.g., reverse phase or normal phase), the make-up of the mobile phase during an analysis (e.g., isocratic or gradient elution) and the type of detector (e.g., UV, fluorescence, and mass spectrometer). The use of mass spectrometers as detectors for the technique are now so common and important that this variant, LC–MS, with optional UV detection *in situ*, is often the instrument of choice and is covered in Section 2.5.5.

2.5.5 High-performance liquid chromatography–mass spectrometry

As mentioned in Section 2.5.4, because of their advantages, these instruments have mainly replaced HPLC instruments and are now found in the majority of commercial laboratories [48–51]. The LC section of LC–MS can be operated in a very similar way, with similar choices of operating mode, to the less sophisticated HPLC technique, although care has to be taken that complex sample media (e.g., salt solutions) do not pose a threat of contamination to the mass spectrometer.

Substances are detected and identified as they elute from the LC column *via* both the mass spectrometer (e.g., a quadrupole type) and the in-line UV detector. It is standard practice to use both detection systems because the UV detector is more sensitive to some chemical compounds and, in other cases, they are not ionised by the mass spectrometer and so are not detected.

There are a number of choices when it comes to the operating mode of the MS section, some options being:
- Atmospheric chemical ionisation – two variants:
  - Negative-ion ionisation (good for compounds with labile hydrogens such as alcohols).
  - Positive-ion ionisation (good for compounds that will accept a hydrogen ion such as amines).
- Electrospray (good for ionic compounds).
- Atmospheric pressure photoionisation (good for aromatic hydrocarbons).

These MS options, which tend to be 'soft techniques' result in little fragmentation of the molecular ion, and obviously extend the scope of the technique. However, one drawback of the lack of standardisation is that is that there are no commercially available LC–MS libraries (unlike GC–MS) and so laboratories have to generate their own in-house libraries to identify unknown samples.

A variant on the LC–MS technique, which can be seen in a number of commercial laboratories, is HPLC–MS/MS. This configuration enables a charged ion from the initial fragmentation of a substance to be selected and then subjected to further fragmentation to produce another mass spectrum. In this way, the technique can improve the specificity available to the analyst as the same initial fragment ion from different substances will not produce the same secondary fragmentation pattern. The technique, therefore, improves the ability to differentiate between the different structural isomers of a compound.

A more recent development in LC–MS that is now accessible to commercial laboratories is high-resolution accurate mass (HRAM) LC–MS instruments that also have the capability of ion mobility. The high-resolution mass spectrometer enables rapid assignment of the molecular formula and structure of compounds having MW up to 50,000 Da. The addition of ion mobility, which provides information on the 3D structure of a molecule, assists data interpretation by contributing an important additional dimension in the confirmation of an assigned molecular structure. The enhanced confidence in assignments that results from this detector combination reduces the need for confirmatory experiments. The performance of these instruments can also be improved by the upgrading of the LC component to one that has an ultra-performance liquid chromatography (UPLC) status, with the acronym then becoming UPLC–HRAM/MS.

2.5.6 Ion chromatography

IC, which can also be referred to as ion-exchange chromatography, separates ions and polar molecules based on their affinity to an ion-exchange resin present in a column. Laboratory-scale instruments usually use chromatographic columns which are designed to separate either cations or anions and the mobile phases that are used are buffered aqueous solutions. The columns are typically resins (e.g., cellulose beads) with covalent-bonded charged functional groups. The target analytes are retained on the column for a time but are displaced and eluted by increasing the concentration of a similarly charged species in the mobile phase (e.g., in cation IC, a positively-charged analyte can be replaced by positively charged sodium ions). Detection of the eluted ions is usually carried out using either a conductivity detector or a UV-Vis detector.

Although not used extensively for the analysis of rubber, one of the applications for IC is to provide accurate quantifications of the amount of sulfur or halogen species (e.g., chlorine) in the reaction mixture than that resulting from an oxygen flask combustion experiment (Section 2.2). It is more accurate and sensitive for these determinations than using a titration-based approach and so it is the preferred route, particularly if the levels of analyte are expected to be low.

2.5.7 Supercritical fluid chromatography

Supercritical fluid chromatography (SFC) uses the unusual properties of supercritical fluids [52]. It offers the dual benefits of the solubility behaviour of liquids and the diffusion and viscosity behaviour of gases. Using mobile phases such as carbon dioxide, it provides unprecedented versatility in obtaining high-resolution separation of mixtures and substances that other chromatographic techniques (e.g., GC and LC) struggle with (e.g., high-MW oligomers and additives).

SFC can be used in combination with the majority of the detectors used in GC and LC in a preparative way or to isolate samples for subsequent offline analysis using techniques such as IR microscopy. Another advantage of the technique is that the loss of the mobile phase due to decompression once it exits the instrument makes deposition onto a substrate or into a sample vial relatively easy and removes this source of potential interference. It is also possible to interface SFC with a number of spectroscopy techniques (e.g., IR and MS).

2.5.8 Gel-permeation chromatography/size exclusion chromatography

Used in its conventional form, with mixed pore-size column packings, gel-permeation chromatography (GPC), or size-exclusion chromatography as it can also be referred to, is used extensively to characterise the MW and molecular distribution of polymers [53, 54]. By using column packings that have small pore sizes (e.g., 50 or 100 Å) it is also able to separate oligomers, although it does not have the same resolving capability in this regard as SFC. Also, by employing large-bore columns, it can be used in a preparative way to accumulate samples for subsequent analysis by GC–MS, LC–MS or IR microscopy.

For GPC to be successful, the sample must be readily soluble in an organic solvent. Hence, by definition, it cannot be carried out on the polymer phase in a rubber vulcanisate and even unvulcanised fully compounded rubber samples can present problems due to filler–rubber interactions. Therefore, it tends to be restricted to the analysis of raw rubbery polymers or thermoplastic rubbers and in this, using solvents such as toluene and tetrahydrofuran, it is usually very successful. However, given that a number of rubbers are free radical-polymerised (e.g., NBR) or have very high MW (e.g., NR), there can be a problem with the insoluble fractions (i.e., 'gel') that result from physical entanglements and chemical interactions.

It is possible to use GPC to obtain information on high-MW additives present in vulcanised rubbers by the analysis of solvent extracts. This approach is particularly useful if it is suspected that oligomeric or polymeric additives may have been used in the compound. Figure 2.4 shows how high-resolution GPC (using 100 A, 10-μm columns) has been used to separate the polymeric and other additives extracted from a NBR sample. Use of an IR detector or LC-Transform® system, as mentioned below, can then assist in identifying these substances.

During the last 30 years, the range of detectors available for use with GPC has increased and a 'state-of-the-art' system now often possesses the following three detectors:

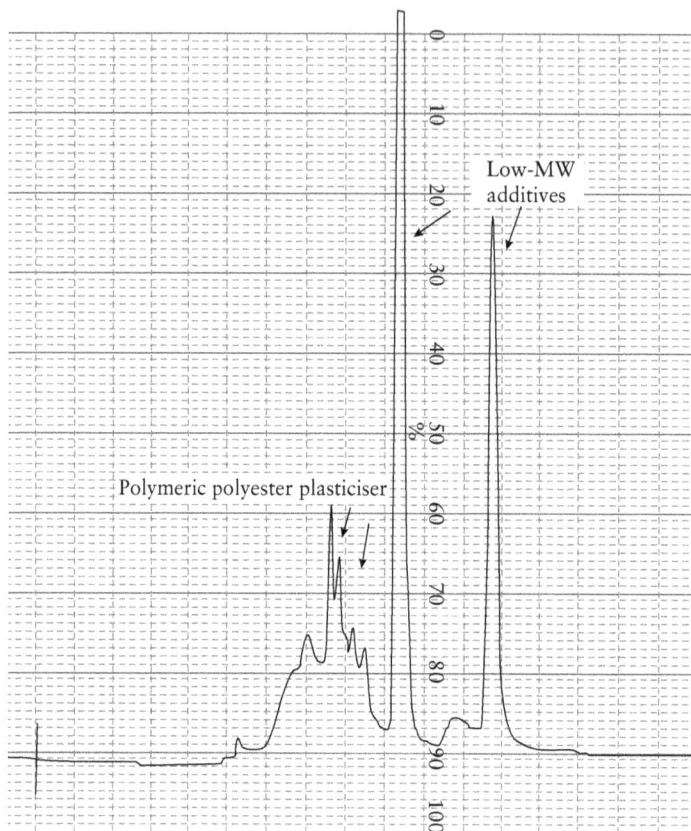

Figure 2.4: Separation of polymeric and monomeric additives extracted from a NBR using high-resolution GPC. Reproduced with permission from M.J. Forrest in *Rubber Analysis – Polymers, Compounds and Products*, Rapra Review Report No.139, Smithers Rapra, Shawbury, UK, 2001, p.10. ©2001, Smithers Rapra [55].

- Refractive index (RI)
- Light scattering
- Viscosity

Another detection system that is becoming available for GPC is IR, and some commercially available systems have these in conjunction with a RI detector. One drawback of IR as an in-line detection system for GPC is that the IR absorption of the mobile phase will reduce the amount of the spectrum that is available for use.

Before the advent of in-line IR detectors for GPC, the development of the Lab Connections LC-Transform® interface offered the capability of obtaining IR spectra on sections or fractions of a GPC chromatogram. The LC-Transform® consisted of two sampling accessories (modules) and operating software. The first module collected the species present in the GPC chromatogram as they eluted from the system.

They were deposited onto a revolving germanium disc with the mobile phase being volatilised by a combination of heat, vacuum and ultrasonic agitation. The other module enabled the prepared disc to be placed into a standard IR bench for analysis. In the case of some rubber systems, by using the correct choice of separating column, polymer blends could be separated and identified, as well as plasticisers and other high-MW additives.

2.5.9 Thin-layer chromatography and paper chromatography

Thin-layer chromatography (TLC) and paper chromatography are relatively simple techniques that have been in existence for many years now and for which little new work has been published since the 1970s; both are rarely used in modern analytical laboratories [56].

Historically, before the advent of modern HPLC-based instruments, TLC and paper chromatography were used for both the separation and identification of additives such as amine and phenolic antioxidants, and UV stabilisers, and they can still fulfil that function today if required. A number of approaches are possible using different developing media, solvents and spraying agents. The standard method of detection is the use of a UV light source.

By the use of plates that have thick layers of the stationary phase, TLC can be used in a preparative way for the isolation of sample fractions for subsequent chromatographic (e.g., GC–MS) or spectroscopic (e.g., IR) analysis.

2.6 Thermal techniques

2.6.1 Differential scanning calorimetry

Differential scanning calorimetry (DSC) can be used for a number of purposes in the analysis of rubbers [57–59]. For example, it is possible to use it to provide quality-control fingerprint thermograms from which deviations in formulation can be ascertained. These will contain both endotherms from the volatilisation of ingredients such as plasticisers, together with exotherms from the curing of the rubber and the degradation of the sample. Analytical work can be performed in both a non-oxidising atmosphere (e.g., nitrogen), which is the most commonly used environment and, if appropriate, such as for investigations involving oxidation stability, an oxidising atmosphere (e.g., air).

A typical two-stage DSC programme used in a nitrogen atmosphere to investigate the composition and properties of a rubber sample is shown below:
1. 2 min at −100 °C.
2. Heat from −100 to +250 °C at 20 °C/min (creates an 'as-received' thermogram).

3. Hold for 2 min at +250 °C.
4. Cool from +250 to −100 °C at 20 °C/min.
5. Hold for 2 min at −100 °C .
6. Heat from −100 to +250 °C at 20 °C/min (creates 'reheat' thermogram).

With rubber samples, the information that is obtained during the initial heating stage (i.e., stage 2 as shown above) is usually the most useful. During this stage, a number of events may be present in the thermogram depending upon the history of the sample and the objective of the analysis, including:
- Volatilisation of absorbed substances (e.g., water or solvents)[*].
- Change in specific heat due to the glass transition temperature (T_g) of the polymer(s).
- Melting of crystalline ingredients[*].
- Loss of plasticiser and other ingredients[*].
- Curing exotherm (if the rubber is compounded but unvulcanised).
- Non-oxidising degradation exotherms.
  [*]endothermic changes

The reheat thermogram (stage 6 as shown above) can sometimes provide evidence of additional compositional changes and reactions, such as further non-oxidising degradation.

An indication of the range of events which can be observed by analysing an uncured compounded rubber by DSC using a nitrogen atmosphere, a temperature ramp of 20 °C/min and a temperature range −100 to around +700 °C, is provided below:
- −100 to 0 °C – T_g.
- 0 to 150 °C – melting of additives and loss of plasticiser/oil.
- 150 to 250 °C – curing.
- >250 °C – degradation and antioxidant function.

Using appropriate standard materials it is possible to employ DSC to quantify the level of certain additives (e.g., peroxide curatives in uncured rubbers) in a sample. DSC can also be used in thermal stability studies of a rubber compound, and to investigate the effectiveness of antidegradants and, to an extent, fire retardants.

DSC can also be used to profile the degree of cure through thick rubber articles, to study cure kinetics, and the compatibility of polymers in a blend. Two examples of the latter studies are those of a blend of EPDM and NR, published by Abou-Helal and El-Sabbagh [60], and that of a blend of polypropylene (PP) and EPDM, a thermoplastic rubber, published by Qixia and co-workers [61].

Crosslink density can influence the T_g of rubber samples. Cook and co-workers [62], using unmodified rubbers (i.e., compounds without plasticisers or fillers), demonstrated that a linear relationship exists between crosslink density and T_g for three types of rubber: NR, polybutadiene (BR) rubber and SBR.

Thermoplastic rubbers, such as those produced by combining EPDM with PP, have a semi-crystalline component (i.e., the PP) and so can be identified and characterised (e.g., determining the proportions of PP and EPDM) by DSC. To assist with this, it is important to have reference spectra that have been recorded under standard conditions (e.g., those in stages 1–6 shown above) and an example of such a library has been published by Price [63].

2.6.1.1 Modulated differential scanning calorimetry

The advent of modulated temperature DSC (also described as alternating or oscillating DSC) over 15 years ago offered advantages to the polymer analyst in that its ability to separate the heat flow into its reversing (heat capacity-related) and non-reversing (kinetic) events improved the quality of the data that could be obtained by DSC and, hence, increased its usefulness [64–66].

The major reversing events that occur in plastics and rubbers include the T_g in both cases, and the melting of the crystalline region in the case of semi-crystalline plastics. There are a number of non-reversing events (including re-crystallisation processes) occurring during melting, and the enthalpic relaxation that occurs in the region of the T_g for amorphous polymers that have undergone physical ageing. In addition, with crystalline plastics, the situation is complicated by the fact that the heating process can alter the crystalline structure of the material, with imperfect crystals melting and becoming absorbed into larger crystals prior to these melting. Conventional DSC cannot separate these two events because they often happen simultaneously, but modulated DSC can resolve the reversing heat flow of crystal melting and the non-reversing heat flow of crystallisation. Modulated DSC is, therefore, capable of providing more accurate data on the latent heat of fusion for the crystal melting event. This is important because the study of the crystallinity of polymers is one of the principal uses of DSC.

Other advantages that have been claimed for the use of modulated DSC in the analysis of plastics and rubbers include:

- Improvements in the accuracy of T_g measurements in plastics due to the greater separation of the T_g (reversing) and enthalpic relaxation event (non-reversing) which occurs just after the T_g [67].
- The inherent greater sensitivity of the technique makes it easier to detect the T_g in highly filled/reinforced polymers, or highly crystalline plastics, where these features reduce the change in the heat capacity that occurs as the materials passes through this transition. HyperDSC™ (Section 2.6.1.2) also offers this advantage over conventional DSC.
- It can be useful in the analysis of certain polymer blends. For example, in the case of a polycarbonate (PC)/polybutylene terephthalate (PBT) blend, the T_g of PC may overlap with the re-crystallisation of parts of the PBT phase.

- It provides, along with power-compensation DSC, a means of directly measuring the heat capacity of a material. It has been demonstrated that the method can also be used to produce thermal conductivity data that agree well with values established by more conventional techniques. This and other applications are covered by a TA Instruments guide to modulated DSC [66].
- Providing that the calibration has taken into account the heat loss through the purge gas surrounding the sample, the thermal conductivity of polymers can also be measured in the range from 0.1 to 1.5 W/°C.
- Its ability to generate an instantaneous heating rate during quasi-isothermal experiments allows cure measurements on rubbers to be made which are not possible by conventional DSC. This can assist in obtaining a greater understanding of the cure mechanism.

Hourston and Song [68] have proposed a modulated DSC method for the measurement of the weight fraction of the interface in a wide range of rubber–rubber blends. A quantitative analysis using the differential of the heat capacity *versus* the temperature signal from the modulated temperature DSC allowed the weight fraction of the interface to be calculated. The data obtained showed how essentially miscible the different rubber types were, as well as revealing the influence of properties such as stereo regularity (with BR blends) and the proportions of comonomers (with SBR blends) on the extent of mixing.

2.6.1.2 HyperDSC™

This variant on DSC involves recording the heat flow cure of a sample as it is heated very rapidly (e.g., at 500 °C/min), as opposed to a standard 20 °C/min heating ramp. This technique can usually be used only with power-compensation systems as an analysis chamber having a low thermal mass is essential; and the ability to record data very quickly is another pre-requisite. Several publications provide information on the use and application of HyperDSC™ [69, 70].

The advantages that this technique offers in the analysis of polymers come from its twin attributes of very low short analysis times and increased accuracy when in comes to measuring crystallinity. In the former case, the instrument has been used effectively on the production lines of semi-crystalline materials to record data, such as T_g and melting temperature (T_m), almost as soon as they are produced; a full characterisation is possible within 60 s. This means that any quality-control problems can be identified very quickly and rectified before too much production has been lost. Although the samples are heated very rapidly, it has been demonstrated that this results in no significant increase in T_g. The data obtained from these HyperDSC™ analyses can, therefore, be compared with library standards that have been run using more traditional conditions.

With respect to the latter case, many semi-crystalline polymers exist in several potentially unstable states. A fast heating rate enables data to be recorded on them before they change due, for example, to microcrystalline regions melting and then re-crystallising during the analysis. If this type of effect occurs under a slow heating rate, it means that the extent of crystallinity of the polymer can change during an experiment, and so the data obtained are not totally representative of the original material.

The two other advantages of HyperDSC™ are:

1. The increased scan rate provides much higher sensitivity due to the resulting higher heat flow, and this enables the analyst to identify transitions which would not be apparent using standard conditions.
2. The ability to have fast heating and cooling rates means that it is possible to closely mimic the conditions that some samples experience in their manufacture and production.

2.6.2 Dynamic mechanical analysis

Dynamic mechanical analysis (DMA) is a very useful technique for rubbers [71]. It is able to generate modulus *versus* temperature data under a range of conditions (e.g., deformation modes and frequencies) and, as a result of this, provide important information on a sample, for example, dynamic mechanical properties, T_g, thermal stability, and cure state.

The effects of temperature on various types of moduli can be obtained over a wide temperature range (typically –150 to +200 °C). A typical set of conditions that can be used to analyse rubber samples by DMA is shown below:

- Deformation mode: double cantilever
- Sample dimensions: 2 × 10 × 25 mm
- Heating rate: 3 °C/min
- Frequency of deformation: 1 Hz
- Start temperature: –150 °C
- Finish temperature: 250 °C
- Amplitude of deformation: 50 µm

The sensitivity of DMA to T_g ($\approx$1,000-times greater than DSC, which measures changes in heat capacity rather than modulus) is very useful in the analysis of unknown samples. For example, it can be difficult to detect the presence of 5–10% of BR in SBR by IR, which would only provide the value of the total butadiene content. However, the large difference in the T_g values of the two polymers makes the presence of BR in such a matrix much easier to detect and, with appropriate standards, quantify.

2.6.3 Microthermal analysis

Microthermal analysis is a characterisation technique that can be regarded as combining the thermal analysis principles of DSC with the high spatial resolution of scanning probe microscopy (SPM). It is possible to create a microthermal analyser by modifying an atomic force microscope by replacing its tip with an ultraminiature-resistive heater that also serves as a temperature sensor. Using this device, it is possible to obtain localised thermal analysis data on the surface of polymer samples, and this feature enabled the technique to be launched commercially in 1998 [72].

An example of the commercial instruments that have been produced is the one marketed by TA Instruments [73, 74]. This device is useful for the analysis of polymers in certain applications because it can obtain DSC and thermal mechanical analysis (TMA) data on very small (i.e., micron-sized) areas. Examples of the applications that the instrument can be used for are presented below.

2.6.3.1 Determination of the homogeneity of polymer blends

The surface of a rubber product can be mapped and occurrence of the events which are characteristic of individual polymers can be obtained. The relative intensity of these will indicate whether some areas of the sample are richer in one polymer than others. For example, in a thermoplastic rubber gasket produced from a mixture of EPDM rubber and PP, the T_g of the EPDM rubber (TMA trace) and the T_m (DSC trace) of the PP could be focused upon.

2.6.3.2 Studying surfaces

Examples of possible applications include rubber products that are modified by processes such as corona discharge or are sterilised by radiation. These processes change the physical properties (e.g., modulus) of the surface layer of the polymer due to crosslinking and/or oxidation reactions. The microthermal analyser can be used to determine how uniform across the surface these changes are.

2.6.3.3 Characterisation of localised thermal history

Although this capability will be more useful for studying plastics, where localised stresses can become 'frozen' into the product if they are cooled too quickly during manufacture, it could have some application for thermoplastic rubbers, and for determining if certain areas of a rubber product are undercured due to cold spots in a mould. In the first case, a microthermal analyser would be able to map the heat capacity of the thermoplastic rubber and see if frozen-in irregularities in the structure can be detected (e.g., by a reduction in crystallinity in certain regions). In the second case, DSC could detect a residual curing agent *via* an exotherm, and TMA could detect differences in the modulus.

A thorough review of the techniques and applications of microthermal analysis has been published by Pollock and Hammiche [75].

2.6.4 Thermal mechanical analysis

TMA is a method for measuring linear or volumetric changes in polymers as a function of temperature, time or force. It is often used in conjunction with DSC to investigate the structure and properties of polymers. This combination is often found in commercial microthermal analysers (Section 2.6.3). While DSC is concerned with the energetics of physical and chemical changes, TMA measures the dimensional effects associated with these changes, and so it can be used to obtain the coefficient of thermal expansion for rubber samples.

TMA can also measure properties such as T_g and T_m, and determine the modulus of rubbers, and so is similar to DMA in this regard.

2.6.5 Thermogravimetric analysis

TGA is extremely useful for the analysis of compounded rubber samples because it can provide an extensive amount of information relating to their bulk composition and other important characteristics (e.g., the absorption of a fluid in service) [9]. It is possible to obtain accurate quantification of the principal ingredients such as plasticisers, polymers, CB and inorganic species. These data can be obtained on small (e.g., 10 mg) samples in a relatively short time ($\approx$45 min). By the use of the temperature maxima at which weight loss events occur, it is also possible to use the technique to obtain some qualitative assignments for the bulk ingredients, particularly the polymer and the CB, and to identify the presence of substances that have reproducible and characteristic thermal-breakdown patterns (e.g., $CaCO_3$).

A typical TGA programme for the analysis of a rubber would be:

1. Heat from 40 to 550 °C in a nitrogen atmosphere at 20 °C/min
2. Hold at 550 °C for 10 min
3. Reduce the temperature to 300 at 20 °C/min
4. Change the atmosphere to air
5. Heat from 300 to 900 °C in an air atmosphere at 20 °C/min

For rubbers that are very thermally stable (e.g., silicone rubbers), it is necessary to modify the programme stated above to ensure that all of the organic material is pyrolysed from the sample before the temperature is reduced to 300 °C. This is done by increasing the final temperature of step 1 to $\approx$700 °C and holding the sample in nitrogen at this temperature for a longer period of time (e.g., 20 min).

It is usually common to plot the weight loss derivative in conjunction with the weight loss. The derivative can be used to detect the presence of a polymer blend of two or more components and to indicate if the sample contains more than one type of CB.

Other uses for TGA include the production of compositional fingerprints for quality-control purposes, and the investigation of thermal stability, including the effects that additives such as flame retardants and antioxidants have upon it.

Modern TGA instruments have the ability to carry out high-resolution TGA experiments (Section 2.6.5.1) and modulated TGA experiments (Section 2.6.5.2). It is also possible to interface modern TGA instruments directly to gas analysers (e.g., FTIR instruments or mass spectrometers) to obtain both quantitative and qualitative data in one analytical step. These combinations can be referred to as 'hyphenated TGA' or 'evolved gas analysis TGA' and add a significant degree of capability to the technique (Section 2.6.5.3).

2.6.5.1 High-resolution thermal mechanical analysis

High-resolution TGA was developed to enhance the resolution of multiple components in complex systems. It, therefore, offers particular potential advantages in the analysis of complex rubber formulations. The principle of its operation is that rather than having a constant heating rate (e.g., 20 °C/min) throughout the analytical run, the TGA furnace is actively controlled by the software and the rate of heating is determined by the rate that the sample loses weight. A review of the use of high-resolution TGA for the analysis of polymer samples has been published by Forrest [76]. This review demonstrates that, in practice, high resolution offers distinct advantages in some applications (e.g., quantification of the plasticiser and polymer components in rubbers), but can cause problems with others (e.g., quantification of two or more rubbers in a blend). For example, Forrest found that NR gave one weight loss in a conventional TGA experiment, but four in a high-resolution operation [76].

2.6.5.2 Modulated thermal mechanical analysis

Modulated TGA was also intended to improve the accuracy of the data resulting from TGA experiments. It was developed by TA Instruments [77, 78] and operates using a similar principle to that of modulated DSC (Section 2.6.1.1), with a sinusoidal modulation overlaid over the linear heating rate so that the average sample temperature changes continuously with time in a non-linear way. Gamlin and co-workers [79] used a combination of high-resolution TGA and modulated TGA to study the effects of the ethylene content and maleated EPDM content on the thermal stability and degradation kinetics of EPDM rubbers. A comparison was made between the data generated from isothermal and non-isothermal TGA experiments, and other values reported in the literature.

2.6.5.3 Thermal mechanical analysis with evolved gas analysis

Combining TGA with an instrument capable of analysing the evolved gaseous products that are emitted from it during an analysis extends and enhances the usefulness and diagnostic power of the technique. Two popular choices of such an instrument are an FTIR or a mass spectrometer, and these are directly coupled to the TGA by the use of a heated interface. The nature of the interface varies according to the instrument being employed. If it is a mass spectrometer, it could be a fused silica capillary tube heated to $\approx 200\ ^{\circ}C$ to prevent condensation of the gaseous products and only a small quantity (e.g., 1%) of the volatiles will be diverted into the instrument. If the instrument is an FTIR, a less sensitive technique, all of the gaseous products in the purge gas (e.g., nitrogen) are transferred through a heated interface line (e.g., glass-coated steel) into a heated gas cell.

In addition to TGA–IR and TGA–MS, other hyphenated TGA techniques that are available to the analyst include TGA–GC, TGA–GC–IR and TGA–GC–MS.

TGA with evolved gas analysis can have a number of applications in the analysis of rubbers, including:
– Thermal degradation processes
– Studies of vulcanisation
– Identification of polymer type
– Identification of additives (e.g., cure system ingredients)
– Analysis of substances (e.g., oils) absorbed in service

2.6.6 Dielectric analysis

Dielectric analysis (DEA) has proved to be very useful over the years. It measures the response of a polymer to an applied alternating voltage signal. It, therefore, measures two fundamental electrical characteristics of a polymer – capacitance and conductance – as a function of time, temperature and frequency. A polymer sample is placed in contact with a sensor (or electronic array for samples in the form of pellets or powders) and the oscillating voltage signal applied at frequencies between 0.001 and 100,000 Hz. Capacitance and conductance can be related to changes in the molecular state of a polymer:
– Capacitance: the ability of a polymer to store an electrical change – dominates at temperatures below T_g or T_m.
– Conductance: the ability of a polymer to transfer an electrical charge – dominates at temperatures above T_g or T_m.

The actual polymeric properties monitored by DEA are:
– E′ (permittivity) – a measure of the degree of alignment of the molecular dipoles to the applied electrical field (analogous to the elastic modulus in DMA).

– E″ (loss factor) – represents the energy required to align the dipoles (analogous to the viscous modulus in DMA).

Tan δ in DEA is the ratio of these two properties and the technique can be used for a similar range of applications as DSC and DMA, but it has a number of practical advantages over those two techniques:
– Data are not affected by evolution of volatile species during such events as curing reactions.
– Data can be obtained on samples in a wide range of physical forms (e.g., solid, liquid, pellets or powder).
– It is more sensitive than DSC and so it can be used to obtain superior data on samples where there is only a small polymer fraction (e.g., highly extended or filled polymer products).
– Data can be obtained by probe samplers situated away from the instrument, and so it can be used for 'in-line' monitoring of events such as curing reactions.

There are a number of application notes on the use of DEA to characterise polymers available from instrument suppliers [80].

2.6.7 Thermal conductivity analysis

Accurate thermal conductivity data are important for assessing the suitability of a particular polymer for a given application. Such data are also being used extensively in the application of computer modelling packages to predict the flow of plastics and rubber compounds during processing operations.

The ability to use modulated DSC to determine such data has been mentioned in Section 2.6.1.1. More traditional methods involve the use of the standard guarded hot plate device that is described in the American Society for Testing of Materials (ASTM) standard, ASTM F-433. In addition to this, Lobo and Cohen [81] described an in-line method that can be applied to polymer melts. Also, Oehmke and Wiegmann [82] described an apparatus that can carry out thermal conductivity measurements as a function of temperature as well as pressure.

2.7 Microscopy techniques

High-magnification (e.g., 150×) optical microscopy can be used to examine the dispersion of fillers, such as CB, within the rubber matrix [83, 84]. The success of this operation is usually dependent on the generation of high-quality microtome sections of the sample. The degree of dispersion can be subjectively compared with photographs of standard examples (e.g., a Cabot Dispersion Classification Chart for CB).

Modern digital microscopes and their associated software have greatly enhanced the capability of the technique, including the measurement of features of interest and the superimposing of multiple images that allow 3D pictures to be created of items such as contaminants and defect structures. Digital imaging also allows the easy and rapid transfer of images from the microscopy to online computers for use in reports and presentations.

SEM has been mentioned in Section 2.3.11 in relation to the elemental analysis of surfaces using EDX. It can also be used in its own right, where its high-magnification powers can be employed to determine the particle size of fillers and in the detailed examination of fracture surfaces. It can also be used to investigate the morphology of rubber compounds, particularly filled rubber compounds [85] and rubber blends [86]. Transmission electron microscopy can also be used to investigate the phase morphology of polymer blends and the dispersion of fillers.

The use of IR microspectrometry to generate IR data on a variety of small samples (e.g., polymers, inclusions, fibres and contaminants) has been covered in Section 2.4.2. Atomic force microscopy [also called scanning force microscopy (SFM)] involves use of a very high-resolution scanning probe microscope capable of resolution of the order of fractions of a nanometre, which is >1,000-times better than the detection limit of an optical microscope. A review of the use of SFM to visualise and study the morphology of polymers has been published by Vancso and Schonherr [87].

There are is wide range of instrumental techniques and methods that can be regarded as being based on SPM. One of these, microthermal analysis using a modified atomic force microscope, has been mentioned in Section 2.6.3. The use of SPM to characterise a variety of polymeric materials has been reviewed by Tsukruk [88].

2.8 Rheological instruments

This class of instrument can be used in a number of ways to characterise rubber samples. One of the most important, which is essential for any quality-control function within a rubber manufacturing environment, is the determination of cure characteristics (e.g., scorch time, cure rate and cure time). The use of these instruments for this type of application is covered within Chapter 5 – curing and cure state studies.

These types of instrument can also be used to obtain other important information on rubber samples, such as those related to their viscosity, viscoelastic properties as well as the interaction between fillers and polymers.

Three of the principal types of instrument that are commercially available include the moving die rheometer, the rubber process analyser, and Mooney viscometer. A publication produced by one of the manufacturers of these types of instruments [89] provides a convenient listing of their capabilities and applications, including:
- Mooney viscosity, Mooney stress relaxation and Mooney scorch determinations.
- Determination of isothermal curing characteristics.

- Strain sweep measurements for filler loading, filler type, filler dispersion and rubber–filler interaction information.
- Monitoring of the breakdown of structure within CB during processing (e.g., milling).
- Studying stress–strain response at high strain to characterise filler content and structure, as well as polymer architecture.
- Monitoring simultaneous curing and blowing agent action during the production of rubber foams.

2.9 Miscellaneous analytical techniques

There are a number of other analytical techniques that can be used to characterise rubber samples. These include specific techniques applied to the analysis of certain additives (e.g., CB), rubber systems (e.g., latex) and to the rubber itself (e.g., viscosity measurement for determining MW). The application and information obtained by such techniques are covered in subsequent sections of this book.

Surface analytical techniques such as SIMS [90, 91], laser ionisation mass analysis [92, 93] and X-ray photoelectron spectroscopy [94–96] can be used for failure diagnosis associated with the poor bonding of rubber products to metal substrates and investigation of surface contaminates.

References

1. T.R. Crompton in *Characterisation of Polymers*, Volume 1 (2008) and Volume 2 (2009), Smithers Rapra, Shawbury, UK.
2. N.P. Cheremisinoff in *Polymer Characterisation*, University Press of Mississippi, Jackson, MS, USA, 1996.
3. B.J. Hunt and M.I. James in *Polymer Characterisation*, Springer Science & Business Media, Berlin, Germany, 2012.
4. T.R. Crompton in *Practical Polymer Analysis*, Springer Science & Business Media, Berlin, Germany, 2012.
5. T.R Crompton in *Polymer Reference Book*, iSmithers, Smithers Rapra Technology Ltd, Shawbury, UK, 2006.
6. T.R. Crompton in *Determination of Additives in Polymers and Rubbers*, Smithers Rapra Technology Ltd, Shawbury, UK, 2007.
7. J.C.J. Bart in *Additives in Polymers*, John Wiley & Sons, New York, NY, USA, 2005.
8. P.P. De, N.R. Choudhury and N.K. Dutta in *Thermal Analysis of Rubbers and Rubbery Materials*, Smithers Rapra Technology Ltd, Shawbury, UK, 2010.
9. M.J. Forrest in *Application to Thermoplastics and Rubbers, Principles and Applications of Thermal Analysis*, Ed., P. Gabbott, Blackwell Publishing, Oxford, UK, 2008, Chapter 6.
10. J.L. Koenig in *Infrared and Raman Spectroscopy of Polymers*, Rapra Review Report No.134, Smithers Rapra Technology Ltd, Shawbury, UK, 2001.

11. Q. Guo in *Polymer Morphology: Principles, Characterisation and Processing*, John Wiley & Sons, New York, NY, USA, 2016.
12. ISO 7269:1995 – Rubber – Determination of free sulfur.
13. ISO 15672:2000 – Rubber and rubber additives – Determination of total nitrogen content using an automatic analyser.
14. ISO 15671:2000 – Rubber and rubber additives – Determination of total sulfur content using an automatic analyser.
15. V.M. Litvinov and P.P. De in *Spectroscopy of Rubbers and Rubbery Materials*, Rapra Technology Ltd, Shawbury, UK, 2002.
16. J.L. Koenig in *Spectroscopy of Polymers*, 2nd Edition, Elsevier, Amsterdam, The Netherlands, 1999.
17. W. Klopffer in *Introduction to Polymer Spectroscopy*, Springer Science & Business Media, Berlin, Germany, 2012.
18. N. Everall, P.R. Griffiths and J.M. Chalmers in *Vibrational Spectroscopy of Polymers: Principals and Practice*, John Wiley & Sons, New York, NY, USA, 2007.
19. M.J. Forrest, Y. Davies and J. Davies in *The Rapra Collection of Infrared Spectra of Rubbers, Plastics and Thermoplastic Elastomers*, 3rd Edition, Rapra Technology Ltd, Shawbury, UK, 2007.
20. M. Jefcoat, H. Tubb and O. Madden in *Proceedings of the 243rd National ACS Meeting*, 25–29th March, San Diego, CA, USA and the Smithsonian Museum Conservation Institute 2012, **53**, 1, 735.
21. D.O. Hummel and F.K Scholl in *Infrared Analysis of Polymers, Resins and Additives: An Atlas*, Volumes 1–3, 2nd Edition, Carl Hanser, Munich, Germany, 1979–1983.
22. M.J. Forrest in *Rubber Analysis – Polymers, Compounds and Products*, Rapra Review Report No.139, Smithers Rapra, Shawbury, UK, 2001, p.7.
23. S. Elliott, *British Plastics and Rubber*, 2013, September, p.43.
24. R.J. Pell, J.B. Callis and B.R. Kowalski, *Applied Spectroscopy*, 1991, **45**, 5, 808.
25. R. Pornprasit, P. Pornprasit, P. Boonma and J. Natwichai, *Journal of Spectroscopy*, 2016, Paper No.4024783, p.7.
26. P.J. Hendra and J.L. Agbenyega in *The Raman Spectra of Polymers*, Wiley, Chichester, UK, 1993.
27. *Fourier Transform Raman Spectroscopy*, Eds., D.B. Chase and J.F. Rabolt, Academic Press, Orlando, FL, USA, 1994.
28. C.G. Smith, P.B. Smith, A.J. Pasztor, M.L. McKelvy, D.M. Meunier and S.W. Froelicher, *Analytical Chemistry*, 1995, **67**, 12, 97R.
29. P.J. Hendra and K.D.O. Jackson in *Proceedings of the 144th IRC Conference*, 26–29th October, Orlando, FL, USA, 1993, Paper No.105.
30. P.J. Hendra, C.H. Jones, P.J. Wallen, G. Ellis, B.J. Kip, M. van Duin, K.D.O. Jackson and M.J.R. Loadman, *Kautchuk und Gummi Kunststoffe*, 1992, **45**, 11, 910.
31. A.H. Kuptsov and G.N. Zhizhin in *Handbook of Fourier Transform Raman and Infrared Spectra of Polymers*, Elsevier, Amsterdam, The Netherlands, 1998.
32. B.J. Clark, T. Frost and M.A. Russell in *UV Spectroscopy*, Springer Science & Business Media, Berlin, Germany, 1993.
33. J. Keeler in *Understanding NMR Spectroscopy*, John Wiley & Sons, New York, NY, USA, 2011.
34. T. Kitayama and K. Hatada in *NMR Spectroscopy of Polymers*, Springer Science & Business Media, Berlin, Germany, 2013.
35. R.N. Ibbett in *NMR Spectroscopy of Polymers*, Springer Science & Business Media, Berlin, Germany, 2012.
36. E. de Hoffmann in *Mass Spectrometry: Principals and Applications*, John Wiley & Sons, New York, NY, USA, 2007.

37. R.P Lattimer and R.E. Harris, *Mass Spectrometry Reviews*, 1985, **4**, 3, 369.
38. B. Welz and M. Sperling in *Atomic Absorption Spectrometry*, 3rd Edition, 1999, Wiley-VCH GmbH, Weinheim, Germany, 1999.
39. J. Nolte in *ICP Emission Spectrometry*, Wiley-VCH, Weinheim, Germany, 2003.
40. R. Jenkins in *Chemical Analysis: A series of Monographs on Analytical Chemistry and its Applications*, Volume 152, 2nd Edition, 1999, Wiley, Hoboken, NJ, USA, 1999.
41. J.C. Russ, M. Ashby Frs, R. Kiessling and J. Charles in *Fundamentals of Energy Dispersive X-ray Analysis*, Elsevier, Amsterdam, The Netherlands, 1984.
42. T. Provder in *Chromatography of Polymers: Hyphenated and Multidimensional Techniques*, Volume 731, American Chemical Society, Washington, DC, USA, 1999.
43. V.G. Berezkin, V.R. Alishoyev and I.B. Nemirovskaya in *Gas Chromatography of Polymers*, Elsevier, Amsterdam, The Netherlands, 1977.
44. H-J. Hubschmann in *Handbook of GC-MS*, John Wiley & Sons, New York, NJ, USA, 2015.
45. S. Tsuge, H. Ohtani and C. Watanabe in *Pyrolysis – GC/MS Data Book of Synthetic Polymers*, 1st Edition, Elsevier, Amsterdam, The Netherlands, 2011.
46. M.J. Forrest, S. Holding and D. Howells, *Polymer Testing*, 2006, **25**, 63.
47. H. Pasch in *HPLC of Polymers*, Springer Science & Business Media, Berlin, Germany, 1999.
48. H. Pasch and B. Trathnigg in *Multidimensional HPLC of Polymers*, Springer Science & Business Media, Berlin, Germany, 2013.
49. W. Radke and J. Falkenhagen in *Liquid Interaction Chromatography of Polymers*, Elsevier, Amsterdam, The Netherlands, 2013, Chapter 5.
50. M.C. McMaster in *LC/MS: A Practical Users Guide*, John Wiley & Sons, New York, NY, USA, 2005.
51. W.M.A. Niessen in *Liquid Chromatography-Mass Spectroscopy*, 3rd Edition, CRC Press, Boca Raton, FL, USA, 2006.
52. K. Ute in *Encyclopedia of Analytical Chemistry*, John Wiley & Sons Ltd, New York, NY, USA, 2006, DOI:10.1002/9780470027318.a2033.
53. S. Mori and H.G. Barth in *Size Exclusion Chromatography*, Springer Science & Business Media, Berlin, Germany, 2013.
54. C-S. Wu in *Handbook of Size Exclusion Chromatography and Related Techniques: Revised and Expanded*, CRC Press, Boca Raton, FL, USA, 2003.
55. M.J. Forrest in *Rubber Analysis – Polymers, Compounds and Products*, Rapra Review Report No.139, Smithers Rapra, Shawbury, UK, 2001, p.10.
56. J.G. Kreiner and W.C. Warner, *Journal of Chromatography A*, 1969, **44**, 315.
57. J.D. Menczei and R.B. Prime in *Thermal Analysis of Polymers: Fundamentals and Applications*, John Wiley & Sons, New York, NY, USA, 2014.
58. W.M. Groenewoud in *Characterisation of Polymers by Thermal Analysis*, Elsevier, Amsterdam, The Netherlands, 2001.
59. V.A. Bershtein and V.M. Egorov in *Differential Scanning Colaorimetry of Polymers: Physics, Chemistry, Analysis, Technology*, Ellis Horwood, Hemstead, UK, 1994.
60. M.O. Abou-Helal and S.H. El-Sabbagh, *Journal of Elastomers and Plastics*, 2005, **37**, 4, 319.
61. Z. Qixia, F. Hong, B. Zhiyang and L. Bogeng, *China Synthetic Rubber Industry*, 2004, **27**, 3, 1000.
62. S. Cook, S. Groves and A.J. Tinker, *Journal Rubber Research*, 2003, **5**, 3, 121.
63. C. Price in *The Rapra Collection of DSC Thermograms of Semi-crystalline Thermoplastic Materials*, Rapra Technology Ltd, Shawbury, UK, 1997.
64. M. Reading and D.J. Hourston in *Modulated Temperature Differential Scanning Colorimetry: Theoretical and Practical Applications in Polymer Characterisation*, Springer Science & Business Media, Berlin, Germany, 2006.

65. *Modulated DSC Compendium; Basic Theory and Experimental Considerations*, TA Instruments Application Note TA210, New Castle, DE, USA, 1996.
66. *Modulated DSC^{TM} Theory & Applications Basic Overview*, TA Instruments Application Note TA211, New Castle, DE, USA, 1995.
67. J.M. Hutchinson, *Journal of Thermal Analysis and Calorimetry*, 2003, **72**, 619.
68. D.J. Hourston and M. Song, *Journal of Applied Polymer Science*, 2000, **76**, 12, 1791.
69. T.F.J. Pijpers, V.B.E. Mathot and B. Goderis, *Macromolecular*, 2002, **35**, 3601.
70. P. Gabbott, P. Clarke, T. Mann, P. Royall and S. Shergill, *American Laboratory*, 2003, **35**, 16, 17.
71. K.P. Menard in *Dynamic Mechanical Analysis: A Practical Introduction*, 2nd Edition, CRC Press, Boca Raton, FL, USA, 2008.
72. H.M. Pollock and A. Hammiche, *Journal of Physics D: Applied Physics*, 2001, **34**, R24.
73. *Polymer Blend Study by Micro-Thermal Analysis*, TA Instruments Application Note TS58, New Castle, DE, USA. *http://www.tainstruments.com/pdf/literature/Hotline_99v1.pdf.*
74. *Micro-Thermal Analysis - An Holistic Approach to Materials Characterization*, TA Instruments Application Note TA257, New Castle, DE, USA.
75. H.M. Pollock and A. Hammiche, *Journal Physics D: Applied Physics*, 2001, **34**, R23.
76. M.J. Forrest in *Proceedings of Polymer Testing 1997*, 7–11th April Rapra Technology Ltd, Shawbury, UK, 1997, Paper No.5.
77. *Modulated Thermogravimetric Analysis: A New Approach for Obtaining Kinetic Parameters*, TA Instruments Application Note TA237, New Castle, DE, USA.
78. *Obtaining Kinetic Parameters by MTGA*, TA Instruments Application Note TA251, New Castle, DE, USA.
79. C. Gamlin, M.G. Markovic, N.K. Datta, N.R. Choudhury and J.G. Matisons, *Journal of Thermal Analysis and Calorimetry*, 2000, **59**, 319.
80. *Thermal Analysis Technical Literature (Theory and Applications)*, TA Instruments, New Castle, DE, USA, 1994.
81. H. Lobo and C. Cohen, *Polymer Engineering and Science*, 1990, **30**, 2, 65.
82. F. Oehmke and T. Wiegmann in *Proceedings of ANTEC 1994*, 1–5th May, San Francisco, CA, USA, 1994, p.2240.
83. G.H. Michler in *Electron Microscopy of Polymers*, Springer Science & Business Media, Berlin, Germany, 2008.
84. L.C. Sawyer, *Polymer Microscopy*, Springer Science & Business Media, Berlin, Germany, 2012.
85. Y. Changjie, Q. Zhang, G. Junwei, Z. Junping, S. Youqiang and W. Yuhang, *Journal of Polymer Research*, 2011, **18**, 6, 2487.
86. D. Ma, J. Tan, Y. Li, C. Bian, G. Wang and S. Feng, *Journal of Applied Polymer Science*, 2014, **131**, 15, 7.
87. G.J. Vancso and H. Schonherr in *Scanning Force Microscopy of Polymers*, *Springer Science & Business Media*, Berlin, Germany, 2010.
88. V.V. Tsukruk in *Scanning Probe Microscopy of Polymers*, Eds., B.D. Ratner and V.V. Tsukruk, ACS – Division of Polymer Chemistry, Washington, DC, USA, 1998.
89. *Rubber Testing*, TA Instruments, New Castle, DE, USA, 2014.
90. P. Albers, B. Freund, G. Prescher, K. Siebold and S. Wolff in *Proceedings of the 2nd International Conference on Carbon Black*, 27–30th September, Mulhouse, France, 1993, p.119.
91. W. van Ooij and M. Nahmias in *Proceedings of the 134th ACS Fall Meeting*, 18–21st October, Cincinnatti, OH, USA, 1988, Paper No.26.
92. K. Harrison and L.B. Hazell, *Materials World*, 1994, **1**, 10, 532.

93. R.S. Lehrle and K.J. Niderost in *Proceedings of the Rubbercon 1992 Conference*, 15–19th June, Brighton, UK, 1992, p.281.

94. M. Dimopoulos, N.R. Choudhury, M. Ginic-Markovic, J. Matisons and D.R.G. Williams, *Journal of Adhesion Science and Technology*, 1998, **12**, 12, 1377.

95. E. Dujko and S.C. Cerelast in *Proceedings of the IRC Conference*, 12–14th May, Paris, France, 1998, p.87.

96. J. Jacks, *Adhesives Age*, 1997, **40**, 11, 36.

3 Principal components and bulk composition

3.1 Introduction

In this section, the principal components, or ingredients, of a rubber compound are defined as being:
- Polymer
- Plasticiser/oil
- Carbon black (CB)
- Inorganic filler

Work of the type described in this section yields important, fundamental information on a rubber sample and so it can be carried out for a number of reasons, including:
- Characterisation of individual component(s) for identification purposes.
- Obtain fundamental information as part of a research and development programme.
- Quality control.
- Defence of a patent.
- Reverse engineering.
- Failure diagnosis.

The information obtained from a bulk compositional analysis obviously plays an important part in a full characterisation (i.e., reverse engineering) study, but there are practical reasons why it can be regarded as having a stand-alone purpose and so be covered in a specific section of this book. This is because for a number of the exercises listed above, it is often convenient to look upon reverse engineering work as comprising of two distinct stages, as defined below:
- Stage 1: Determination of bulk composition and the identity of the principal components.
- Stage 2: Determination of additives, such as antidegradants, curatives and accelerators.

This chapter covers stage 1, Chapter 4 deals with stage 2, and Chapter 5 presents guidance and examples of how the various approaches and techniques can be brought together to achieve the best results in reverse engineering and product deformulation studies.

One of the practical reasons why splitting the work into two stages is useful is because failure diagnosis work can often reach a successful conclusion by carrying out the relatively cost effective and quick tests that comprise stage 1, and to have set off a comprehensive analysis programme would have wasted time and money.

https://doi.org/10.1515/9783110640281-003

Unfortunately, it is often the case that rubber products fail in service because the wrong type of rubber has been used or the level of one of the principal components is not correct. If the initial work shows that nothing obvious is wrong with the bulk composition, then a more in-depth analysis can proceed to target those aspects of the formulation that are more likely to provide a solution to the problem given the nature of the failure.

It will be evident that splitting the work into two stages, rather than setting off a comprehensive analysis programme right from the start, can also be advantageous in other instances, particularly if a large number of samples is concerned and some 'screening' has to take place to make the task more manageable.

The work that can be carried out to identify, quantify and characterise the principal components in a rubber is described in Sections 3.2–3.7 and then the approaches that can be used to obtain an overall measure of the bulk composition of a rubber sample by a single operation [e.g., thermogravimetric analysis (TGA) analysis], are brought together in Section 3.8.

3.2 Identification and quantification of the polymer in rubber compounds

One of the most basic and useful tests to carry out on a rubber sample is to determine the generic type of polymer (or blend of polymers) that is present. This has resulted in some useful textbooks being published that address this particular aim [1].

Because a typical rubber compound contains only ≈50% polymer, a direct analysis by the most convenient technique, attenuated total reflectance (ATR) Fourier–Transform infrared (FTIR) spectroscopy, rarely provides a definitive answer. It is also the case that often the majority of the work carried out by laboratories is on samples that have been vulcanised and so are insoluble in solvents, meaning that transmission FTIR work cannot be carried out.

In order to obtain an FTIR spectrum, a certain amount of preparation work is required to remove the interference of the organic additives. After milling or commuting the sample to increase the surface area, a solvent extraction is performed to remove organic additives such as plasticisers/oils (Section 3.4.1) and the species related to the cure system and antidegradant systems. The extracted portion of the sample can then be analysed by an ATR route (e.g., using a single bounce ATR accessory having a diamond window), although the data obtained will still be compromised by the absorption of the infrared (IR) light by any CB present, and any inorganic fillers will contribute IR absorption bands to the sample spectrum.

The usual practice is, therefore, to carry out an additional preparation step to remove the influence of the CB and the inorganic fillers and additives, such as zinc oxide (ZnO). This is achieved by pyrolysing the solvent extracted portion (e.g., in a modified ignition tube) and to record a FTIR spectrum of the

RP28.0501 Urethane castable ether

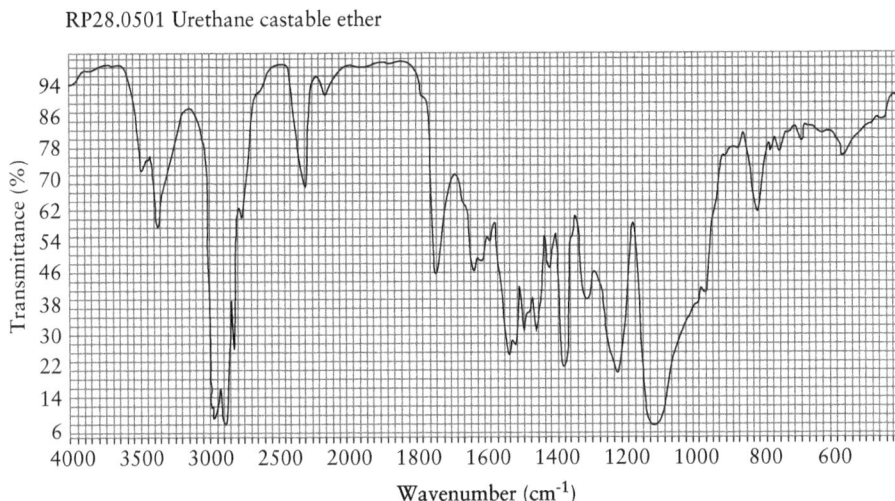

Figure 3.1: Pyrolysate IR spectrum of a polyether-type castable elastomer. Reproduced with permission from M.J. Forrest in *Chemical Characterisation of Polyurethanes*, Rapra Review Report No.108, Smithers Rapra, Shawbury, UK, 1999, p.19. ©2001, Smithers Rapra [3].

collected pyrolysis condensates. One example of a pyrolysate FTIR spectra of a rubber, a natural rubber (NR) is provided in Chapter 2 (Figure 2.1), and another, that of a castable polyurethane (PU) rubber, is shown in Figure 3.1. Providing that an extensive database of pyrolysate spectra is available, for example, the one published by Forrest and co-workers [2], this method can be very successful in both differentiating between generic polymer types [e.g., NR and styrene-butadiene rubber (SBR)]. Also, in some cases, it can show the differences between the grades of a polymer, such as the copolymer, terpolymer, and tetrapolymer versions of fluorocarbon rubber. It is also possible to use the technique to identify certain rubber blends, particularly blends of diene rubbers [e.g., SBR and polybutadiene (BR)], and the results can be expressed semi-quantitatively if the appropriate standards are available.

In cases where the rubber is not vulcanised or compounded, the situation is obviously much simpler and either the sample can be analysed directly using an ATR technique or by transmission IR if a suitably thin film can be prepared. It may be necessary to dissolve an amount of the sample in a suitable solvent (e.g., chloroform) and obtain a transmission IR spectrum from a cast film. Identifications can then be made by reference to FTIR libraries containing transmission and ATR spectra [2]. The transmission approach can also be used to obtain information on the microstructure of a rubber (Section 3.3.2). An example of a transmission FTIR spectrum of a rubber, NR, is shown in Figure 2.2 in Chapter 2, and that of a millable-type polyether-type PU rubber, is shown in Figure 3.2.

RT28.0801 Urethane millable

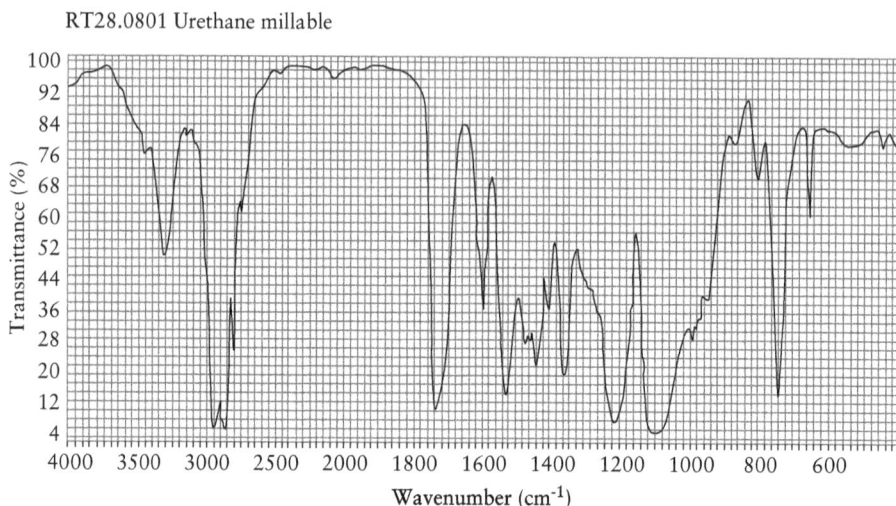

Figure 3.2: Millable polyether-type PU rubber. Reproduced with permission from M.J. Forrest in *Chemical Characterisation of Polyurethanes*, Rapra Review Report No.108, Smithers Rapra, Shawbury, UK, 1999, p.15. ©2001, Smithers Rapra [4].

Another method which enables the polymer type within a rubber sample to be identified is pyrolysis gas chromatography (GC). This method can be applied to either the extracted portion of the sample or the sample as-received. In the latter, the relatively low-molecular weight(s) (MW) oils elute early on and do not tend to have a significant influence on the pyrogramme as a whole. However, to avoid complications it is advisable to use samples that have been solvent-extracted. As with pyrolysis IR spectroscopy, a database of standards needs to be put together and then the pyrogram fingerprints of unknowns can be matched.

The temperature that the pyrolysis unit is set at, and the design of the unit (e.g., Curie point or microfurnace), will affect the pyrograms produced from a particular sample. As with pyrolysis FTIR, it is possible to identify and quantify blends and blend ratios. As an illustration of the data produced by this approach, reference fingerprint GC pyrograms for SBR and NR obtained at 750 °C are shown in Figures 3.3 and 3.4, respectively, and typical experimental conditions for pyrolysis GC experiments are shown below.

General requirements for pyrolysis GC are:
- Sample weights: 0.1–5.0 mg.
- Pyrolysis temperatures: 400–800 °C (microfurnace and Curie point pyrolysers).
- 800–1,200 °C (platinum filament pyrolyser).

Microfurnace-type pyrolyser – specific conditions using a commercial example: a SGE™ Pyrojector:

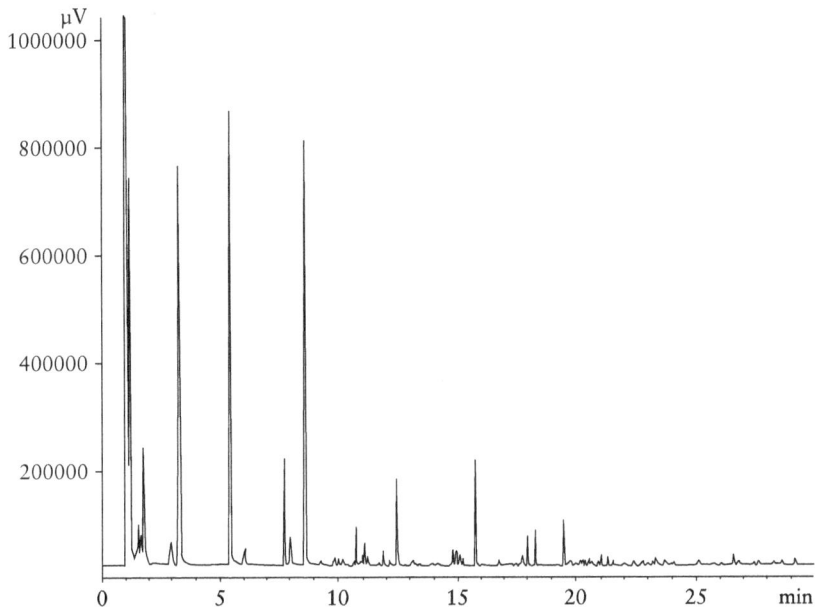

Figure 3.3: GC pyrogram obtained at 750 °C for SBR. Reproduced with permission from M.J. Forrest in *Rubber Analysis – Polymers, Compounds and Products*, Rapra Review Report No.139, Smithers Rapra, Shawbury, UK, 2001, p.16. ©2001, Smithers Rapra [5].

– Weight of sample: 0.4 mg.
– Pyrolysis temperature: 600 °C.
– Pyrolysis gas: helium at 15 psi.
– GC column: 5% diphenyl siloxane 30 m × 0.25 mm, 1.0-μm film.
– GC temperature programme: 2 mins at 50 °C followed by 10 °C/min to 280 °C and then 10 min at 280 °C.
– GC carrier gas: helium at 10 psi.
– GC injector temperature: 320 °C.
– Detector: flame ionisation detector at 350 °C.

Adding a mass spectrometer to the system so that the technique becomes pyrolysis GC–mass spectrometry (MS) further increases the usefulness of the method and the information that can be obtained. For example, Table 3.1 lists the principal diagnostic pyrolysis species that can be detected in the pyrograms of some common rubbers.

It can be seen from Table 3.1 that rubbers will often regenerate monomers and yield dimers and other oligomers upon pyrolysis under non-oxidising conditions (e.g., under helium). However, care needs to be taken in the interpretation of

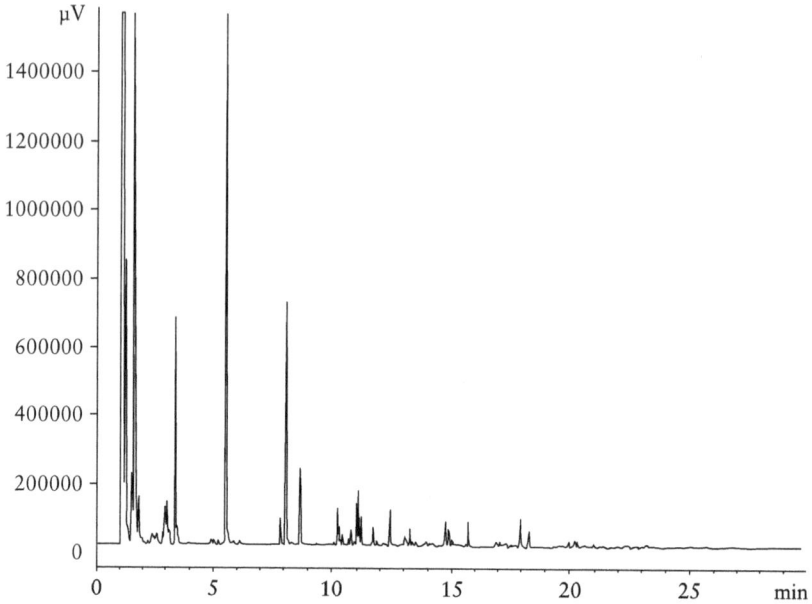

Figure 3.4: GC pyrogram obtained at 750 °C for NR. Reproduced with permission from M.J. Forrest in *Rubber Analysis – Polymers, Compounds and Products*, Rapra Review Report No.139, Smithers Rapra, Shawbury, UK, 2001, p.16. ©2001, Smithers Rapra [5].

Table 3.1: Principal diagnostic pyrolysis species for common rubbers.

Rubber type	Diagnostic pyrolysis species
PI rubber and NR	Isoprene and dipentene
SBR	Butadiene, 4-vinyl-1-cyclohexene and styrene
BR	Butadiene and 4-vinyl-1-cyclohexene
Isobutylene-isoprene rubber	Isobutene and isoprene
NBR	Butadiene, acrylonitrile and 4-vinyl-1-cyclohexene

NBR: Nitrile rubber PI: Polyisoprene
Reproduced with permission from M.J. Forrest in *Rubber Analysis – Polymers, Compounds and Products*, Rapra Review Report No.139, Smithers Rapra, Shawbury, UK, 2001, p.17. ©2001, Smithers Rapra [6]

the results, as the presence of various 'monomer type' substances in the following list of the pyrolysis products of a polychloroprene (CR) rubber demonstrate:

- Chloroethene
- Butadiene
- Chloroprene
- Toluene

- Styrene
- Chlorobenzene
- 1-Chloro-3-methylbenzene
- 4-Chloro-1,2-dimethylbenzene

In the case of the aromatic compounds in this list, these non-chlorinated species originate from the loss of hydrogen chloride (HCl) followed by cyclisation. Common pyrolysis products can cause problems in trying to detect amounts of, for example, SBR, in CR. Hence, it is important to target 'secondary' diagnostic pyrolysis products as well as 'primary'.

In addition to identifying the polymer type of a rubber, pyrolysis GC can also be used to quantify the polymers in a blend. Fuh and Wang [7] used the peak ratio of 1-chloro-4-(1-chloroethenyl)-cyclohexene from CR, and benzonitrile from NBR, to accurately quantify the two polymers in standard samples having blend compositions varying from 80:20 to 20:80.

A review covering the use of pyrolysis IR spectroscopy, pyrolysis GC, solid-state magic-angle spinning ^{13}C-nuclear magnetic resonance (NMR) and 300-MHz proton NMR to identify polymers in rubber samples was presented at the International Tyre Conference in Ohio in 1996 [8]. In addition to reviewing the subject, the analytical results obtained using the four techniques on six prepared rubber samples by twelve laboratories were detailed.

In addition to being used to identify the polymer type, solid-state ^{13}C-NMR can also be used in conjunction with physical testing data to characterise crosslinked rubber samples [9]. The data obtained at a constant cure temperature and time, but with a varying amount of sulfur curative, was found to correlate well. However, keeping the sulfur level constant but varying the cure temperature and time was found to produce large changes in the results obtained by physical testing (e.g., tensile strength, elongation and modulus), but led to only small differences in the ^{13}C-NMR data.

It is also possible to obtain an indication of the type of rubber in a sample using dynamic mechanical analysis (DMA) and utilising its high sensitivity with regard to recording the glass transition temperature (T_g). As the rubber sample is heated from well below sub-ambient (e.g., −150 °C) to above ambient (e.g., +200 °C) temperatures, a major change in its modulus occurs as it passes through its T_g due to segmental rotation taking place within its polymer molecules. This change is most easily detected by recording the change in the tan δ plot (i.e., loss modulus/ elastic modulus) and the peak in this plot is a measure of the T_g of the rubber. This is demonstrated in Figure 3.5, which shows the DMA thermogram of a CR-based rubber sample.

To carry out this type of work, it is necessary to have a reference list of the T_g of a range of rubbers. An example of such a list, obtained using differential scanning calorimetry (DSC) data, is provided in Table 3.2. However, care needs to be taken

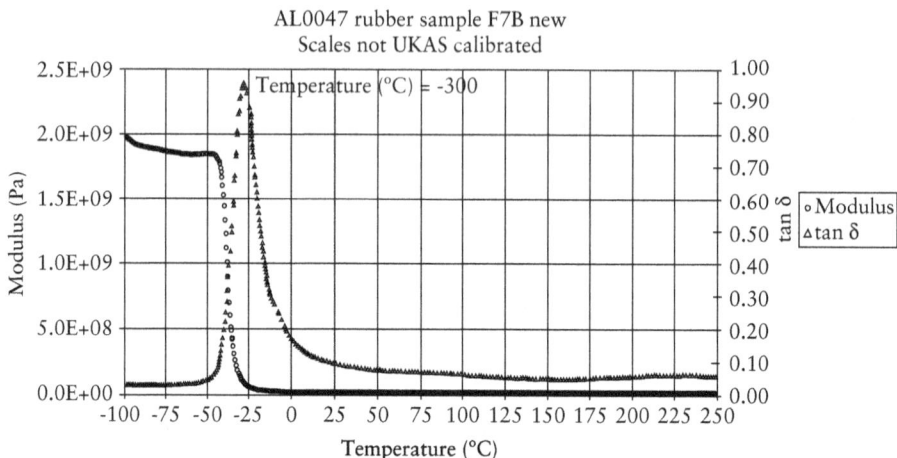

Figure 3.5: Typical DMA trace for a rubber sample (a CR rubber). The modulus shows a significant reduction at T_g, while tan δ shows a large peak. Reproduced with permission from M.J. Forrest in *Principles and Applications of Thermal Analysis*, Ed., P. Gabbot, Blackwell Publishing, Oxford, UK, 2008, p.214. ©2008, Blackwell Publishing [10].

because there are no definitive values for T_g because the value obtained for a given rubber or plastic is dependent on both the type of test that is performed (e.g., DMA, DSC, torsion pendulum) and the precise operating conditions of the instrument (e.g., the heating rate, rate of deformation). It is, therefore, important, as with so many other analytical procedures, to standardise the DMA conditions and run standard samples in order to make the fullest use of this capability.

The value for a T_g recorded by DMA will also be dependent to varying degrees on sample-related factors, such as the presence and level of a plasticiser, because the value reflects the T_g of the complete sample and not just the polymer that is present. However, even so, it can often provide a useful indication of polymer type, and is particularly valuable in detecting if a blend of polymers is present. An example of this type of application is shown in Figure 3.6, in which the DMA thermogram of a blend of NBR and SBR is shown. The ratio of the two rubbers in this blend was NBR (85%):SBR (15%) and so, as mentioned above, by the use of such standard materials, it is possible to use DMA to obtain the approximate blend ratio of an unknown sample. Depending upon the particular blend of polymers present, in can be possible to detect the presence of a polymer in a blend down to ≈5% of the total polymer fraction. Due to this capability, carrying out DMA can often corroborate and help to reinforce the data generated by FTIR.

One specific area where DMA is also very useful is in differentiating between polymeric materials and samples that have the same monomers in their polymer

Table 3.2: Approximate T_g values for a range of commercial rubbers in the uncompounded, uncured state obtained using DSC.

Rubber Type	Approximate values for T_g (°C)
Polyisobutylene	−70
EPM and EPDM	−65
Fluorocarbon	−20
Hydrin	−50
Polydimethyl siloxane	−120
BR	−100
CR	−45
Butyl rubber	−70
Halogenated butyl rubber	−70
Synthetic polyisoprene	−70
NR	−70
NBR	−65 to 0*
SBR	−50
Polysulfide	−60
PU (ether and ester types)	−40**

* Depends upon level of acrylonitrile monomer
** Varies with chemical structure EPM: Ethylene propylene monomer
EPDM: Ethylene propylene diene monomer
Reproduced with permission from M.J. Forrest in *Rubber Analysis – Polymers, Compounds and Products*, Rapra Review Report No.139, Smithers Rapra, Shawbury, UK, 2001, p.54. ©2001, Smithers Rapra [16]

molecules, for example, between vulcanisable SBR and styrene-butadiene-styrene (SBS) thermoplastic rubbers. FTIR can struggle in such cases, both in terms of differentiating between these two distinct types of rubbers and if the characteristic absorption bands for styrene and butadiene units are present for other reasons (e.g., a blend of SBR and BR) or the presence of a styrene-butadiene thermoplastic resin in either SBR or BR. The differences in the T_g between these different types of styrene-butadiene based polymers (e.g., −40 °C for the SBR; −80 and +100 °C for the styrene and butadiene units in the SBS thermoplastic rubber, respectively) means that DMA can provide answers.

As indicated above with the example of styrene-butadiene resin, it is also possible to use DMA to detect the presence of a rubber within another matrix, such as a plastic. A common example of this situation is when BR has been mixed with polystyrene (PS) to produce a high-impact grade. The large difference between the T_g of the two polymers (−80 and +100 °C, respectively) and its high sensitivity makes the technique ideal for this type of investigation. Another example of DMA being used to detect the presence of a rubber within a rubber–plastic blend is shown in the DMA thermogram in Figure 3.7. In this case, butyl rubber and a semi-crystalline

Mix C-NBR 85%: SBR 15%
Scales not UKAS calibrated

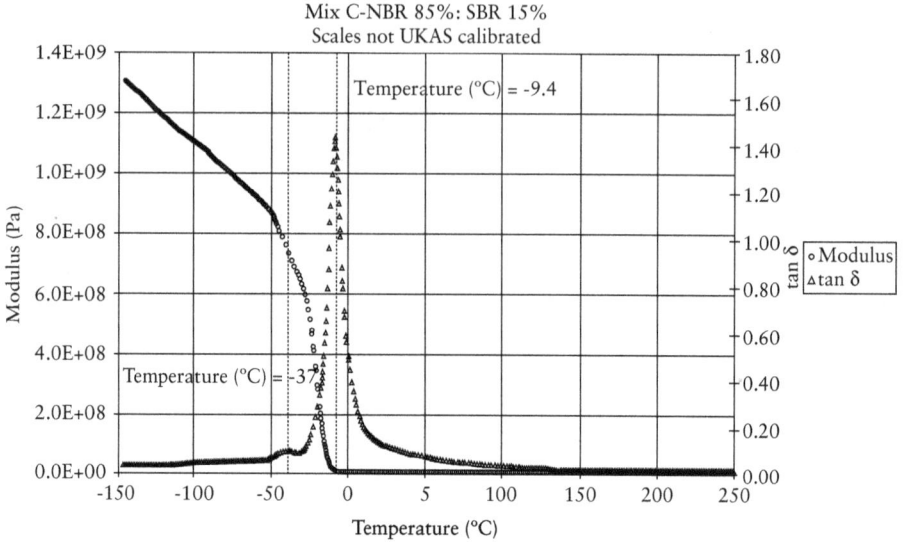

Figure 3.6: DMA trace showing the presence of 15% SBR within a NBR matrix – see the small peak in the tan δ just above –50 °C. Reproduced with permission from M.J. Forrest in *Principles and Applications of Thermal Analysis*, Ed., P. Gabbot, Blackwell Publishing, Oxford, UK, 2008, p.218. ©2008, Blackwell Publishing [11].

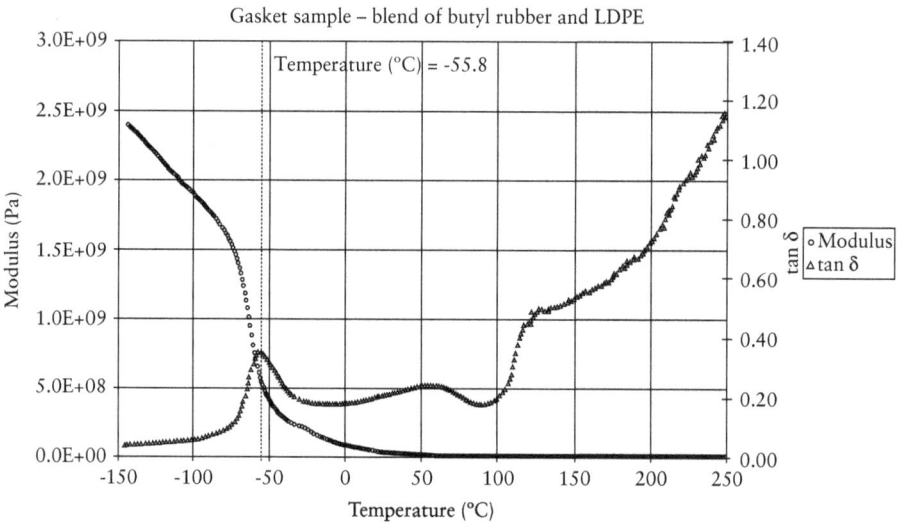

Gasket sample – blend of butyl rubber and LDPE

Figure 3.7: DMA thermogram of a 50:50 blend of butyl rubber and LDPE. The butyl rubber supports the material through the melting range of the LDPE. Reproduced with permission from M.J. Forrest in *Principles and Applications of Thermal Analysis*, Ed., P. Gabbot, Blackwell Publishing, Oxford, UK, 2008, p.222. ©2008, Blackwell Publishing [12].

thermoplastic, low-density polyethylene (LDPE), have been mixed in a ratio of 50:50. The data plotted in the tan δ curve show very clearly the T_g of the butyl rubber phase (about −56 °C) and the softening and melting of the LDPE phase (≈60 to 120 °C). It was possible to continue to record data on the sample after the LDPE phase had melted due to the presence of the butyl rubber; a sample of pure LDPE would have lost dimensional stability once past its melting point and dropped from the instruments grips.

DMA can also be used as a means of characterising blends of rubbers in terms of their compatibility, degree of mixing, and of assessing which rubber is present as the dispersed phase and which as the continuous phase [13, 14]. Another application involves its use to investigate the reinforcement of rubbers by different fillers and the influence of coupling agents in this process [15].

Another common thermal technique that can also be used to detect the T_g is DSC (Table 3.2). This is done by looking for a characteristic inflexion in the heat capacity curve (i.e., specific heat curve) within the thermogram. A similar temperature range is employed as that used for DMA and again standard materials should be analysed to obtain in-house reference data. Because the change in the heat capacity of a sample as it moves through its T_g is far less pronounced than the change in its modulus, the DSC is a much less sensitive technique than DMA for this type of work. This makes the identification of polymers in blends much harder, particularly if one of the polymers is only present at the relatively low level (e.g., ≤20%). A guide to the relative sensitivity of different thermal analytical techniques to detect the T_g is shown in Table 3.3.

Table 3.3: Relative sensitivity of a range of thermal analytical techniques for the detection of T_g.

Technique	Measured parameter	Relative sensitivity*
DSC	Heat capacity	1
Thermal mechanical analysis	Coefficient of expansion	5
DMA	Modulus	10
Dielectric analysis	Permittivity	10

* where a value of 10 represents the most sensitive
Reproduced with permission from M.J. Forrest in *Principles and Applications of Thermal Analysis*, Ed., P. Gabbot, Blackwell Publishing, Oxford, UK, 2008, p.236. ©2008, Blackwell Publishing [17]

With respect to the quantification of the polymer component within a rubber, the easiest technique to use is TGA. In a number of cases it is also possible to detect a blend of polymers using TGA and to obtain a semi-quantification of the blend ratio by reference to the splitting of the weight loss derivative. Unfortunately, some rubber types (e.g., CR) undergo two-stage weight loss when heated and the first stage (loss of HCl in the case of CR) will overlap with the weight loss due to the volatilisation of any low-MW organic substances in the rubber (e.g., plasticiser). It is, therefore, difficult to use TGA alone to determine the polymer content of these samples and other analytical techniques (e.g., elemental analysis) often have to be used as well.

Another complicating factor is that a number of rubbers (e.g., halogenated rubbers, NBR, acrylic rubbers) do not completely pyrolyse in the nitrogen atmosphere region and leave a carbonaceous residue (Table 3.4) which is lost due to oxidisation in the air atmosphere region. This carbonaceous residue, which may also overlap to a degree with the CB weight loss in this region, has to be included in order to obtain an accurate value for the polymer content. It is, however, possible, in some limited, specific cases involving co-polymers, to use the fact that some polymers give a carbonaceous residue during TGA run to obtain some approximate quantitative information on the monomer ratio and, hence, the generic 'grade' of rubber is present. One classic example of this is NBR, a copolymer of acrylonitrile (ACN) and butadiene. NBR is available in a range of grades in which the ACN level can vary from ≈20 to ≈50% according to the level of oil resistance, low-temperature flexibility and other properties that are desired. Any carbonaceous residue obtained in the TGA trace is only derived from the ACN units in the NBR and so the amount, adjusted for the total amount of polymer in the rubber compound, is indicative of the level of this monomer (Table 3.4).

Table 3.4: Carbonaceous residues obtained from a range of commercial rubbers by TGA.

Polymer type	Carbonaceous residue (%)*
Chlorosulfonated PE rubber	2–4
Fluorocarbon rubber	3–10
Ethyl acrylate rubber	6–8
NBR (20–50% ACN)	2–12
Epichlorohydrin rubber	5–15
CR rubber	20–25
PVC (used in blends with NBR)	15–20

* Carbonaceous residue obtained from 100% polymer and a range is given because the exact amount obtained depends upon the actual grade of polymer
PE: Polyethylene
PVC: Polyvinyl chloride
Reproduced with permission from M.J. Forrest in *Principles and Applications of Thermal Analysis*, Ed., P. Gabbot, Blackwell Publishing, Oxford, UK, 2008, p.204. ©2008, Blackwell Publishing [18]

The use of TGA to determine the polymer content of a rubber sample and provide an indication of the type of polymer that is present within it is described in more detail in Section 3.8. This section also describes how TGA can be used to determine the bulk composition of rubber samples.

3.3 Characterisation of rubbery polymers

Section 3.2 covered the determination of polymer type, one of the most basic requirements of rubber analysis, and this section will cover the analytical techniques that can be applied both to raw, uncompounded polymers and fully compounded materials in order to obtain a fuller characterisation (e.g., MW, ratio of monomers) of the polymer.

3.3.1 Determination of molecular weight

In common with other polymers, rubbery polymers have a distribution of MW [19]. This distribution will be widest for polymers produced by free-radical polymerisation (e.g., NBR) and narrowest for ionic polymers such as EPDM rubbers.

It is possible to work with both raw (i.e., uncompounded) rubbers, as well as compounded, unvulcanised ones, although the production of an insoluble, infinite network by vulcanisation obviously excludes the analysis of cured rubbers.

There are a number of ways in which the molecular weight distribution (MWD) of a polymer can be expressed:

$$\text{Mn} = \frac{\sum \text{NM}}{\sum \text{N}} \tag{3.1}$$

$$\text{Mw} = \frac{\sum \text{NM}^2}{\sum \text{NM}} \tag{3.2}$$

$$\text{Mz} = \frac{\sum \text{NM}^3}{\sum \text{NM}^2} \tag{3.3}$$

where 'N' is the total number of molecules of molecular mass 'M' in all cases; Mn is the number average molecular weight, Mw is the weight average molecular weight, and Mz is the Z average molecular weight.

The 'Z' average is very sensitive to high-MW molecules within the polymer. The weight average is always greater than the number average. A measure of the breadth of the molecular distribution [i.e., polydispersity (PDI)] of the polymer can be obtained by dividing the weight average by the number average:

$$\text{PDI} = \text{Mw}/\text{Mn} \tag{3.4}$$

As a polymer tends towards a uniform MWD, its PDI approaches unity and it is said to be 'monodisperse'.

There are a wide variety of methods that can be used to obtain the MW. The principal ones are described in Sections 3.3.1.1–3.3.1.4.

3.3.1.1 Gel permeation chromatography/size exclusion chromatography

This is the pre-eminent technique and has been refined and developed extensively over the past 45 years. It is possible to use gel permeation chromatography (GPC) for both raw polymers and compounded ones, and Mn, Mw and Mz values can be obtained. GPC is essentially a liquid chromatography (LC) technique with the sample being dissolved in a suitable solvent [e.g., toluene for SBR; tetrahydrofuran (THF) for NBR] and injected into the system where the molecules are separated according to their molecular size by a column that contains a swollen, particulate, porous gel packing.

The standard detector for GPC systems is the refractive index (RI) detector, but double (i.e., RI and viscosity) or triple (i.e., RI, viscosity and light scattering) detection systems have been developed recently. These multiple detection systems provide more accurate measurements of the MW by obtaining actual Mark–Houwink parameters, *via* the viscosity detector (Section 3.3.1.2), for the polymer being analysed, as opposed to PS-equivalent MW from the sole use of monodisperse PS calibrants. In addition to the advantage that the viscosity detector brings to GPC, an inherent property of the light-scattering detector is that it provides 'true' MW data. Equipping a GPC system with triple detection also has other advantages, such as the capability to provide structural information on polymers.

It is also possible to use spectroscopic detectors such as IR and ultraviolet (UV) with GPC systems. One practical use of the latter is to determine if chemical modifiers (e.g., silane coupling agents) are bound to polymer molecules. This is done by measuring the MWD of the modified rubber at a UV wavelength where the rubber itself does not contribute to the absorbance and comparing this with the distribution of the unmodified rubber.

There can be two problems encountered with the GPC of rubbers. The first is that rubbers can have relatively high MW (>1,000,000) and care has to be taken that the flow rates used with modern small particle size packings (i.e., <10-μm diameter) do not cause the rubber molecules to break up under the imposed shear forces. Flow rates <0.5 ml/min may have to be used to minimise back pressure. The other problem is that free radical-polymerised rubbers, and NR, can have a degree of insoluble-crosslinked gel associated with them.

GPC is often carried out on a comparative basis and Figure 3.8 illustrates this, where the GPC chromatograms of two NBR samples produced using THF as the solvent for the samples and the mobile phase have been overlaid. In this case, the work was carried out because it was thought that differences in processing performance were related to MW differences. However, the MW data computed from

Figure 3.8 are shown in Table 3.5 and it can be seen that the difference between the samples is relatively insignificant.

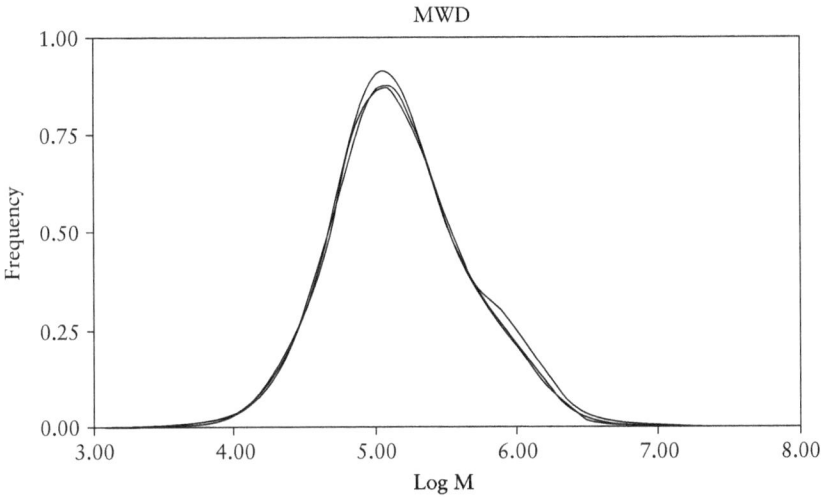

Figure 3.8: Overlaid GPC chromatogram of two NBR. Reproduced with permission from M.J. Forrest in *Rubber Analysis – Polymers, Compounds and Products*, Rapra Review Report No.139, Smithers Rapra, Shawbury, UK, 2001, p.13. ©2001, Smithers Rapra [20].

Table 3.5: MW data for two NBR samples (Figure 3.8).

	Mw	Mn	PDI
Sample A	281,000	89,800	3.1
Sample B	331,000	92,000	3.6

Reproduced with permission from M.J. Forrest in *Rubber Analysis – Polymers, Compounds and Products*, Rapra Review Report No.139, Smithers Rapra, Shawbury, UK, 2001, p.13. ©2001, Smithers Rapra [20]

Some of the typical applications that can be conducted on rubbers by GPC/size exclusion chromatography (SEC) are described by Agilent Technologies in a recent Application Compendium [21]. This publication describes applications in which various GPC/SEC systems are used to obtain the MW and, in some cases, structural information, on various rubber samples. The applications that are covered include:

- High-sensitivity analysis of NR with evaporative light scattering (ELS) detection – This example demonstrates how ELS detection can improve the data that are obtained on NR samples because it eliminates the interference of the 'system peaks' that occur with standard RI detection.

- The analysis of NR and synthetic BR with triple detection – The example provided show how both MW data and structural information can be obtained on NR and BR samples using systems that are equipped with triple detection capability.
- General analysis of synthetic elastomers – Use of GPC/SEC systems with columns capable of resolving MW up to 10,000,000 Da to provide MW data on PI, BR, SBR and butyl rubbers is demonstrated.
- Polydimethylsiloxane analysis using GPC/viscometry – Illustrates how two silicone rubbers, although quite different in MW, are structurally similar and so their viscoelastic behaviour as a function of MW would be expected to be comparable.
- Analysis of SBS block copolymers – Demonstrates how 'conventional' GPC (i.e., THF solvent at 40 °C using RI detection) can be used to detect the presence of homopolymers, which can significantly affect the end properties in SBS thermoplastic rubbers.

Prior to the advent of advanced detection systems, the use of modular accessories like the LC–Transform with GPC extended its capabilities by allowing IR spectroscopy data to be generated on the molecules eluted during an analysis. The LC–Transform system worked by impinging the solute that elutes from a GPC column onto a germanium disc by removing the solvent using a combination vacuum and high temperature. The germanium disc is then placed into a module fitted into the sample compartment of an IR bench and IR data collected on the entire chromatogram using continuous collection software. This procedure, in conjunction with elution data obtained from a GPC chromatogram recorded on the same sample with an RI detector, enables the IR spectrum of any particular retention time, or range of retention times, to be viewed. This technique is very useful for compounded rubbers because it allows polymer blends to be identified, as well as relatively high MW additives such as plasticisers, and its full capabilities have yet to be replaced by a commercial in-line detection system.

3.3.1.2 Viscosity

Viscosity is a standard procedure for MW determinations and involves the use of specially designed viscometers to accurately measure the viscosity of a polymer solution. From this the intrinsic viscosity is determined and, hence, the MW. The time taken for the polymer solution to pass between two marks on the viscometer is compared with that of pure solvent and the ratio is the viscosity of the solution. Successive dilutions give a range of concentrations and times from which the intrinsic viscosity can be calculated. The value for this is then entered into the Mark–Houwink–Sakwada equation:

$$\text{Intrinsic viscosity} = KM^X \qquad (3.5)$$

where 'K' and 'X' are constants for the particular solvent being used and are referred to as the 'Mark–Houwink parameters'.

Depending on the source of the Mark–Houwink parameters, the MW can be expressed as either the number average or weight average.

3.3.1.3 Osmometry

There are two principal osmometry techniques [22, 23]: vapour pressure osmometry and membrane osmometry.

Vapour pressure osmometry involves the indirect measurement of the lowering of the vapour pressure of a solvent due to the presence of a solute. It is based on the measurement of the temperature difference between droplets of pure solvent and of the polymer solution maintained in an isothermal atmosphere saturated with the solvent vapour. Calibration involves the analysis of standards of known MW and should be over the entire MW of interest to ensure the best results. The technique is useful for polymers that have MW in the range 500–50,000 Da.

Membrane osmometry involves the use of a membrane between the solvent and the polymer solution that is permeable to the solvent molecules, but not to the solvated polymer molecules. This is the major restriction on the technique and means that it cannot be used for low-MW polymers (i.e., those below ≈50,000 Da).

3.3.1.4 Light scattering

Recent developments in instrumentation have led to a resurgence of this technique: the Mw is obtained from the light-scattering behaviour of polymer solutions. These developments have centred on low-angle laser light scattering [24], multi-angle laser light scattering (MALLS) [25, 26], and ELS. As well as being stand-alone instruments, light scattering detectors have been developed for GPC systems (Section 3.3.1.1). MALLS detectors have the additional benefit of being able to provide direct information on the branching ratio of a polymer [27].

To reduce variability, care has to be taken to ensure that the polymer solution is clear and low cell volumes (e.g., 0.1 µl) and dilute solutions help with this. Solvent choice is also critical, with the best results being obtained with solvents that have a very different RI to the polymer and have low-light scattering properties.

ELS involves the removal of the solvent from the solvated polymer as it elutes down a drift tube. The isolated polymer particles then scatter light from the light source.

Light scattering can produce precise data and has been used successfully on polymers with MW ranging from 10,000 to 10,000,000.

3.3.1.5 Miscellaneous techniques

Other methods of determining or investigating MW include:
- End group analysis – provides Mn data.
- Ultra-centrifugation – provides Mw data.
- Comparative 'indirect' techniques (e.g., DSC or DMA) that work by monitoring properties that can be influenced by MW (e.g., T_g and crystallinity).

3.3.2 Monomer types and microstructure

The use of pyrolysis FTIR spectroscopy and pyrolysis GC to determine the generic polymer type of a rubber, and hence information on the monomers used to produce it, is covered in Section 3.2. This section also mentions the use of FTIR in the transmission mode for identification of a polymer where the rubber is uncompounded or unvulcanised. As with all FTIR work, access to a comprehensive FTIR library is important and the example published by Forrest and co-workers [2], already mentioned in Section 3.2 in relation to its collection of pyrolysate spectra, also contains an extensive collection of transmission spectra. With respect to quantitative work, where appropriate standards are available it is also possible to use FTIR spectroscopy to quantify the amounts of monomers in copolymers and terpolymers, providing a reasonably large amount of each monomer is present.

Regarding the minor monomers (e.g., those at <10% *w/w*) FTIR in the transmission mode with uncompounded and unvulcanised rubber samples can, in some cases, enable qualitative information to be obtained. For example, in the case of the terpolymer EPDM, transmission FTIR can be used on the pure polymer to identify the diene monomer. However, if this rubber is compounded, it may not be possible to remove enough of the additives to stop them masking the diene monomer in the rubber and, if the sample is vulcanised, additional detection difficulties are encountered as a large proportion of the diene monomer will have reacted with the cure system components.

Other analytical techniques which are more sensitive than FTIR can be used for the determination of minor monomers, even in rubber samples that have been vulcanised. For example, Yamada and co-workers showed that high-resolution pyrolysis GC can be used to both identify and quantify ethylidene norbornene diene monomer in vulcanised EPDM rubbers [28].

To assist workers in their efforts to characterise the structural features present in *cis*- and *trans*-1,4-polybutadiene, Nallasamy and co-workers [29] analysed both these polymers using Raman spectroscopy and FTIR. The analyses were performed over the following frequency ranges:
- Raman spectroscopy: 100–4,000 cm^{-1}
- FTIR spectroscopy: 200–4,000 cm^{-1}

Tables are provided in this reference which list probable assignments for the absorption bands that the research team observed in the data, made using the help of intensities and normal co-ordinate analyses. The research team found that the calculated normal frequencies were in good agreement with those determined experimentally.

Another technique that is very effective at identifying the exact monomers present in a rubber is NMR. Although the best results are obtained on a solution of the pure polymer, it is possible to use the solid-state technique on vulcanised samples and also to analyse solvent extracts of these by ^{13}C-NMR; the chemical composition of the extracted high-MW oligomers usually provide an accurate enough reflection of the original polymer. The NMR technique will also provide information on the proportions of the various monomers in copolymers and terpolymers. In contrast to IR spectroscopy, this information can be obtained without having to analyse or having available polymer standards of known monomer composition. It is possible for experienced specialists to interpret the NMR data using fundamental principles and the areas under the diagnostic chemical shifts.

NMR instruments can also be used to obtain fundamental structural information on polymers such as end group chemistry, branch points and structural isomerism (e.g., *cis*- and *trans*-1,4-isomer ratios in diene rubbers). It is also possible to use FTIR spectroscopy for the latter example, although some of the isomeric structures have relatively weak absorptions which makes detection difficult if they are at a low concentration.

Two sets of workers who have carried out a lot of work on the use of solid-state ^{13}C-NMR to investigate the structure of vulcanised rubbers are Koenig and his team at Case Western Reserve University [30–32] and Gronski and co-workers at the University of Freiberg [33–35]. The low density of the crosslinks present in a vulcanisate challenge the sensitivity of the technique, but it has still been possible to distinguish modifications to the main chain (isomerism and cyclic sulfides), the crosslinks themselves, and the reactions of cure and reversion [36].

Sung-Seen Choi and co-workers [37] at Sejong University used both FTIR and H-NMR to determine the microstructural composition of unbound polymer (i.e., solvent extractable) that was present in filled SBR/BR blends. They concluded that H-NMR had fewer experimental errors than transmission FTIR. The research team used H-NMR to determine the 1,2- and 1,4-butadiene contents of the unbound rubber to obtain average SBR/BR ratios. The bound rubber microstructural composition in the filled SBR/BR blends was then obtained using a combination of the bound rubber content and the average unbound microstructural compositions. The results showed that the bound rubber had a higher proportion of SBR to BR than that present in the rubber formulation. Three of the authors of this work also carried out a study on sulfur-cured solution SBR, the aim of which was to obtain information on their microstructure [38]. The cured rubbers were analysed by both solution H-NMR and transmission FTIR and the results obtained were compared with that of raw solution SBR. The comparison revealed that the microstructures were different because of the pendent group formed by the cure accelerator and *cis–trans* isomerism. Also, the styrene and *trans*-1,4 unit contents of the cured SBR were larger than those of the raw SBR, whereas the 1,2 unit contents of the cured SBR tended to be smaller than those of the raw SBR. The research team also compared the H-NMR and FTIR results, and found that the NMR data had narrower error ranges and better correlation than that obtained by FTIR.

Shield and Ghebremeskel [39] investigated the possibility of using TGA to determine the styrene content of SBR. An alternative method to the traditional methods for determining the styrene content of SBR in tyre treads was thought desirable because these could be hampered by the presence of additives and other polymers in the samples. The approach that they used involved monitoring the shifts in the decomposition temperature of SBR as a function of the styrene content. It was found that the magnitude of the shift could be used to determine the styrene content of an SBR. The effect of emulsifiers, CB filler, curing agents and other polymers on this shift was investigated and ways of minimising their influence on the data were suggested.

For further information in this area, the reader is directed to the work of Brame, who provides a review on the use of spectroscopic techniques (NMR, FTIR and MS) for the determination of polymer structure [40].

3.3.3 Specific heat values

DSC can be used to determine the specific heat of polymers and the effect that bulk ingredients (e.g., CB) and the vulcanisation process have on these values. Such a study has been published by Changwoon Nah and co-workers [41] and encompassed a number of rubbers, including NR, SBR, BR and butyl rubber.

3.3.4 Crystallinity

Conventional rubbers are, to a very large extent, amorphous materials. This feature is essential if they are to exhibit the properties that define them as a class of material (e.g., very high elasticity and high resilience). Hence, analyses to acquire information on their crystallinity are relatively rare, but can be important in some cases. One of these cases is in the analysis of EPDM, both the conventional rubber and when it is blended with polypropylene (PP) to produce a thermoplastic elastomer (TPE). A commercial example of an EPDM/PP blend is marketed under the Santoprene™ tradename, and a DSC thermogram of a Santoprene™ TPE (grade 103–40) in the 'as-received' state (i.e., without any annealing) is shown in Figure 3.9. In this thermogram, the endotherm due to the semi-crystalline PP phase is very obvious and has a melting point maximum at 154.27 °C.

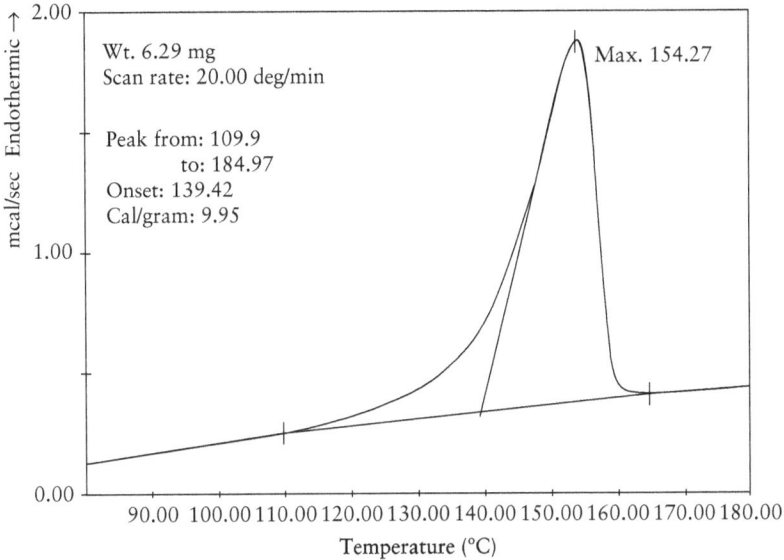

Figure 3.9: DSC thermogram of Santoprene™ 103-40 in the 'as-received' state. Reproduced with permission from M.J. Forrest, Y. Davies and J. Davies in *The Rapra Collection of Infrared Spectra of Rubbers, Plastics and Thermoplastic Elastomers*, 3rd Edition, Rapra Technology Ltd, Shawbury, UK, 2007, p.74. ©2007, Rapra Technology Ltd [2].

Within the structure of EPDM rubber there can be crystalline regions due to the ethylene component. The ratio of ethylene to propylene in EPDM can vary between the different grades and, as the amount of ethylene increases, the degree of crystallinity within the rubber increases and this increases its T_g. An application note published by PerkinElmer [42] describes how DSC can be used to characterise EPDM elastomers. In particular, it demonstrates how DSC can be used to show distinct and significant differences between two EPDM TPE samples in terms of crystalline content and their T_g values. The results displayed show how the EPDM with the greatest concentration of ethylene has a higher T_g, melting temperature and heat of melting; this is valuable information because the thermo-physical properties of the EPDM will affect the end use characteristics of the TPE.

Other rubbers that exhibit crystalline regions include members of the PU group. Figures 3.10–3.12 show the DSC thermograms obtained on a PU rubber in three distinct ways in order to characterise its morphology and obtain information on its 'heat history'. The following information is present in these three DSC thermograms:

- Figure 3.10 – 'As-received' DSC thermogram of the PU rubber sample providing information how its crystalline phase has been influenced by its processing (i.e., its heat history).

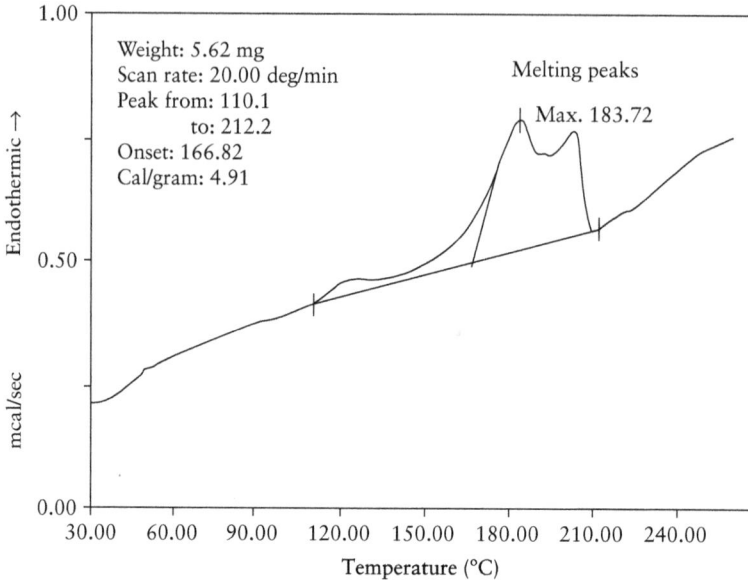

Figure 3.10: As-received DSC thermogram of a PU elastomer. Reproduced with permission from M.J. Forrest in *Chemical Characterisation of Polyurethanes*, Rapra Review Report No.108, Smithers Rapra, Shawbury, UK, 1999, p.23. ©2001, Smithers Rapra [43].

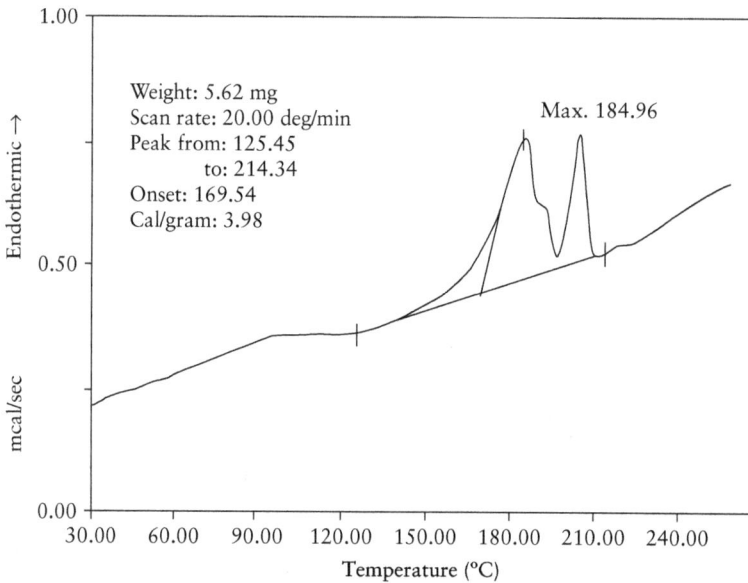

Figure 3.11: Reheat DSC thermogram of the PU elastomer (Figure 3.10) after slow cooling from the melt. Reproduced with permission from M.J. Forrest in *Chemical Characterisation of Polyurethanes*, Rapra Review Report No.108, Smithers Rapra, Shawbury, UK, 1999, p.23. ©2001, Smithers Rapra [43].

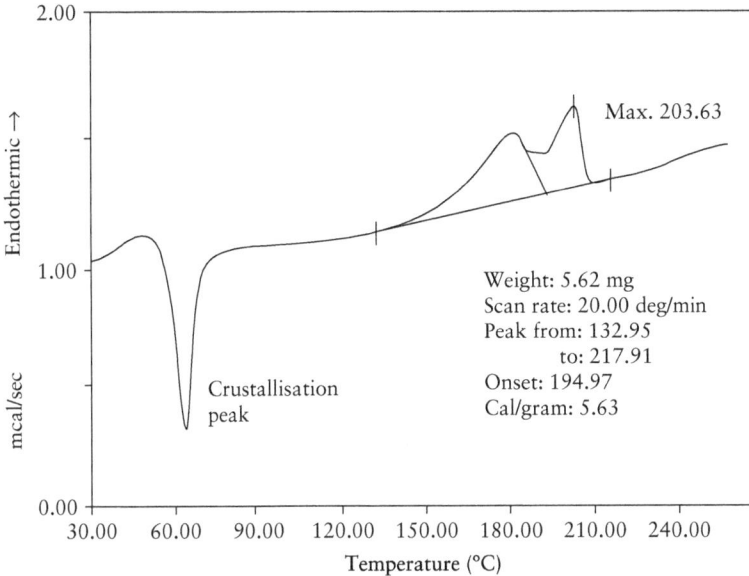

Figure 3.12: Reheat DSC thermogram of the PU elastomer (**Figure 3.10**) after quenching from the melt. Reproduced with permission from M.J. Forrest in *Chemical Characterisation of Polyurethanes*, Rapra Review Report No.108, Smithers Rapra, Shawbury, UK, 1999, p.23. ©2001, Smithers Rapra [43].

- Figure 3.11 – 'Reheat' DSC thermogram obtained after the PU rubber had been slowly cooled from the melt state to obtain information on its true melting characteristics.
- Figure 3.12 – Reheat DSC thermogram obtained on the PU rubber after it had been quenched from the melt state to obtain information on the formation of its crystalline phase (e.g., crystallisation temperature) and the subsequent melting behaviour of this phase.

3.3.5 Analysis of rubber blends

When two or more rubbers are blended together, in order to obtain the best properties from the resulting compound it is important to ensure that they are well mixed and analytical techniques can assist with monitoring this. For example, as the degree of mixing of two rubbers increases, this will influence a number of physical properties, one of which will be the T_g of the blend. Due to its high sensitivity to this property, this change can usually be detected by DMA and an assessment as to the degree of mixing can be obtained (Section 3.2).

Another important consideration when blending rubbers is to ensure that they are compatible with one another. Theraftil and co-workers [44] assessed the compatibility of blends of EPDM and chlorobutyl rubber (CIIR) in sulfur-cured compounds. The research team used a range of physico-chemical analytical techniques in their study, including tensile tests, DSC and DMA, and they also analysed the fracture surfaces of test pieces using scanning electron microscopy (SEM). The effects of adding chlorosulfonated PE as a compatibilisation agent on the phase separation of EPDM and CIIR was also evaluated by mechanical testing (e.g., tensile testing and tear testing) and by an examination of the fracture surface morphology by SEM.

DeCastro and co-workers [45] investigated the effect on dynamic mechanical properties that results from varying the sequence that additives are incorporated into a 50:50 blend of NR/BR. In their work, four addition sequences were used, the curing conditions were kept constant, and the blend samples were analysed by both DMA and SEM. Their work revealed that a selective partition of the additives within each rubber phase, and that this resulted in a correlation between mechanical performance and blend morphology.

It is common practice to blend rubbers with other polymers (e.g., plastics and resins) in order to create a material with enhanced properties (e.g., toughness). One typical example of such a blend is when rubber is blended with an epoxy resin. Nigram and co-workers [46] used a combination of thermal techniques (DSC, TGA and DMA) to study a blend of an epoxy cresol novolac resin that had been toughened using carboxyl-terminated nitrile rubber (CTNR). The research team used these techniques to obtain data on several properties of the blend. For example, they found that the exothermic heat of reaction due to crosslinking of the resin using diaminodiphenyl methane as the curing agent showed a decreasing trend as the rubber content of the material increased, and an increase in rubber content was also found to provide an increase in thermal stability and a decrease in the storage modulus. The loss modulus and loss tangent values of the material were found to initially increase as the temperature was increased, but then fell away, and addition of 10% of CTNR was found to give the maximum values for both the loss modulus and tan δ.

Jana and co-workers [47] at the Indian Institute of Technology used TGA to evaluate the thermal stability of blends of silicone rubber and LDPE that contained four ethylene copolymer compatibilisers. The TGA data showed that the blends underwent degradation in two stages and that it followed first-order reaction kinetics. The activation energy at 10% degradation, determined using the Freeman–Carroll method, was at a maximum for the 25:75 LDPE/silicone rubber blend, and this blend also yielded the maximum half-life at 200 °C (812.5 days) calculated using the Flynn–Wall method. Of the four compatibilisation agents evaluated, ethylene methyl acrylate copolymer (EMA) was found to function the most effectively. In a related project by members of the same research team [48] TGA and derivative TGA were used to study compatibilised 50:50 blends of silicone rubber and LDPE that contained different levels (0 to 15%) of EMA copolymer compatibiliser. The TGA and derivative TGA data of the different blend samples showed that thermal degradation took place in two stages, whereas pure LDPE showed a three-stage degradation, and a mixture of only EMA and silicone rubber degraded in a single step. The research team calculated the activation energies of degradation at the 10% point using the Freeman–Carroll method and found that the maximum value (171 kJ/mol) was obtained with the blend that contained 6% of EMA. This level of EMA was also found to give the maximum half-life period at 200 °C using the Flynn–Wall method of calculation.

Amash and co-workers [49] investigated the solubility behaviour of various elastomeric processing promoter additives in EPDM, NBR and PP using thermal analysis (e.g., DMA). The research team then prepared binary blends of rubber/PP-containing selected promoter additives and examined their phase morphology, compatibilisation and certain final properties that related to these characteristics. The results showed that the promoter additives acted as effective homogenising agents and contributed to a significant improvement in the dispersion and compatibility of the two blend components.

Blends are used extensively in the tyre industry (e.g., for tread compounds) and researchers are continually looking at how improvements in performance can be achieved. A research team from South Korea used DMA to characterise rubber compounds that contained 5 phr of various phenolic resins, some of which were modified resins that had been prepared specifically for this research work. Overall, the investigation showed that because alkyl phenolic resins have a higher softening point, phenolic resins that had been modified with either terpene or rosin showed potential as new additives for improving the grip properties of tyres, although not all the resins could enhance their rolling resistance, which is another important property.

Varghese and co-workers [50] used TGA to study the effect that a number of compounding variations had on the thermal behaviour of NBR/ethylene-vinyl acetate (EVA) blends. The following aspects of the formulation were altered during the course of their work:
- Ratio of NBR to EVA
- Filler type (e.g., silica, clay and CB)
- Amount of filler
- Cure system type (e.g., sulfur and peroxide)

With respect to the initial decomposition temperature, it was found to increase as the amount of NBR increased in the blend and the peroxide cure system provided the highest value, which was thought to be due to the high-bond dissociation energy of the carbon–carbon crosslinks. The peroxide-crosslinking system was also found to exhibit the best retention of properties upon heat ageing and the addition of fillers improved thermal stability. The data obtained by the research team on the blend samples included the activation energies of degradation and mass loss at different temperatures.

3.3.6 Degradation studies and changes in chemical structure

Aoyagi and Sano [51] analysed a fluorocarbon rubber shaft seal after 300,000 km of service by micro-hardness and ATR–FTIR. The workers found that the hardness of the lip seal increased due to both a structural change within the polymer and an ageing reaction within the material. The ATR–FTIR results showed that the peak width of the CF_2 absorption band had increased due to the ageing and the structural changes that had resulted. Simulation work to create the in-service environment showed that it was the additives within the engine oil that were mainly responsible for the degradation observed.

Under certain conditions, rubber products can fail in service due to a phenomenon known as 'blow-out' whereby gaseous substances produced within the product can escape in an explosive manner, causing it to rupture. A research team at Sejong University [52] used a microwave oven, along with GC–MS analysis of the volatile substances generated, to investigate the causes of blow-out from CB-filled NR and SBR samples. The species that were identified by GC–MS in the blow-out gas from the NR sample included 1-methyl-4-(1-methylethenyl) cyclohexene, and 4-vinylcyclohexene and styrene were detected from the SBR sample. These three substances were considered to be the thermal degradation products of the NR and SBR, respectively. Upon examination of the burst region of the sample, it was not possible to find evidence of any organic additives that had been used in the compounding stage, but a large number of hydrocarbon substances were detected for both samples. The research team concluded that the principal cause of blow-out was the decomposition of the polymer, which caused cavities to form in the samples, thereby reducing the resistance to the increased internal pressure.

Pappa and co-workers [53] analysed the volatile products that result from the thermal decomposition of NBR/PVC blends using TGA–MS. Such blends are used to manufacture products, such as forest fire hoses, and experimental compounds containing various inorganic fillers (e.g., magnesium hydroxide, organic-modified kaolin and a nanoclay material) were prepared to determine how they affected the nature of the decomposition products. The TGA–MS data revealed no significant differences between these experimental compounds and an unfilled NBR/PVC control compound, apart from the absence of HCl in the gaseous products from the blend that contained the magnesium hydroxide filler.

3.4 Determination of plasticisers or process oils

3.4.1 Quantification

Mention has been made in Section 3.2 of the isolation of plasticisers from a rubber matrix by solvent extraction. In that case, the objective was to clean up the polymer sample for further work (e.g., FTIR analysis for polymer identification). However, it can be an end in its own right and it is the subject of the International Organization for Standardization (ISO) standard, ISO 1407, which shows how it is possible to obtain accurate quantification of the plasticiser present in a rubber compound.

The choice of solvent is an important factor in obtaining accurate quantification of the plasticiser or oil and a good-quality extract for identification work (Section 3.4.2). Ideally a solvent should be used that is a good solvent for the plasticiser/oil to facilitate efficient removal, but a poor solvent for the vulcanised rubber polymer to avoid a high proportion of low-MW polymer being extracted. In reality, due to the similarities in the solubility parameter values of the plasticiser/oil and the polymer (essential if they are to be compatible) a compromise often has to be reached and some contribution from the polymer in the extract cannot be avoided completely, particularly as vulcanised rubbers are only lightly cross-linked systems.

The solubility parameters of a range of rubbers and common solvents are shown in Tables 3.6 and 3.7, respectively. In general, the closer the solubility parameter values of a rubber and solvent are, the more compatible they will be and the better the solvent will be for the rubber. These tables can also be used, therefore, to select a solvent to dissolve an unvulcanised rubber to create a solution (e.g., to enable a film to be cast for a transmission IR analysis).

Because they are viscous liquids with relatively low MW and are volatile within the temperature range 150–300 °C, the TGA technique can usually provide a good quantification of plasticiser or process oil content, particularly with synthetic plasticisers, such as phthalates and adipates. These substances are lost from a rubber sample during a TGA experiment by volatilisation and this is true whether or not an oxidising atmosphere is used in the experiment.

3.4.2 Identification

Having quantified the plasticiser/oil present by solvent extraction, it is then possible to obtain an identification by use of transmission FTIR spectroscopy. To assist with this process, there are extensive IR spectral databases available to the polymer analyst [42]. For example, the data shown in the spectra in Figures 3.13 and 3.14 are taken from standard spectra present in the in-house

Table 3.6: Solubility parameters of common solvents.

Solvent	Solubility parameter ($cal^{1/2}\,cm^{-3/2}$)
n-Hexane	7.0
Diethyl ether	7.4
Cyclohexane	8.2
Carbon tetrachloride	8.6
Xylene	8.8
Toluene	8.9
THF	9.1
Chloroform	9.3
Methyl ethyl ketone	9.3
Trichloroethane	9.6
Dichloromethane	9.7
Cyclohexanone	9.7
N,N-diethylacetamide	9.9
Acetone	9.9
o-Dichlorobenzene	10.0
Carbon disulfide	10.0
N,N-diethylformamide	10.6
n-Butanol	11.4
Cyclohexanol	11.4
ACN	11.9
N,N-Dimethylformamide	12.1
Ethanol	12.7
Dimethylsulfoxide	12.9
Methanol	14.5
Ethylene glycol	14.6
Water	23.4

Reproduced with permission from M.J. Forrest in *Rubber Analysis – Polymers, Compounds and Products*, Rapra Review Report No.139, Smithers Rapra, Shawbury, UK, 2001, p.56. ©2001, Smithers Rapra [54]

collection at Smithers Rapra and are consistent for the plasticisers dioctyl phthalate and dibutyl sebacate, respectively. Also, where the plasticisers have distinct IR spectra (e.g., phthalates and sulfonamide) it is possible to detect the presence of a blend of plasticisers. If quantification of the blend composition is required, though, this is usually easier to achieve by high-performance LC or LC–MS, than it is by FTIR.

The use of TGA to quantify the amount of a plasticiser or oil in a rubber sample has been mentioned in Section 3.4.1. It is also possible to use the technique to obtain some qualitative information. For example, by the use of standardised conditions and control samples, the TGA derivative peak temperature at which the maximum rate of volatilisation occurs can be used to obtain some qualitative

Table 3.7: Solubility parameters of typical rubbers.

Rubber type	Solubility parameter (cal$^{1/2}$ cm$^{-3/2}$)
NR	7.9–8.5
Polyisoprene	7.9–8.5
SBR (4 to 40% styrene content)	8.1–8.6
Butadiene rubber	8.0–8.6
Butyl rubber	7.5–8.0
EPDM	7.5–8.6
Chloroprene rubber	8.1–9.4
NBR (18 to 30% ACN)	8.7–9.3
NBR (40% ACN)	10.4–10.5
Silicone rubbers	7.0–11.0
Polysulfide rubbers	9.0–10.0
PU rubbers	9.8–10.3
PVC (plastic used in blends with NBR)	8.5–11.0

Reproduced with permission from M.J. Forrest in *Rubber Analysis – Polymers, Compounds and Products*, Rapra Review Report No.139, Smithers Rapra, Shawbury, UK, 2001, p.55. ©2001, Smithers Rapra [55]

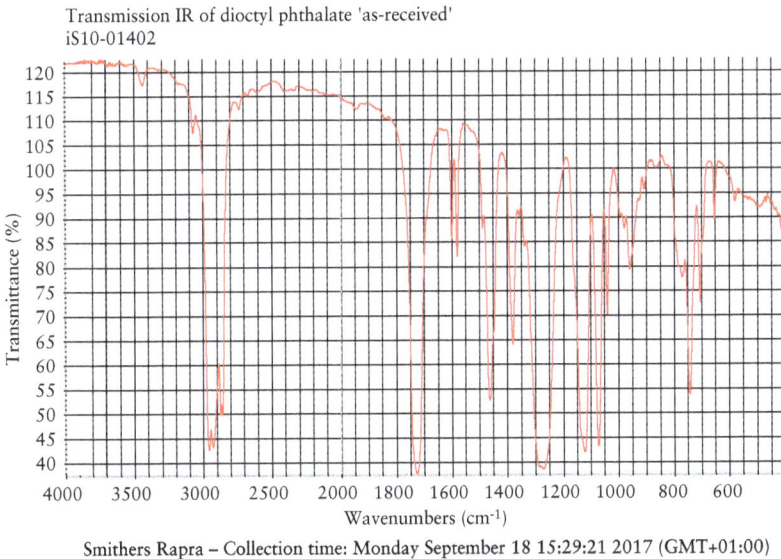

Transmission IR of dioctyl phthalate 'as-received'
iS10-01402

Smithers Rapra – Collection time: Monday September 18 15:29:21 2017 (GMT+01:00)

Figure 3.13: Transmission FTIR spectrum of dioctyl phthalate.

information on the type of plasticiser present in a sample. For example, under the same analytical conditions, dibutyl phthalate (DBP) will have a derivative peak temperature of 220 °C, whereas di-(2-ethylhexyl)phthalate will have a peak

Transmission IR of dibutyl sebacate 'as-received'
iS10-01403

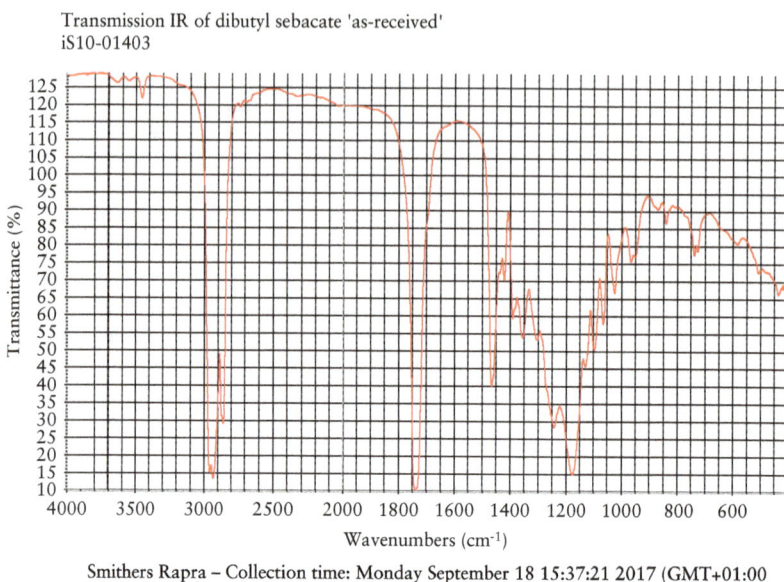

Smithers Rapra – Collection time: Monday September 18 15:37:21 2017 (GMT+01:00

Figure 3.14: Transmission FTIR spectrum of dibutyl sebacate.

at 265 °C. The chemical structure of the plasticiser also has an influence, with di-(2-ethylhexyl)adipate and di-(2-ethylhexyl) sebacate having peak temperatures of 250 and 280 °C, respectively.

Some of the mineral process oils and plasticisers that have higher MW (e.g., polymeric plasticisers) can give problems because the weight loss event due to their volatilisation from the sample can merge with the pyrolysis weight loss event of the polymer. However, the relative positions of the derivative peak temperature, and the temperature at which volatilisation begins, can continue to be used to obtain comparative data.

Modifying the conditions under which a TGA experiment is performed can improve the quality of the data obtained. Knappe [57] described how operating a TGA under vacuum can improve the separation between plasticisers and the polymer component in a rubber sample, and also how coupling a TGA to an FTIR instrument can assist in the identification of additives such as plasticisers. Interfacing a TGA to an FTIR is covered in more detail in Section 3.8.

As mentioned in Section 3.2, rubber compounds containing polymers that lose small molecules by heat degradation (e.g., the HCl lost by CR) give more complicated TGA traces in which the early weight loss events overlap. These samples require a solvent extraction step to determine their plasticiser level. It is also harder to obtain qualitative information on the type of plasticiser; often an IR analysis of the solvent extract is required.

3.4.3 Determination of polyaromatic hydrocarbons in hydrocarbon oils and rubber products

For many years it was a common and widespread practice to use aromatic mineral oils, also called 'extender oils', in large tonnage products like passenger tyres. In fact, the incorporation of relatively large amounts of these types of mineral oil in tyres was beneficial to certain prime properties of the final product (e.g., road holding and good grip).

However, around 20 years ago, health and safety concerns began to grow over the polyaromatic hydrocarbons (PAH) present in these types of mineral oils [e.g., the distillate aromatic (DAE) types]. These concerns included both their potential threat to the environment, as a result of tyre rubber being deposited onto road surfaces and substances leaching from it, and to the workers who came into contact with them in industry. This pressure on the industry to substitute these types of aromatic oils with safer mineral oils led to the development of low-PAH content extender oils (e.g., mild-extraction solvate types) for use in tyres and other rubber products. The European Commission (EC) also published a regulation addressing the use of these oils: Directive (EC) 2005/69. This Directive came into force on 1st January 2010 and prohibited the use of DAE-type oils.

Another approach to the regulatory problems associated with highly-aromatic oils was to look for renewable alternatives. One such alternative is cardanol (*m*-pentadecenyl phenol), a cheap and abundant byproduct of the cashew nut industry, which has been shown to be capable of acting as a plasticiser and multifunctional additive in polymers. Researchers in India [58] carried out comparative analyses using a range of techniques (e.g., thermal, physical and microscopic) on NR that had been chemically grafted with cardonol (CGNR) and an oil-plasticised NR compound. The compound based on CGNR was found to be superior to the NR compound in the following ways:

- Physico-mechanical properties
- Crosslink density
- Bound rubber content
- Lower Payne effect
- Better and more uniform dispersion of CB

Another possible source of PAH in rubbers is CB, a material that can also be used in other products, such as plastics. As a consequence, the EC published another regulation in 2013 [59] which restricted PAH in several consumer articles that were manufactured using both rubber and plastics materials. The presence of PAH in extender oils and tyres has also been addressed by the Registration, Evaluation, Authorisation and Restriction of CHemicals (REACH) regulation [60]. REACH, in Annex XVII (i.e., EC 1907/2006 XVII amendment), placed a specific restriction on the amount of benzo-alpha-pyrene (a common PAH compound) that could be present

(1 mg/kg) and a limit (10 mg/kg) on eight specified PAH compounds, one of which is benzo-alpha-pyrene. It also designated ISO 21461 as the method that should be used to determine the combined level of these eight PAH compounds.

Because of the specific concern over extender oils and tyres, and the general concern over PAH, work was carried out to develop analytical methods that were capable of accurately determining the amount of specific PAH compounds in products and in determining the overall level of PAH compounds. One example of a specific method is the method set by the American Society for Testing of Materials (ASTM) for quantifying benzo-alpha-pyrene in CB [61]. Important general methods for the determination of PAH include:

- IP 346 – total amount of PAH in extender oils.
- ISO 21461 – NMR-based method for determining the PAH level of tyres (eight PAH compounds are targeted).
- German ZEK method (16 PAH compounds).
- US Environmental Protection Agency (18 PAH compounds).

With regard to ISO 21461, the NMR instrument determines the level of PAH in an extender oil or the extract of a rubber sample by measuring the amount of Bay protons that are present. According to REACH (EC) 552/2009, extender oils and tyres are compliant with the restriction on PAH (re: the limit introduced on 1st January 2010) if they do not exceed 0.35 Bay proton as measured and calculated by ISO 21461.

Hamm and co-workers [62] investigated the extraction and migration behaviour of PAH from cured, CB-filled rubber compounds. The research team used standard PI-based formulations to develop and validate a GC–MS method for the identification and quantification of PAH compounds that had migrated into various aqueous media. The objective was to develop a method that could be used to assess the exposure to PAH compounds from rubber products.

3.5 Determination of carbon black

CB is by far the most commonly used filler for rubber compounds and has been in constant use in formulations since just before the First World War. Although CB can be added to rubber compounds to make them less expensive by reducing the proportion of the more expensive polymer component (e.g., in highly 'extended' EPM and EPDM compounds), it is more often added to improve the physical properties (e.g., tensile strength and tear strength) of a rubber compound, something which few other filler materials can do. In common with other fillers, it also increases the hardness and viscosity of a compound.

Historically, there have been three main types of CB. Each type is designated by reference to the manufacturing technique used to produce it from oil, namely:

- Furnace black
- Thermal black
- Channel black

The first of these is the most common type of CB (95% of the market). It is available in a large range of grades (more than 20), and is used in a wide range of rubber compounds. The second type is available in only a few grades and tends to be used in a limited range of compounds (e.g., fluorocarbon rubbers). The third type is rarely seen today, although it has some specific uses due to its relatively high purity (e.g., food-grade rubbers for approval by the US Food and Drug Administration).

In addition to the three types shown above, two specialist CB exist. These are lampblacks (which have been developed by the Chinese for ink and lacquer formulations) and acetylene black (which is conductive and used to formulate conducting rubber compounds). As mentioned in Section 3.5.1, graphite can also be added to increase conductivity and carbon fibres can be present in some compounds to alter physical properties. Another carbonaceous material that may become more common in rubber compounds in the future are carbon nanotubes, which are showing some promise in improving the electrical conductivity of plastics [63].

3.5.1 Quantification of carbon black in rubber vulcanisates

It is possible to quantify the amount of black in a vulcanisate by using a pyrolysis approach in a tube furnace in a way that is analogous to that described in Section 2.1 for the determination of halogens. The initial, non-oxidative pyrolysis stage involves heating an accurately weighed sample in a stream of nitrogen and then weighing the resulting residue. The final stage involves heating this residue in an oxidising atmosphere (either air or oxygen) to oxidise the CB to carbon dioxide (CO_2). The final inorganic residue is then weighed and the amount of black in the sample calculated using Equation 3.6:

$$\text{Amount of CB (\%)} = \frac{A - B \times 100}{\text{original weight}} \tag{3.6}$$

where 'A' is the residue after non-oxidative pyrolysis stage and 'B' is the final residue.

Carrying out the above experiment in a tube furnace has the advantage that relatively large (e.g., 1–2 g) samples can be used, and so a relatively accurate result is obtained.

It is more usual, however, to use the automated and sensitive technique of TGA and this instrument can determine the amount of CB in a sample, along with the amount of the other principal ingredients (e.g., plasticiser and polymer) in one seamless operation with automatic gas switching and data collection (Sections 3.8).

Compared with the tube furnace, small samples are analysed (1–10 mg), but this has the advantage that the technique can be used on small samples, such as contaminants or, if the sample is large enough, in a virtually non-destructive way.

A TGA trace of a NBR formulation showing the quantification of carbonaceous residue and CB is shown in Figure 3.15.

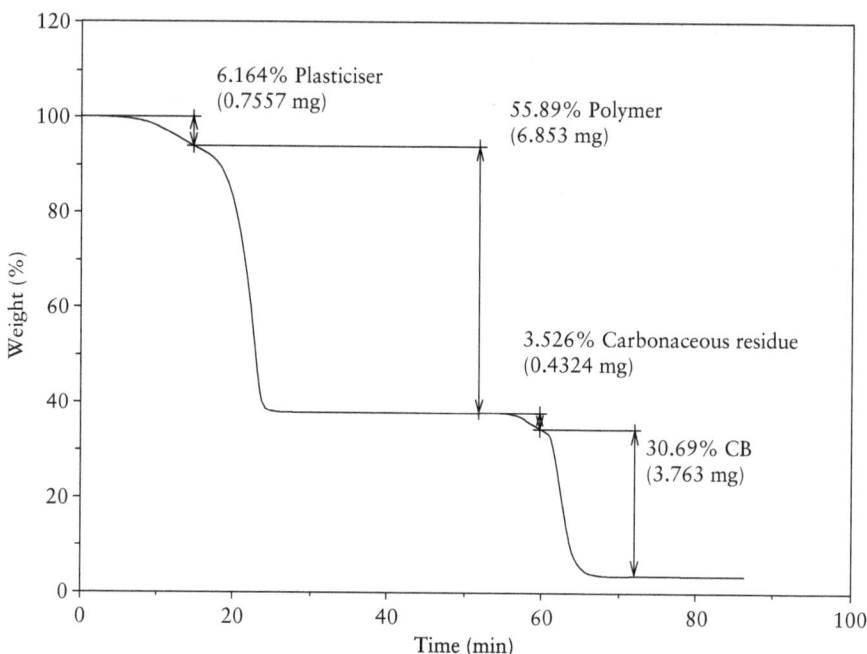

Figure 3.15: TGA trace of NBR. Reproduced with permission from M.J. Forrest in *Rubber Analysis – Polymers, Compounds and Products*, Rapra Review Report No.139, Smithers Rapra, Shawbury, UK, 2001, p.16. ©2001, Smithers Rapra [5].

Its inherent advantages mean that TGA is an excellent technique for the quantification of CB in rubber compounds, and the results obtained have a reasonable degree of accuracy, as shown by the data generated using a range of CB in different rubber compounds by Pautrat and co-workers [64]. In general, the research team found that, in all cases, the amount of CB recorded by the TGA was slightly above ($\approx$5%) the quantity that had been added to the rubber compounds at the mixing stage.

Although TGA is useful for quantifying the amount of CB, it has a limited capability when it comes to distinguishing between the different types of CB. This is because there is no clear and definitive relationship between particle size and/or particle structure and the oxidation temperature of a CB. There is, however, a loose relationship in that the higher surface areas of the small particle size blacks (e.g., super abrasion furnace) results in lower oxidation temperatures than the larger particle size blacks. The temperature at which a CB oxidises also varies between its virgin form and when it has been incorporated into a rubber compound. For example, the oxidation temperature of a virgin semi-reinforcing furnace black has been reported as being ≈670 °C, as opposed to ≈580 °C when it is present in a cured butyl rubber compound [65]. Such a result shows that it is important to have access to a number of appropriate rubber compound reference standards.

For reference, some indications of the oxidation temperatures for a selected range of CB, when they are present in a rubber compound, is provided in Table 3.8.

Table 3.8: Indicative-TGA oxidation temperatures of a selection of CB when present in rubber compounds.

Type of CB	Indicative-oxidation temperature (°C)
Medium thermal	600
Semi-reinforcing furnace	575
High-abrasion furnace	550
Medium process channel	520

Reproduced with permission from M.J. Forrest in *Principles and Applications of Thermal Analysis*, Ed., P. Gabbot, Blackwell Publishing, Oxford, UK, 2008, p.196. ©2008, Blackwell Publishing [66]

A number of workers, for example Pautrat and co-workers [64] and Maurer [67], used the time taken for 15% of the weight of the CB to oxidise (T_{15}) to investigate if the type of CB in a rubber compound could be ascertained by TGA. Again, a loose relationship has been found but no absolute correlation exists, and some of the problems that are encountered in assigning CB type by TGA can be summarised as shown below:
- Blacks vary in structure as well as particle size and this can blur the distinction between different grades resulting in overlap within a TGA experiment.
- Analytical conditions (e.g., temperature ramp) can affect the data.
- The cure system (i.e., sulfur or peroxide) used in the compound can influence the data.
- The polymer can affect the data if it generates a carbonaceous residue that overlaps with the black weight loss.

The last point is important as a number of commercially important rubbers (e.g., halogenated, nitrile and PU) produce carbonaceous residues during the nitrogen atmosphere region and care must be taken to ensure that these do not affect the quantification of CB in the air atmosphere region. Use of a high-resolution TGA instrument may assist in this because it should improve the separation of the two weight loss events. However, one must be beware of the difficulties that can be encountered with the operation of certain high-resolution TGA instruments because an incorrect choice of programme can cause greater problems, as described in Section 3.8.

As mentioned previously, in addition to CB, rubbers can contain other types of carbonaceous fillers, for example:
- Acetylene black
- Graphite
- Carbon fibres

The first two products are used to create a rubber that is conductive and the third, which is found relatively rarely in rubber compounds, is a filler to improve strength and increase hardness. All three products are purer forms of carbon than CB so they have higher oxidation temperatures. CB, which contains various trace substances (e.g., sulfur compounds) that originate from the oil which it is derived from, usually oxidises in rubber compounds over the range 500–600 °C, whereas the three purer forms of carbon all oxidise ≈700 °C in rubber compounds, and this additional stability makes it possible to differentiate between the two classes in a TGA experiment. In both cases, the temperatures cited are for a TGA programme in which the temperature of the rubber sample is increased at a rate of 20 °C/min.

3.5.2 Dispersion of carbon black within rubber products

To achieve optimum properties, it is important that the CB within a rubber is uniformly dispersed. There are a number of tests that have been developed to assess the degree of dispersion.

3.5.2.1 Cabot dispersion test

This test [68] involves the cold-stage microtoming of thin (≈2-µm thick) sections from the rubber, which are then immersed in xylene and examined between glass slides using a light transmission microscope fitted with a Cabot graticule. The degree of dispersion is graded using a Cabot Dispersion Classification Chart.

3.5.2.2 Cut surface and torn surface methods

The surfaces of the rubber are viewed using incident illumination, thereby avoiding the need to prepare microtome sections, and a degree of automation is provided by equipment such as the Optigrade Dispergrader. The torn surface method has been described by Sweitzer and co-workers [69] and Stumpe and Railsback [70] and is, in effect, a method for the determination of the agglomeration of CB within the sample. Grading is possible and a scale of 1 (very poor dispersion) to 10 (excellent dispersion) is used.

3.5.2.3 Transmission electron microscopy

Very thin sections are required (<100 nm) and the small area examined makes it possible to evaluate the dispersion of black within phases of polymer blends. Care has to be taken though that the small area analysed does not lead to unrepresentative results. A detailed description of the microscopy of rubber samples has been provided by Kruse [71].

3.5.2.4 Online measurement during processing

Le and co-workers [72] used an online measurement method to investigate the effect of curing additives on the dispersion kinetics of CB within SBR compounds. The online method used involved the measurement of electrical conductance during mixing and the data obtained showed that curing agents such as stearic acid and diphenyl guanidine (DPG) greatly accelerated the dispersion of CB within the rubber compound. The research team postulated that stearic acid and DPG might alter filler–filler interactions and that this subsequently accelerated the dispersion process. Their work also revealed that ZnO was not capable of improving dispersion because of its limited amount of interaction with the CB, that sulfur and N-cyclohexyl benzothiazole-2-sulfenamide decelerated the dispersion process and that, as the amount of styrene in the SBR increased, the influence that the curing additives had on dispersion decreased.

3.5.3 Isolation of carbon black from a vulcanisate

Historically, there are two main routes for the recovery of CB from rubber vulcanisates.

3.5.3.1 Digestion of the matrix followed by filtration

In this method, the polymer in the rubber vulcanisate is destroyed and solubilised by a chemical such as a strong (e.g., nitric) acid in the presence of an organic solvent (e.g., nitrobenzene) and heat. The CB is then recovered by filtering the mixture. The drawback in the method is the difficulty of achieving a sufficient degree of degradation, especially with the more chemically resistant rubbers.

3.5.3.2 Isolation by non-oxidative pyrolysis

In a non-oxidising atmosphere, heating of the rubber compound to $\approx$500 °C results in the pyrolysis and loss of the polymer fraction as volatile products. This leaves a residue comprising CB and inorganic species. CB is then isolated by an acid wash, a water wash and then low-temperature drying ($\approx$105 °C), with light grinding to break up aggregates.

3.5.4 Surface area tests on recovered black

All three test methods to obtain total surface area (Section 3.6.2) have been applied to CB that has been recovered from vulcanisates. The results obtained suggest that with the Brunauer–Emmett–Teller method (BET) the type of rubber that the black has been recovered from can influence the data (Table 3.9). The results obtained using the cetyltrimethylammonium bromide (CTAB) method were more independent of polymer type (Table 3.10).

Table 3.9: Total surface area (m^2/g) of CB determined by the BET method.

CB type	Original black	Recovered from NR	Recovered from SBR	Recovered from SBR/BR
N110	144.6	132.5	127.5	123.9
N220	124.2	124.3	112.9	110.0
N330	82.9	86.4	78.2	74.5

Adapted from data published in *European Rubber Journal*

Table 3.10: Total surface area (m^2/g) of CB determined by the CTAB method.

CB type	Original black	Recovered from NR	Recovered from SBR	Recovered from SBR/BR
N110	103.9	98.2	99.1	100.1
N220	95.9	96.1	92.0	91.0
N330	65.0	71.5	62.0	68.9

Adapted from data published in *European Rubber Journal*

The DBP method to obtain the external surface area has been carried out on CB samples recovered from vulcanised rubber using the preferred pyrolysis route and, as with the total surface area tests, reasonable results have been obtained enabling the type of black used to be determined.

3.6 Characterisation of carbon black

CB is an extremely pure form of carbon which consists of very small, mostly spherical particles which fuse together in clusters referred to as 'aggregates'. The latter group together as 'agglomerates' which break up during the mixing process. The aggregates tend to remain intact in the rubber matrix and the type of aggregate defines the structure of a CB; the higher the structure the greater the number of particles in an aggregate. The greater the number of particles in an aggregate, the larger will be the volume within it that is not accessible to rubber molecules.

The properties that define a particular CB are as follows:
- Sphere size
- Structure (aggregate size and shape)
- Total surface area
- Surface area available to rubber molecules.

3.6.1 Particulate and aggregate size

The most popular technique for the examination of the particles of CB is SEM. However, other techniques have been used, such as X-ray diffraction (XRD), transmission electron microscopy, atomic force microscopy and scanning tunneling microscopy, and these feature in a review of the subject [73]. These last two techniques can be used to examine the surface structure of CB [74]. In addition to the examination of individual particles, high-resolution electron microscopy is a useful tool for the studying particle aggregates [75].

A team drawn from industry and academia in Japan [76] used contrast variation small-angle neutron scattering (SANS) to study the surface of CB dispersed within SBR samples. The team prepared the SBR samples by swelling them in solvents (mixtures of toluene and deuterated toluene) that had different scattering length densities. Analysis of the swollen vulcanisates by SANS revealed that there was an adsorption layer of SBR, of ≈10 nm in thickness, on the surface of the CB, and the results also provided information on the CB aggregate structure. One of this group of researchers, Takenaka, also published a review of the use of SANS to analyse the structure of sulfur-crosslinked rubber and used the technique to characterise the structure of PI rubber and SBR, with a non-uniform structure being observed in both cases [77].

3.6.2 Total surface area

The total surface area of a CB sample (i.e., the area that can be accessed by rubber molecules and that within pores and between particles) provides an indication of particle size. There are three main methods for the determination of total surface area.

3.6.2.1 Brunauer–Emmett–Teller method (nitrogen adsorption)
Named after Brunauer and co-workers, who developed it in 1938, this method is described in ISO 4652 and is carried out using a Ni-Count-1 instrument. It also features in ASTM D3037, in which there are four accepted procedures, one of which uses the Ni-Count-1.

3.6.2.2 Iodine adsorption

This method uses simple laboratory equipment and produces results that correlate well with the nitrogen adsorption method. The procedure to be followed is described in ISO 1304 (ASTM D1510). Reinforcing grades of CB have iodine adsorption numbers typically in the range 70–160 mg/g whereas semi-reinforcing grades are in the range 30–45 mg/g.

3.6.2.3 Cetyltrimethylammonium bromide adsorption

This method, in which a large molecule is used as an adsorption species, has been described by Lamond and Gillingham [78]. It is thought to be more accurate than the other two methods because it is less influenced by the chemical nature of blacks and is less sensitive to particle porosity. A similar procedure to that suggested by Lamond and Gillingham is presented in ASTM D3765. Reinforcing grades of CB have CTAB values in the range 80–140 m^2/g, whereas semi-reinforcing grades are in the range 30–45 m^2/g.

3.6.3 External surface area

This test enables a measurement of the amount of structure in a CB to be established.

The DBP test is described in ASTM D2414 and involves adding phthalate from a burette into a weighed portion of CB which is stirred constantly. As DBP is added, the CB powder changes to a semi-liquid mass with an increase in torque. The torque peaks at the limit of DBP absorption. The result is expressed as ml of DBP per 100 g of CB. Low-structure CB have values in the range 60–80 ml/100 g, with high-structure types in excess of 120 ml/100 g.

3.6.4 Determination of polyaromatic hydrocarbons in carbon black

Due to their method of manufacture (i.e., combustion of mineral oil) it is possible for CB to contain some PAH compounds. A statement published by the International Carbon Black Association (ICBA) stated that, based on laboratory assessments involving vigorous Soxhlet extraction using solvents such as toluene, most grades of CB typically have PAH levels not exceeding 0.1%.

The ICBA contributed to the development of an ASTM method [61] for the determination of a particular PAH compound, benzo-alpha-pyrene, in CB. This method, and other analytical methods to determine PAH compounds in extender oils and final rubber products, such as tyres, have been covered in Section 3.4.3.

3.6.5 Miscellaneous analytical techniques

A number of analytical techniques have been used to characterise CB, and a representative selection of these, along with the references in which information can be sourced, are shown in Table 3.11.

Table 3.11: Examples of analytical techniques used to characterise CB.

Technique	Application	References
IGC	Surface energies and thermodynamic parameters	Wang, Wolff and Donnet [79]; Wang and Wolff [80]
Neutron scattering	Particle structure	Hjelm and co-workers [81]
Raman spectroscopy and X-ray scattering	Particle microstructure	Gruber, Zerda and Gerspacher [82]; Gruber, Zerda and Gerspacher [83]; and Gerspacher and Lasinger [84]
GC–MS, secondary ion mass spectrometry, X-ray photoelectron spectroscopy and IGC	Surface chemistry	Ayala and co-workers [85]

IGC: Inverse gas chromatography
Reproduced with permission from M.J. Forrest in *Rubber Analysis – Polymers, Compounds and Products*, Rapra Review Report No.139, Smithers Rapra, Shawbury, UK, 2001, p.22. ©2001, Smithers Rapra [86]

3.7 Determination of inorganic fillers

In addition to being extremely useful for the determination of polymers, plasticiser/oil and CB, TGA is also a very rapid and convenient technique for the quantification of inorganic fillers, such as barium sulfate ($BaSO_4$) (i.e., barytes), silica, silicates (e.g., clays) and calcium carbonate ($CaCO_3$), in rubber compounds and products.

One limitation of the TGA technique is that the total inorganic content of the rubber is obtained, with no indication of the relative proportion if a blend of inorganic fillers or a mixture of inorganic filler and other inorganic additives is present. Most rubber compounds contain ZnO as part of the cure system (sulfur systems) or as an acid acceptor (peroxide cure systems), so it is common practice to complement the TGA data with a semi-quantitative elemental technique, such as X-ray fluorescence spectroscopy (XRF), in order to obtain qualitative as well as quantitative data. Although it is possible to use the ash obtained by the TGA experiment for this analysis, it is more common to use the ash that has been obtained by a qualitative or quantitative ashing process, such as the one described below.

$CaCO_3$ can be detected and quantified more effectively by TGA than most of the other common inorganic fillers, which are stable up to 1,000 °C, because of the quantitative way that it decomposes upon heating. It will undergo a characteristic quantitative degradation at around 680–700 °C (at a heating rate of 20 °C/min) into CO_2 and calcium oxide, which remains as a stable residue up to 1,000 °C. It is, therefore, possible to both identify the presence of $CaCO_3$ in a sample by this characteristic weight loss, and to quantify it because the CO_2 weight loss represents 44% of the $CaCO_3$ by mass.

The value of the residue that remains will also show if any other inorganic compounds are present in the sample. Information can then be obtained on these using the analytical techniques (e.g., XRF) described elsewhere in this section.

The other technique commonly used for the quantification and isolation of inorganic substances from a rubber sample is furnace ashing. One common method used for this process is described in ISO 247 and two procedures are provided. The first, dry ashing, may be unsuitable for rubbers containing halogens due to the loss of volatile halides, such as zinc chloride, formed by the reaction of ZnO with the HCl liberated from the polymer. The second, acid ashing, is not recommended for raw rubbery polymers. There are two temperature options in the standard: 550 and 950 °C. Care has to be taken with the higher temperature if knowledge of the types of inorganic compounds present is not available. For example, if $CaCO_3$ is present, as mentioned above, it will breakdown at ≈700 °C. With this in mind, 550 °C is the more common ashing temperature used in commercial laboratories.

Although it is rarely used for this purpose because of the practical and time-efficient benefits of TGA and standardised ashing techniques, it is possible to use the tube furnace approach if relatively large samples (e.g., 2–5 g) are available. In order to separate the inorganic species from the rubber sample, the tube furnace is employed using a similar set of operating conditions to the TGA. Values for total organics, total carbonaceous material and total inorganics are obtained by the experiment and the isolated ash can be analysed to obtain qualitative and quantitative information on the inorganic compounds present as described below.

Once an ash has been obtained, by either TGA, tube furnace, or a standard ashing process, qualitative information on the inorganic compounds present can be obtained by a variety of analytical techniques. One of the most common and convenient is FTIR, and it is standard practice to use a liquid paraffin mull technique. If only a small amount of ash is available and the quantity of ash is too small to use the paraffin mull technique effectively, other FTIR techniques can be used such as 'single bounce' ATR or FTIR microscopy. This eventuality is very rare in the case of inorganic filler additives which by nature are bulk additives. Hence, these alternatives are usually required only for additives such as inorganic pigments (Chapter 4).

The use of FTIR alone to identify inorganic fillers in a sample ash can be problematic, particularly if a mixture is present, so it is often necessary to compliment this work with an elemental technique. For qualitative or semi-quantitative work, XFR is very useful in providing the elements present which, in conjunction with the FTIR spectrum, will usually enable specific compounds to be identified. If it is suspected that a compound, such as a silicate is present, use of XRD can often provide an assignment of the particular type by virtue of its characteristic crystalline fragmentation pattern. If quantification data are obtained on a particular element, or if it is thought that a semi-quantitative multi-element scan would be useful, then inductively coupled plasma can be employed.

The use of FTIR to assist in the identification of inorganic fillers in a rubber sample, *via* analysis of its ash, is mentioned above and some of the characteristic, diagnostic IR absorption bands that can be employed in this task are shown in Table 3.12.

Table 3.12: Examples of characteristic IR absorption bands for some typical inorganic fillers.

Inorganic filler	Diagnostic IR absorption bands
Silica	Broad band at 950–1,330 cm^{-1} peaking at 1,050–1,100 cm^{-1}
Silicates	Broad band at 850–1,300 cm^{-1} peaking at 950–1,100 cm^{-1}
$CaCO_3$	1,420, 870 and 710 cm^{-1}
$BaSO_4$	1,080 and 610 cm^{-1}

3.8 Determination of the bulk composition of a rubber

Sections 3.2–3.7 have shown how analytical work can be carried out on a rubber sample to obtain information on the principal components which make up its bulk composition. In addition to obtaining this information, it is also important in a number of cases (e.g., checking a formulation against a specification) to obtain information on the overall composition of a rubber sample (i.e., the proportions of principal components). This kind of information can be referred to as the 'bulk composition' because it addresses only the major components and does not provide information on the many types of additives that can be added in relatively small quantities; obtaining this type of information is the subject of Chapter 4.

It is possible to obtain this information by combining the qualitative and quantitative data obtained using particular techniques and approaches in ections 3.2–3.7. However, it is often the case that a simplified approach is preferred, and one of the most useful and effective techniques for obtaining bulk compositional data on a rubber sample is TGA. This is because the technique has a number of inherent advantages:

- It only requires a small sample (e.g., 10 mg).
- It is relatively quick – a complete analysis can be completed within 90 min.

- It is relatively accurate (e.g., ±0.5% w/w).
- It is relatively sensitive (e.g., it can indicate the presence of two polymers or two types of CB).

A typical TGA analysis will involve two stages:
- Stage 1 – The sample is heated from 40 °C under a nitrogen atmosphere at a heating rate of 20 °C/min to 550 °C and held at that temperature for 10 min to ensure that constant weight is achieved.
- Stage 2 – The temperature is reduced from 550 to 300 °C, the atmosphere changed to air, and then the sample heated to 850 °C.

Because of its effectiveness and popularity for a range of important applications, including quality control, failure diagnosis, and reverse engineering, the use of TGA to determine bulk composition has been incorporated into international standards. An example is ISO 9924-1, which describes the use of TGA for the determination of the composition of butadiene, ethylene propylene, butyl, isoprene and SBR. The technique also lends itself to being standardised because, apart from differences in optional features (e.g., high-resolution programmes), the majority of the commercial instruments use the same fundamental design.

Because of the small sample size that is used, and the practical options that are available (e.g., heating rate), it is best to standardise the analytical conditions to ensure that meaningful comparisons between samples can be made, even if there is a time delay in their analysis.

A representative TGA trace of a rubber sample, in this case a NBR, recorded using the analysis programme shown above, is shown in Figure 3.16.

The data in Figure 3.16, with both the weight loss and the derivative weight loss curves displayed, show the main weight loss events present in such traces. From this TGA trace it is possible to obtain the following bulk compositional information:
- Plasticiser content – 5.7%
- Polymer content – 67.3%
- Carbonaceous residue* – 3.2%
- CB – 10.1%
- Inorganic material – 13.7%

  *Polymers that contain atoms such as nitrogen, oxygen, sulfur or a halogen produce a carbonaceous residue during reactions such as cyclisation during stage 1 (see the programme above). This residue oxidises to CO_2 during stage 2 (see above).

The general application of TGA to determine the bulk composition of rubber samples is also covered in the section on reverse engineering (Chapter 5) and the data generated on tyre samples using the AutoStepwise method is included.

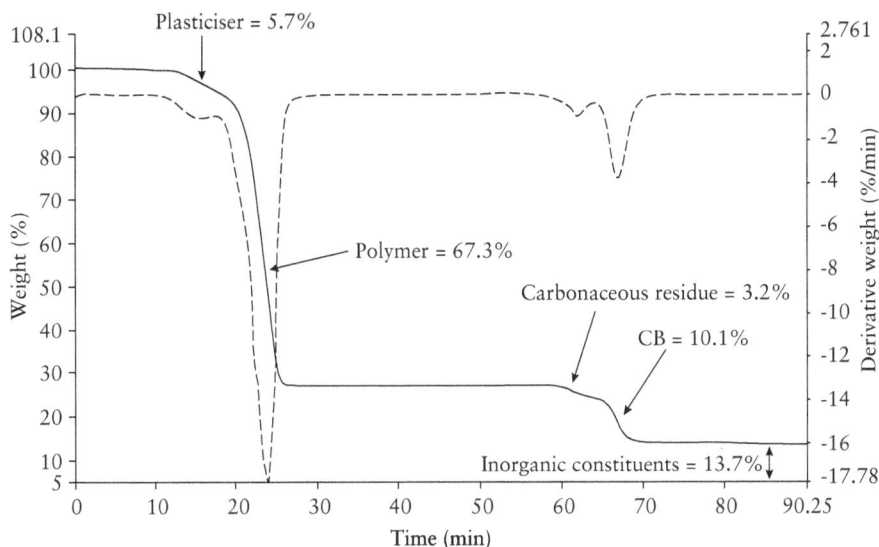

Figure 3.16: Typical TGA trace for a rubber (NBR) compound. The weight loss is shown as the full curve and the derivative curve is dashed. Reproduced with permission from M.J. Forrest in *Principles and Applications of Thermal Analysis*, Ed., P. Gabbot, Blackwell Publishing, Oxford, UK, 2008, p.193. ©2008, Blackwell Publishing [87].

An important point to make regarding the carbonaceous residues referred to above is that, because their oxidation temperature is usually less than that of CB, this helps to resolve these events within the TGA trace and so aids their respective quantifications. Also, the ability to detect the presence of a carbonaceous residue in a TGA trace and its size can, in a limited sense, assist in the diagnostic value of the technique to identify the generic polymer type of the rubber present in a sample. This information can then be used to corroborate data obtained by the more standard approaches, such as the use of FTIR (Section 3.2). In addition to the example of NBR (which is discussed in more detail in Section 3.2), there are a number of other polymers which give reproducible amounts of carbonaceous residue that can be used to assist with generic polymer identifications. The most popular commercial polymers used in rubbers and rubber blends and which yield carbonaceous residues are shown in Table 3.4 (Section 3.2).

Figure 3.16 shows the TGA trace plotted in the form of weight loss against time. This is the standard format because it makes the information presented easier to see by eliminating the influence of the two independent heating stages. However, it is often useful to obtain temperature data on weight loss events, for example, the onset temperature of the polymer weight loss, to see if a reduction in thermal stability can be detected (e.g., due to conditions in service). The temperature at the maximum rate of weight loss (i.e., the peak in the derivative curve) is also useful for

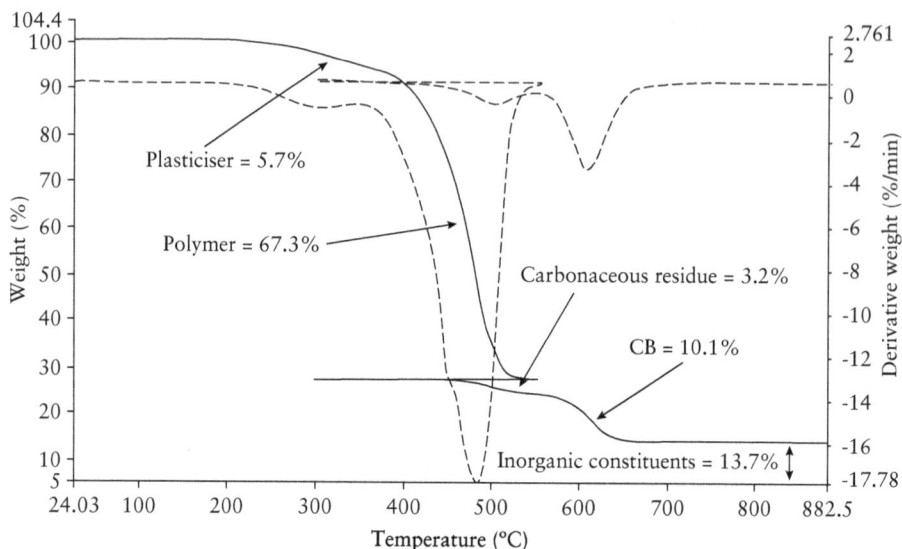

Figure 3.17: The TGA data presented in Figure 3.16 shown in the form of weight loss against temperature. Reproduced with permission from M.J. Forrest in *Principles and Applications of Thermal Analysis*, Ed., P. Gabbot, Blackwell Publishing, Oxford, UK, 2008, p.194. ©2008, Blackwell Publishing [88].

using the TGA trace to provide some qualitative information on a rubber compound. Examples of this particular application include indications to the polymer type and CB type present by reference to data obtained on rubber compounds of known composition analysed under standardised conditions (Sections 3.2 and 3.5). Figure 3.17 shows how the data in Figure 3.16 would look if displayed as weight loss *versus* temperature.

Polymers in rubber compounds, due to their high MW, will be lost during a TGA experiment by a pyrolysis-type reaction and this is true whether a non-oxidising (e.g., nitrogen) or oxidising (e.g., air) atmosphere is used. By using a standard TGA programme, for example, the one shown above, sufficient data can been generated on reference rubber compounds in the nitrogen atmosphere region to enable some limited, qualitative information on the types of polymers present in 'unknown samples' to be obtained. This task is performed using the temperature at the peak of the derivative curve. Indicative examples of the data obtained are provided below:

- NR – peak maximum 370 °C
- Butyl rubber – peak maximum 390 °C
- SBR, BR or EPDM – peak maximum 460 °C
- Fluorocarbon rubber – peak maximum 480 °C

The list shown above illustrates that the presence of polymers that have relatively low amounts of thermal stability (e.g., NR) can be detected quite readily in a compound. However, for a number of polymers, this approach has little or no value and does not begin to replace the more effective methods of polymer identification reviewed in Section 3.2. However, in those cases, if a rubber sample contains more than one polymer (i.e., a blend of polymers) and one of those is NR, these temperature differences can result in TGA being be a useful tool for, firstly, identifying that a blend exists and, secondly, in obtaining an approximate measure of the different proportions of each polymer. This type of work is illustrated in Figure 3.18, where the TGA trace for a tyre tread compound that contains a blend of NR and SBR is shown. In this trace, the weight loss events for the two polymers are partially resolved due to the differences in their respective pyrolysis peak maximum temperatures (i.e., 360 and 460 °C) and so it is possible, by using the inflexion on the derivative weight loss curve, to obtain an approximate quantification of their amounts in the compound: 44.6% for the NR and 14.0% for the SBR.

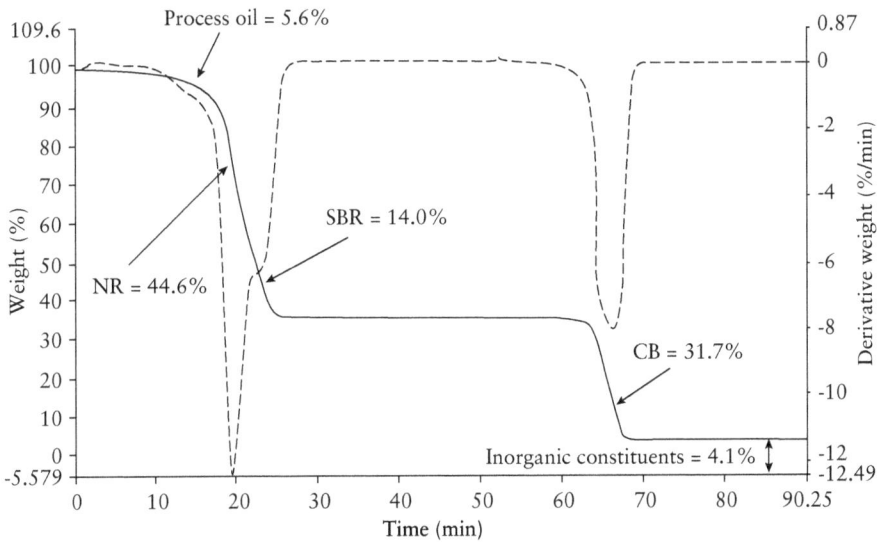

Figure 3.18: TGA trace for a tyre tread compound containing both NR and SBR. The weight loss derivative curve has been crucial in this application to determine the crossover point of the weight losses. Reproduced with permission from M.J. Forrest in *Principles and Applications of Thermal Analysis*, Ed., P. Gabbot, Blackwell Publishing, Oxford, UK, 2008, p.202. ©2008, Blackwell Publishing [89].

It might be thought that the use of a high-resolution TGA programme might assist with this kind of analysis. However, some experiences have shown that because these can result in additional weight loss events for the individual polymers, causing additional overlapping within the TGA trace, their use does not always result in more accurate quantifications. An example of this effect has been presented by Forrest [90], who showed that whereas in a conventional TGA experiment only one weight loss event is obtained for NR, analysing the same sample by using a high-resolution TGA programme resulted in four weight loss events being observed. Another example in this published work that indicated that use of a high-resolution TGA programme can cause problems, was based on its ability to detect a blend of CB in a compound. Operating the instrument using this mode showed that additional weight loss events could be created, which gave the impression that more than one type of black was present in a rubber sample, when in fact only one had been used in the compound.

The examples shown in Figures 3.16 and 3.18 show that if a sample has a composition that is typical for rubber compounds, the resulting TGA trace is relatively simple to interpret. In some cases, however, a rubber compound contains a larger number of components, some with multiple weight loss events, and the situation can be much more complex. Such a case can be encountered with a flame-retardant rubber compound and, in these cases, knowledge of the behaviour of the components within the compound and of the technique itself is crucial to understanding the data. To illustrate this point, the TGA of a flame-retardant rubber compound based on a blend of NBR and PVC, and which contains both organic and inorganic flame-retardant additives in addition to other ingredients, would be expected to exhibit the weight loss events shown in Table 3.13.

Table 3.13: Weight loss position in the TGA trace for the components present in a flame-retardant nitrile/PVC compound.

Ingredient/function	Position in the TGA trace*
Pentabromodiphenylether	1st weight loss
Phosphate ester	1st weight loss
Phthalate ester	1st weight loss
NBR	2nd and 3rd weight losses
PVC	1st, 2nd and 3rd weight losses
CB	4th weight loss
Hydrated alumina	1st weight loss and inorganic residue
Antimony trioxide	Inorganic residue
ZnO	Inorganic residue
Cure system, antidegradant system, pre-vulcanisation inhibitor and so on	Obscured by weight losses of principal components due to low level of addition

* TGA experiment run according to the stage 1 and 2 conditions shown above
Reproduced with permission from M.J. Forrest in *Principles and Applications of Thermal Analysis*, Ed., P. Gabbot, Blackwell Publishing, Oxford, UK, 2008, p.206. ©2008, Blackwell Publishing [91]

TGA is not capable of detecting and quantifying a number of common additives and ingredients due to their relatively low level in a rubber compound and the relatively poor detection limit of conventional TGA. Another reason why it cannot detect and quantify certain ingredients is that they are reactive towards the polymer and each other (e.g., cure system and antidegradant system). A summary of the most important additives that cannot usually be determined directly by conventional TGA is:
- Curatives and accelerators
- Antioxidants and antiozonants
- Process aids
- Friction reducing agents
- Pre-vulcanisation inhibitors and cure retarders

The analytical techniques and methods that can be used to detect and quantify these types of additives are reviewed in Chapter 4. However, the increasing use of TGA instruments that are interfaced to either an FTIR spectrometer or mass spectrometer has improved the chances of rubber analysts obtaining at least some of this information. These combinations can also be used to obtain qualitative information on the plasticiser/oil and the polymer fractions in a rubber sample (Section 2.6.5.3).

It is also possible to use the tube furnace method to obtain bulk compositional information on a rubber sample, but this has a number of disadvantages over TGA because it:
- Is often more time-consuming
- Requires a larger sample
- Is not as informative*

    *Because there is no record of the weight loss, only values for the total amount of organic material (i.e., plasticiser and polymer), the total amount of carbonaceous material (i.e., all CB and any carbonaceous material), and the total amount of inorganic material are obtained.

The advantage that the tube furnace method has over TGA is that, because a larger sample size is used, it is usually more accurate, particularly if quantifying small amounts (e.g., 1–5%) of CB, or other carbonaceous additives, in rubbers that do not produce carbonaceous residues.

References

1. G.A.L. Verleye, N.P.G. Roeges and M.O. De Moor in *Easy Identification of Plastics and Rubbers*, Smithers Rapra Publishing, Shawbury, UK, 2001.
2. M.J. Forrest, Y. Davies and J. Davies in *The Rapra Collection of Infrared Spectra of Rubbers, Plastics and Thermoplastic Elastomers*, 3rd Edition, Rapra Technology Ltd, Shawbury, UK, 2007.

3.   M.J. Forrest in *Chemical Characterisation of Polyurethanes*, Rapra Review Report No.108, Smithers Rapra, Shawbury, UK, 1999, p.19.

4.   M.J. Forrest in *Chemical Characterisation of Polyurethanes*, Rapra Review Report No.108, Smithers Rapra, Shawbury, UK, 1999, p.15.

5.   M.J. Forrest in *Rubber Analysis – Polymers, Compounds and Products*, Rapra Review Report No.139, Smithers Rapra, Shawbury, UK, 2001, p.16.

6.   M.J. Forrest in *Rubber Analysis – Polymers, Compounds and Products*, Rapra Review Report No.139, Smithers Rapra, Shawbury, UK, 2001, p.17.

7.   M.R.S. Fuh and G-Y. Wang, *Analytica Chimica Acta*, 1998, **371**, 1, 89.

8.   G.J. Frisone, D.L. Schwarz and R.A. Ludwigsen in *Proceedings of the International Tire Exhibition & Conference (ITEC) 1996 Select*, Akron, OH, USA, 1996, p.21.

9.   D.D. Parker, J.L. Koenig and M. Mori, *Rubber Chemistry and Technology*, 2003, **76**, 1, 212.

10.   M.J. Forrest in *Principles and Applications of Thermal Analysis*, Ed., P. Gabbot, Blackwell Publishing, Oxford, UK, 2008, p.214.

11.   M.J. Forrest in *Principles and Applications of Thermal Analysis*, Ed., P. Gabbot, Blackwell Publishing, Oxford, UK, 2008, p.218.

12.   M.J. Forrest in *Principles and Applications of Thermal Analysis*, Ed., P. Gabbot, Blackwell Publishing, Oxford, UK, 2008, p.222.

13.   K.N. Pandey, D.K. Setua and G.N. Mathur, *Polymer Engineering Science*, 2005, **45**, 9, 1265.

14.   M. Carlberg, D. Colombini and F.H.J. Maurer, *Journal of Applied Polymer Science*, 2004, **94**, 5, 2240.

15.   B.L. Lopez, L.D. Perez and M. Mesa, *E-Polymer*, 2005, **18**, 1.

16.   M.J. Forrest in *Rubber Analysis – Polymers, Compounds and Products*, Rapra Review Report No.139, Smithers Rapra, Shawbury, UK, 2001, p.54.

17.   M.J. Forrest in *Principles and Applications of Thermal Analysis*, Ed., P. Gabbot, Blackwell Publishing, Oxford, UK, 2008, p.236.

18.   M.J. Forrest in *Principles and Applications of Thermal Analysis*, Ed., P. Gabbot, Blackwell Publishing, Oxford, UK, 2008, p.204.

19.   S.R. Holding and E. Meehan in *Molecular Weight Characterisation of Synthetic Polymers*, Rapra Review Report No.83, Rapra Technology Ltd, Shawbury, UK, 1995.

20.   M.J. Forrest in *Rubber Analysis – Polymers, Compounds and Products*, Rapra Review Report No.139, Smithers Rapra, Shawbury, UK, 2001, p.13.

21.   G. Cleaver in *Application Compendium: Analysis of elastomers by GPC/SEC*, Agilent Technologies Inc., Santa Clara, CA, USA, 1st August 2015.

22.   E. Aik-Hwee, S. Ejiri, S. Kawahara and Y. Tanaka in *Proceedings of the International Seminar on Elastomers: Applied Polymer Symposium*, 14–16th July 1993, Akron, OH, USA, 1993, p.5.

23.   M.R. Ambler, *Journal of Applied Polymer Science*, 1980, **25**, 5, 901.

24.   L. Mrkvickova, *Macromolecules*, 1997, **30**, 17, 5175.

25.   W.S. Fulton, W.M.H. Thrope and R.J. White in *Proceedings of the International Rubber Conference (IRC) 1996*, 17–21st June, Manchester, UK, 1996.

26.   W.S. Fulton, W.M.H. Thrope and R.J. White, *European Rubber Journal*, 1996, **178**, 9, 30.

27.   P.J. Wyatt, *Analytica Chimica Acta*, 1992, **272**, 1.

28.   T. Yamada, T. Okumoto, H. Ohtani and S. Tsuge, *Rubber Chemistry and Technology*, 1991, **64**, 5, 708.

29.   P. Nallasamy, P.M. Anbarasan and S. Mohan, *Turkish Journal of Chemistry*, 2002, **26**, 105.

30.   D.D. Parker and J.L. Koenig, *Journal of Applied Polymer Science*, 1998, **70**, 7, 1371.

31. M. Mori and J.L. Koenig, *Rubber Chemistry and Technology*, 1997, **70**, 4, 671.
32. D.J. Patterson, J.L. Koenig and J.R. Shelton in *Proceedings of the ACS 123rd Spring Meeting*, 9–13th May, Toronto, Canada, American Chemical Society, Washington, DC, USA, 1983, Paper No.67, p.30.
33. U. Hoffmann, W. Gronski, G. Simon and A. Wutzler, *Die Angewandte Makromolekulare Chemie*, 1992, **202–203**, 283.
34. W. Gronski, U. Hoffmann, G. Simon, A. Wutzler and E. Straube, *Rubber Chemistry and Technology*, 1992, **65**, 1, 63.
35. W. Gronski, U. Hoffmann, B. Freund and S. Wolff in *Proceedings of the International Rubber Conference (IRC) 1991*, 24–27th June 1991, Essen, Germany, 1991, p.311.
36. M.R. Krejsa and J.L. Koenig, *Rubber Chemistry and Technology*, 1991, **64**, 1, 40.
37. S-S. Choi, H-M. Kwon, Y. Kim, E. Ko and E. Kim, *Polymer Testing*, 2017, **59**, May, 414.
38. S-S. Choi, Y. Kim and H-M. Kwon, *Polymer International*, 2107, **66**, 6, 803.
39. S.R. Shield and G.N. Ghebremeskel, *Rubber World*, 2000, **223**, 2, 24.
40. E.G. Brame in *Applications of Polymer Spectroscopy*, Elsevier, Amsterdam, The Netherlands, 2012.
41. C. Nah, J.H. Park, C.T. Cho, Y-W. Chang and S. Kaang, *Journal of Applied Polymer Science*, 1999, **72**, 12, 1513.
42. W.J. Sichina, *Thermal Analysis*, Application Note, Perkin Elmer, PETech-45, 2000.
43. M.J. Forrest in *Chemical Characterisation of Polyurethanes*, Rapra Review Report, No.108, Smithers Rapra, Shawbury, UK, 1999, p.23.
44. S.J. Theraftil, A.A. Kuzhuppully and R. Joseph, *Iranian Polymer Journal*, 2008, **17**, 6, 419.
45. D.F. DeCastro, A.F. Martins, J.C.M. Suarez and R.C. Reis Nunes, *Kautschuk, Gummi, Kunststoffe*, 2003, **56**, 1–2, 49.
46. V. Nigam, D.K. Setua and G.N. Mathur, *Journal of Thermal Analysis and Calorimetry*, 2001, **64**, 2, 521.
47. R.N. Jana and G.B. Nando, *Journal of Applied Polymer Science*, 2003, **90**, 3, 635.
48. R.N. Jana, P.G. Mukunda and G.B. Nando, *Polymer Degradation and Stabilisation*, 2003, **80**, 1, 75.
49. A. Amash, R.H. Schuster and T. Frueh, *Kautschuk und Gummi Kunststoffe*, 2001, **54**, 6, 315.
50. H. Varghese, S.S. Bhagawan and S. Thomas, *Journal of Thermal Analysis and Colorimetry*, 2001, **63**, 3, 749.
51. Yuichi Aoyagi and H. Sano in the *Proceedings of the ACS Rubber Division 184th Technical Fall Meeting and Education Symposium*, 8–10th October, Cleveland, OH, USA, American Chemical Society: Rubber Division, Washington, DC, USA, 2013, Paper No.89, p.14.
52. S-S. Choi, J-C. Kim and H-M. Lee, *Journal of Applied Polymer Science*, 2008, **110**, 5, 3068.
53. A. Pappa, K. Mikedi, A. Agapiou, S. Karma and G.C. Pallis, *Journal of Analytical and Applied Pyrolysis*, 2011, **92**, 1, 106.
54. M.J. Forrest in *Rubber Analysis – Polymers, Compounds and Products*, Rapra Review Report No.139, Smithers Rapra, Shawbury, UK, 2001, p.56.
55. M.J. Forrest in *Rubber Analysis – Polymers, Compounds and Products*, Rapra Review Report No.139, Smithers Rapra, Shawbury, UK, 2001, p.55.
56. Hummel and F. Scholl in *Atlas of Polymer and Plastics Analysis*, Volume 3, 2nd Revised Edition, Verlag Chemie GmbH, Weinheim, Germany, 1981.
57. S. Knappe, *Rubber World*, 2002, January, 33.
58. S. Mohapatra and G.B. Nando, *Rubber Chemistry and Technology*, 2015, **88**, 2, 289.
59. Commission Regulation (EU) 1272/2013: Amending Annex XVII to Regulation (EC) No 1907/2006 of the European Parliament and of the Council on the Registration, Evaluation,

Authorisation and Restriction of Chemicals (REACH) as regards polycyclic aromatic hydrocarbons Text with EEA relevance.

60. EC REACH Regulation (EC) 552/2009: Amending Regulation (EC) No 1907/2006 of the European Parliament and of the Council on the Registration, Evaluation, Authorisation and Restriction of Chemicals (REACH) as regards Annex XVII (Text with EEA relevance).

61. ASTM D7771-11 – Standard test method for determination of benzo-alpha-pyrene (BaP) content in carbon black.

62. S. Hamm, T. Frey, R. Weinand, G. Moninot and N. Petiniot, *Rubber Chemistry and Technology*, 2009, **82**, 2, 214.

63. Polycond – An EU Integrated Project funded under FP6-NMP. *http://www.polycond.info.*

64. R. Pautrat, B. Metivier and J. Marteau, *Rubber Chemistry and Technology*, 1976, **49**, 4, 1060.

65. M.J.R. Loadman in *Analysis of Rubber and Rubber-like Polymers*, 4th Edition, Kulwer Academic Publishers, Dordrecht, The Netherlands, 1998, **Chapter 11**.

66. M.J. Forrest in *Principles and Applications of Thermal Analysis*, Ed., P. Gabbot, Blackwell Publishing, Oxford, UK, 2008, p.196.

67. J.J. Maurer, *Rubber Age*, 1970, **102**, 47.

68. A.I. Medalia and D.F. Walker, Technical Report RG-124, Revision 2, Cabot Corporation: Carbon Black Division, Boston, MA, USA, 1970.

69. C.W. Sweitzer, W.M. Hess and J.E. Callan, *Rubber World*, 1958, **138**, 6, 869.

70. N.A. Stumpe and H.E. Railsback, *Rubber World*, 1964, **151**, 3, 41.

71. J. Kruse, *Rubber Chemistry and Technology*, 1973, **46**, 3, 653.

72. H.H. Le, S. Ilisch, E. Hamann, M. Keller and H-J. Radusch, *Rubber Chemistry and Technology*, 2011, **84**, 3, 415.

73. W.M. Hess and C.R. Herd in *Carbon Black Science and Technology*, 2nd Edition, Marcel Dekker, New York, NY, USA, 1993, p.89.

74. H. Raab, J. Froehlich and D. Goeritz in *Proceedings of the International Rubber Conference (IRC) 1997*, 6–9th October, Kuala Lumpur, Malaysia, 1997, p.171.

75. W.M. Hess, L.L. Ban and G.C. McDonald, *Rubber Chemistry and Technology*, 1969, **42**, 4, 1209.

76. M. Takenaka, S. Nishitsuji, S. Fujii, N. Amino and Y. Ishika, *International Polymer Science and Technology*, 2011, **38**, 5, T/1.

77. M. Takenaka, *International Polymer Science and Technology*, 2015, **42**, 3, pT/43.

78. T.G. Lamond and C.R. Gillingham, *Rubber Journal*, 1970, **152**, 65.

79. M.J. Wang, S. Wolff and J-B. Donnet, *Rubber Chemistry and Technology*, 1991, **64**, 714.

80. M.J. Wang and S. Wolff, *Rubber Chemistry and Technology*, 1992, **65**, 5, 890.

81. R.P. Hjelm, W.A. Wampler, P.A. Seeger and M. Gerspacher, *Journal Material Research*, 1994, **9**, 3210.

82. T.C. Gruber, T.C. Zerda and M. Gerspacher, *Carbon*, 1993, **31**, 1209.

83. T.C. Gruber, T.C. Zerda and M. Gerspacher, *Carbon*, 1994, **32**, 1377.

84. M. Gerspacher and C.M Lansinger in *Proceedings of the 133rd ACS Spring Meeting*, 19–22nd April, Dallas, TX, USA, American Chemical Society, Washington, DC, USA, 1988, Paper No.7.

85. J.A. Ayala, W.M. Hess, A.O. Dotsan and G.A. Joyce, *Rubber Chemistry and Technology*, 1990, **63**, 747.

86. M.J. Forrest in *Rubber Analysis – Polymers, Compounds and Products*, Rapra Review Report No.139, Smithers Rapra, Shawbury, UK, 2001, p.22.

87. M.J. Forrest in *Principles and Applications of Thermal Analysis*, Ed., P. Gabbot, Blackwell Publishing, Oxford, UK, 2008, p.193.

88. M.J. Forrest in *Principles and Applications of Thermal Analysis*, Ed., P. Gabbot, Blackwell Publishing, Oxford, UK, 2008, p.194.

89. M.J. Forrest in *Principles and Applications of Thermal Analysis*, Ed., P. Gabbot, Blackwell Publishing, Oxford, UK, 2008, p.202.

90. M.J. Forrest in *Proceedings of the Polymer Testing '97 Conference*, 7–11th April, Rapra Technology Ltd, Shawbury, UK, 1997, Paper No.5.

91. M.J. Forrest in *Principles and Applications of Thermal Analysis*, Ed., P. Gabbot, Blackwell Publishing, Oxford, UK, 2008, p.206.

4 Additives

4.1 Introduction

The determination of the principal components of a rubber sample has been covered in Chapter 3. Two of these components, fillers and plasticisers/oils, can also be considered, quite rightly, additives. However, due to levels that they are added to a rubber compound and, as described in Chapter 3 because it is often not necessary in practice to delve further into the details of a formulation to solve problems (e.g., failures in service), it was thought better to cover these particular additives in their own section, along with the investigations that take place into the nature of the polymer.

This chapter includes two very important groups of additives, antidegradants and curatives, both of which contain many members. Members of the antidegradant group can also be referred to using other names (e.g., stabilisers, protection agents) and it can also include additives that have designations that relate to their functions (e.g., antioxidants and antiozonants). Curing agents are also referred to as 'vulcanisation agents' and 'crosslinking agents'.

With the exception of those that have inherently excellent resistance to degrading agencies such as high temperatures and ozone (e.g., silicones and fluorocarbons), rubbers need additives (i.e., antidegradants) that prevent them from suffering degradation during processing and in service. In addition to heat and ozone, degradation is also possible from other sources (e.g., radiation and chemicals).

As mentioned above, different rubbers have differing degrees of inherent stability to these degrading agencies. Tables 4.1–4.3 provide some comparative data on the resistance of rubbers to degradation due to gamma radiation (Table 4.1), ozone (Table 4.2) and provide an approximate guide to their maximum service temperature when they have been compounded with antidegradants (Table 4.3).

In addition to the need for antidegradants to protect them, rubbers also need curatives to generate the three-dimensional (3D) network which provides them with the inherent properties (e.g., high elasticity) that are such important and defining characteristics.

There have been many technological developments associated with these two classes of additive over the last 80 years. These developments, together with the development of new rubbers, often requiring novel curing and antidegradant systems, means that having a good technological background in rubber technology is of great assistance to the rubber analyst. This is a recurring theme throughout this book and its importance cannot be overestimated.

In addition to antidegradants and curatives, there are many classes of additives that may, or may not, be used in a particular rubber compound depending upon the properties that it needs to perform satisfactorily during processing and in

https://doi.org/10.1515/9783110640281-004

Table 4.1: Gamma radiation resistance of a range of rubbers.

Rubber type	Insignificant damage [radiation dose (Gy)]
Butyl rubber	Up to 10,000
Acrylic rubber	Up to 100,000
Silicone rubber	Up to 100,000
Chlorosulfonated PE	Up to 100,000
NBR	Up to 100,000
Fluorocarbon rubber	Up to 100,000
CR	Up to 100,000
SBR	Up to 500,000
EPM and EPDM	Up to 500,000
PU rubber	Up to 500,000

CR: Polychloroprene
EPDM: Ethylene propylene diene monomer EPM: Ethylene propylene monomer
NBR: Nitrile rubber PE: Polyethylene PU: Polyurethane
SBR: Styrene-butadiene rubber
Reproduced with permission from M.J. Forrest in *Rubber Analysis – Polymers, Compounds and Products*, Rapra Review Report No.139, Smithers Rapra, Shawbury, UK, 2001, p.55. ©2001, Smithers Rapra [1]

Table 4.2: Relative resistance of a range of rubbers in the unstabilised state to ozone.

Rubbers with very poor resistance	Diene rubbers (e.g., NR, SBR, NBR, butadiene rubber)
Rubbers having some resistance	Butyl rubbers, hydrin rubber, hydrogenated nitrile, polythioethers, CR
Rubbers having good resistance	Acrylics, chlorosulfonated PE, EPR, EPDM, fluorocarbon rubbers, silicone rubbers

NR: Natural rubber
Reproduced with permission from M.J. Forrest in *Rubber Analysis – Polymers, Compounds and Products*, Rapra Review Report No.139, Smithers Rapra, Shawbury, UK, 2001, p.55. ©2001, Smithers Rapra [1]

service. Some of the more commonly encountered classes of additive that are used in this way are:

- Fire retardants
- Blowing agents
- Process aids
- Resins

Table 4.3: Approximate maximum service temperatures of a range of rubbers.

Rubber type	Approximate maximum service temperature (°C)
PU rubber	80
NR and polyisoprene rubber	80
Butadiene rubber and SBR	100
NBR and CR rubber	120
Butyl rubber	130
EPR and EPDM rubbers	140
Chlorosulfonated PE	150
Hydrogenated NBR	160
Acrylic rubber	160
Silicone rubber	225
Fluorocarbon rubber	250

Reproduced with permission from M.J. Forrest in *Rubber Analysis – Polymers, Compounds and Products*, Rapra Review Report No.139, Smithers Rapra, Shawbury, UK, 2001, p.55. ©2001, Smithers Rapra [1]

The analytical techniques and approaches that can be used to determine antidegradants, curatives and these other types of additives will be reviewed in this section. For this type of work, because of the relatively low levels and complex mixtures involved, it is combinations involving chromatographic techniques and certain spectroscopic techniques [e.g., gas chromatography (GC)–mass spectrometry (MS) and liquid chromatography (LC)–MS] that are usually the most effective due to their sensitivity, specificity and diagnostic power. However, it is also possible to use other many approaches and techniques. For example, the relatively volatile nature of a many rubber additives enabled Lattimer [2] to use a direct probe method MS to analyse the additives present in a vulcanised rubber sample.

The use of the thermal technique thermogravimetric analysis (TGA) to identify and quantify principal additives (i.e., plasticisers and fillers) has been addressed in Chapter 3, and it has a part to play in investigative work of this type (e.g., stability studies), but unless it is coupled with an infrared (IR) spectrometer or MS it does not have the sensitivity or diagnostic power to be capable of providing specific information on many of the additives mentioned above. Similar comments apply to other popular thermal analysis techniques, such as differential scanning calorimetry (DSC) and dynamic mechanical analysis (DMA).

As with all the sections in this book, this is a large subject in its own right and a number of specialist texts exists, for example, the book written by Crompton [3]. There are also published articles which provide useful background information and look to the challenges and needs of the future [4].

In addition to the information in this section, there are also many other examples in this book in which additives have been determined in rubber products. This work has been carried out for a wide range of purposes, including reverse engineering (Chapter 5), curing studies (Chapter 6), rubber fume studies (Chapter 7) food-contact work (Chapter 8) and extractables and leachables (E&L) work (Chapter 9), but often the analytical methodologies and approaches that have been used are flexible enough to be adapted and applied to other applications and objectives.

4.2 Determination of antidegradants in a rubber compound

In common with other polymer systems, to maintain the optimal physical properties imparted to a rubber compound by a judicious choice of polymer type, filler, plasticiser and cure system, stabilisers need to be added to stop, or at least retard, degradation by a number of agencies [e.g., heat and ultraviolet (UV) light].

The use of an initial solvent extraction step is common in the analysis of a rubber matrix for antidegradants, and other types of additives because these species may not be volatile enough to be removed by heating the sample. The use of organic solvents to remove additives (e.g., plasticisers) from rubber samples was introduced in Section 3.2 and, particularly when quantification as well as identification is required, similar considerations apply in these studies in that the solvent should be compatible with the additive of interest. Table 3.6 provides, as a guide, a list of the solubility parameters of a range of solvents, a number of which can be used for this type of work.

In the case of some relatively volatile antidegradants [e.g., butylated hydroxytoluene (BHT) and isopropyl p-phenylenediamine (IPPD)], heating a sample liberates enough of the species to enable an identification to be made using GC–MS. However, quantification of an antidegradant, even volatile examples, will usually require extraction by a solvent to as near to 100% efficiency as possible. Analysis of the extract can then be carried out by high-performance liquid chromatography (HPLC) or spectroscopy (UV or IR) or, providing the species has sufficient thermal stability and volatility, GC. A chromatographic approach to prevent 'masking' by other co-extractants will usually be required if the extract is relatively complex. For this application, relatively non-specific solvents (i.e., those having mid-range solubility parameters), such as chloroform and acetone, are good extracting solvents. If a high proportion of oligomeric material is present in the extract (sometimes a problem with unvulcanised rubber samples), it can be necessary to add a precipitating solvent (e.g., methanol or heptane) to prevent this from interfering with the chromatographic or spectroscopic step.

If the type of organic antidegradant in a compound is completely unknown, analysis of the solvent extract by GC–MS is an excellent method for identification. Once the identification has been achieved, quantification can be carried out using a preferred method and a validated procedure (e.g., a published standard or in-house method) if one is available. To ensure confidence in the data, a validated method will usually have to be developed if one is not already available.

If the antidegradant is not volatile enough to be identified using a GC-based technique, direct probe MS can be used and the antidegradant identified by reference to its molecular ion and a database of molecular weights (MW). This technique can be carried out on samples in the as-received state as well as dried sample extracts.

4.2.1 Ultraviolet stabilisers

The widespread use of carbon black (CB) in rubber compounds gives the additional advantage of stabilisation against UV light, with the filler preferentially absorbing the energy and so protecting the polymer. The determination and quantification of CB (usually regarded correctly as a filler) in a rubber is described in Section 3.5.

The white pigment titanium dioxide (TiO_2) also provides protection against UV light and can be incorporated into non-black articles. The amount of TiO_2 in a rubber can be determined after acid digestion of the matrix by the use of an elemental technique such as atomic absorption spectroscopy or inductively coupled plasma.

Benzotriazoles and hindered amines can be used as UV stabilisers. The UV-absorbing qualities of these species can be employed for their detection and quantification by UV spectroscopic analysis of solvent extracts of rubber samples.

4.2.2 Antioxidants

The International Organization for Standardization (ISO) standard, ISO 11089, deals with the determination of N-phenyl-β-naphthylamine and poly-2,2,4-trimethyl-1,2-dihydroquinoline (TMQ) as well as two generic types of antiozonant (Section 4.2.3). This method has been tested on CR, NBR and SBR and involves an initial extraction step using a 2:1 mixture of isopropanol and dichloromethane (DCM). The extract obtained is then analysed by reverse-phase HPLC using a mixed solvent programme with UV detection.

In addition to the above, HPLC has been widely used for many years by analysts for the determination of antioxidants in rubbers, and reviews have been provided by Sidwell [5] and Sullivan and co-workers [6].

A useful HPLC method, which is a modification of a method developed by an antidegradant manufacturer, can be used on rubbers for the detection and quantification of a wide range (i.e., 20 species) of antidegradants. The sample is cryogenically ground up into a fine powder and then extracted with diethyl ether for 30 min with ultrasonic agitation. The extract is then analysed by HPLC using the following conditions:
- Injection volume: 25 μl
- HPLC column: aqua 5 C18 ODS reverse-phase column 150 × 3.0 mm (or similar)
- Mobile phase: A = 75/25 methanol/water and B = 50/50 ethyl acetate/acetonitrile:

Gradient	%A	%B
Initial	90	10
After 25 min	0	100

- Flow rate: 2.0 ml/min
- Detection: UV detector set at 270 nm
- Temperature: 40 °C

The following antioxidants can be determined using this technique:
- Irganox® 1010
- Irganox® 1098
- Irganox® 1076
- Irganox® 1330
- BHT

The wide applicability of this HPLC method is due to the non-specific nature of the mobile-phase mixture.

The capability of HPLC has been greatly extended by the commercialisation of affordable LC–MS instruments. The mass selective detector enables unknown samples to be analysed and the antioxidants present to be identified. The LC–MS technique, therefore, compliments GC–MS and, with it, enables an extensive MW/volatility range to be covered.

Thin-layer chromatography (TLC) is comparatively easy and cheap to use and it is both quick and accurate for the identification of antioxidants in rubber extracts. Irrespective of the complexity of the extract composition and the amount used, this method can be used to give a sharp separation for identification purposes. Distinctive retardation factor values and colours can be seen for each antioxidant depending on the stationary phase, the developing solvent used and the detection agent [7, 8]. Normal silica gel has proved effective as a stationary phase, as have others such as alumina starch, gypsum–silica gel and starch–silica gel, and where identification is difficult because of the interferences of co-extracted additives, two-dimensional development methods using two types of solvent system have been employed. [9].

The TLC technique for antidegradants is also described in the American Society for Testing of Materials (ASTM) standard, ASTM D3156 and ISO 4645. The rubber extract is taken up in DCM and then a spot of this solution developed in the developing tank using the nominated solvent. In addition to being used for identification purposes, TLC can be used in a semi-quantitative way; the logarithm of the spot area being proportional to the quantity of the analyte in the spot, and this value being compared with spots containing standard amounts of the analyte. The complexity of rubber extracts usually limits the accuracy of the quantifications obtained.

GC has been used for the analysis of phenolic and amine-type antioxidants. Antioxidants which have high boiling points cannot be directly analysed by GC but they can be analysed in the form of derivatives, such as acetates, trifluoroacetates, trimethyl silylethers, and methyl ethers. Trimethyl silane-based antioxidants, for example, give good separations on standard siloxane-based stationary-phase columns [10].

The newer, more sophisticated variants of GC–MS-type instruments also offer great potential in this area. An example is GC×GC–time-of-flight (ToF)–MS and this type of instrument has been used to analyse rubber samples by Forrest and co-workers [11]. Data were obtained by this research team from an acetone extract of a cured butyl rubber sample using the analytical conditions shown below:

- Sample for analysis:
    - Extract obtained from 0.3 g of finely divided sample which was ultrasonicated with acetone for 30 min.
- Instrument details and analysis conditions:
    - Instrument: Agilent 6890 gas chromatograph with LECO Pegasus III GC×GC–ToF–MS.
    - Injection: programmed temperature vaporiser injection, 10 °C above primary oven temperature; 1 μl of the acetone extract.
    - Primary column: J&W scientific HP-5MS (30 m × 0.250 mm with film thickness of 0.25 μm).
    - Secondary column: SGE BPX-50 (1.8 m × 0.100 mm with film thickness of 0.10 μm).
    - Carrier gas: helium at 1.0 ml/min with constant flow.
    - Primary oven program: 40 °C for 10 min then 10 °C/min to 320 °C and hold for 15 min.
    - Secondary oven program: 50 °C for 10 min then 10 °C/min to 330 °C and hold for 15 min.
    - Modulator offset: 30 °C.
    - Modulator frequency: 4 s.
    - Hot time of the modulator: 0.30 s.
    - Mass spectrometer: scanning 30 to 650 Da at 76 spectra/s.

The data generated by a GC×GC–ToF–MS instrument can be presented in a number of ways (Section 5.2.1). For example, Figure 4.1 shows the 3D chromatogram total ion current (TIC) that resulted from the analysis of the butyl rubber acetone extract.

The information shown in Figure 4.1 can be contrasted with the chromatogram obtained on the same butyl rubber extract by conventional GC–MS (Figure 4.2) using the analytical conditions shown below:

- Sample for analysis:
    - Extract obtained from 0.3 g of finely divided and ultrasonicated with acetone for 30 min.

- Instrument details and analysis conditions:
  - Instrument: Agilent 6890/5973 GC–MS.
  - Injection: 1 μl splitless at 310 °C.
  - Carrier: helium at 1.0 ml/min.
  - Column: Restek RTX-5 amine 30 m × 0.32 mm and film of 1.0 μm.
  - Column oven: 40 °C for 5 min; 20 °C/min increase up to 300 °C held for 15 min (30-min run time).
  - MS: scanning 35 up to 650 Da every 0.25 s.

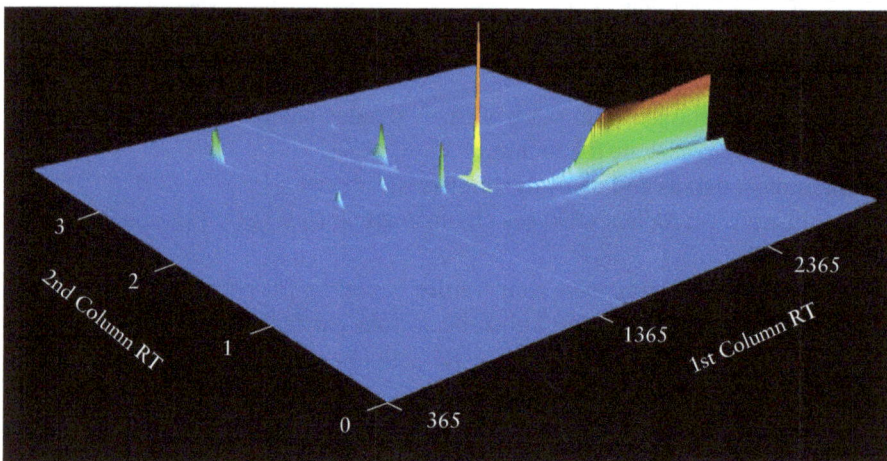

Figure 4.1: 3D GC×GC–ToF–MS TIC chromatogram for the butyl rubber acetone extract (RT: retention time). Reproduced with permission from M.J. Forrest, S. Holding and D. Howells, *Polymer Testing*, 2006, 25, 63. ©2006, Elsevier [11] and M.J. Forrest, S. Holding and D. Howells in *Proceedings of the High Performance and Specialty Elastomers*, 20–21st April, Geneva, Switzerland, Rapra Technology Ltd, Shawbury, UK, 2005, Paper No.2. ©2005, Rapra Technology Ltd [12].

In the case of this sample, both the GC×GC–ToF–MS and conventional GC–MS analyses were successful and by analysing and interpreting the data it was possible to identify the antidegradant that was present in the butyl rubber: 6-ethoxy-1,2-dihydro-2,2,4-trimethylquinoline. If the rubber sample, and hence its solvent extract, had been more complex, then the GC×GC–ToF–MS system would have provided the analyst with a better chance of success due to its superior resolving power.

4.2.3 Antiozonants

To date, the only organic compounds shown to be very effective in the protection of diene rubbers against chain scission by ozone (*via* the cleavage of the carbon–carbon double bond) are *p*-phenylenediamine compounds.

Abundance

TIC: 0804031.D

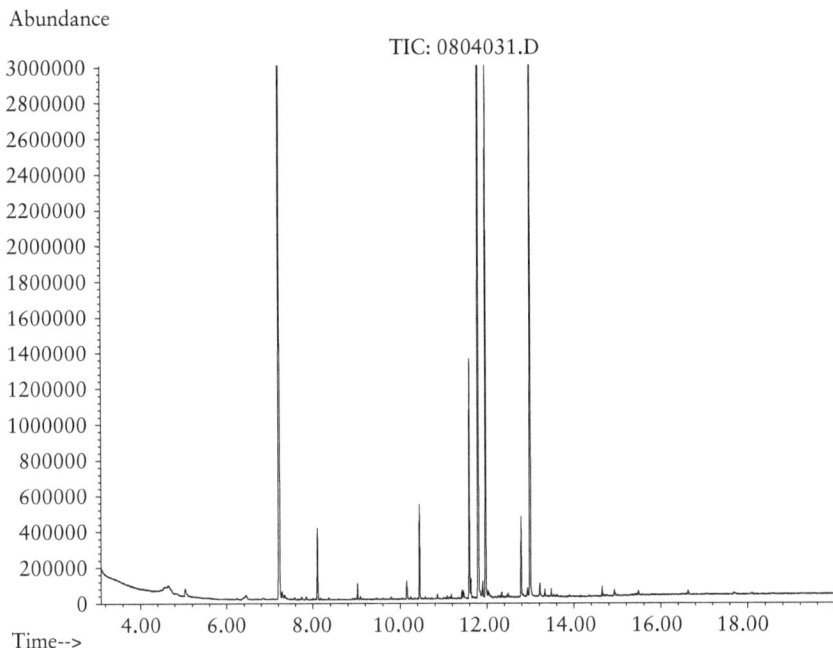

Time-->

Figure 4.2: Conventional GC–MS TIC chromatogram for the butyl rubber acetone extract. Reproduced with permission from M.J. Forrest, S. Holding and D. Howells, *Polymer Testing*, 2006, 25, 63. ©2006, Elsevier [11] and M.J. Forrest, S. Holding and D. Howells in *Proceedings of the High Performance and Specialty Elastomers*, 20–21st April, Geneva, Switzerland, Rapra Technology Ltd, Shawbury, UK, 2005, Paper No.2. ©2005, Rapra Technology Ltd [12].

In addition to the analysis of antioxidants (Section 4.2.2), the standard ISO 11089 can be used to determine N-alkyl-N'-phenyl-p-phenylenediamines [e.g., IPPD and N-(1,3-dimethyl butyl)-N'-phenyl-p-phenylenediamine (6PPD)] and N-aryl-N'-aryl-p-phenylenediamines (e.g., diphenyl-p-phenylenediamine) by HPLC. The HPLC chromatogram of a methanol extract from a CR compound, obtained using a UV detector, is shown in Figure 4.3. From the data in Figure 4.3 it is possible to identify and quantify a number of peaks due to diaryl-p-phenylenediamine antiozonants (highlighted in the chromatogram) in the sample.

A microcrystalline PE wax is often added to a rubber formulation along with a p-phenylene-type antiozonant. These types of wax have a limited solubility in the rubber matrix and so bloom to the surface. In doing so they help the antiozonant migrate to the surface where it is needed to function but, more importantly, they provide a physical barrier to ozone. This approach is most successful in static applications, but does not work for dynamic applications because as the rubber product is stressed the protective film will be ruptured. The fact that these waxes bloom to the surface can be used for their detection by surface analysis techniques such as IR microspectrometry.

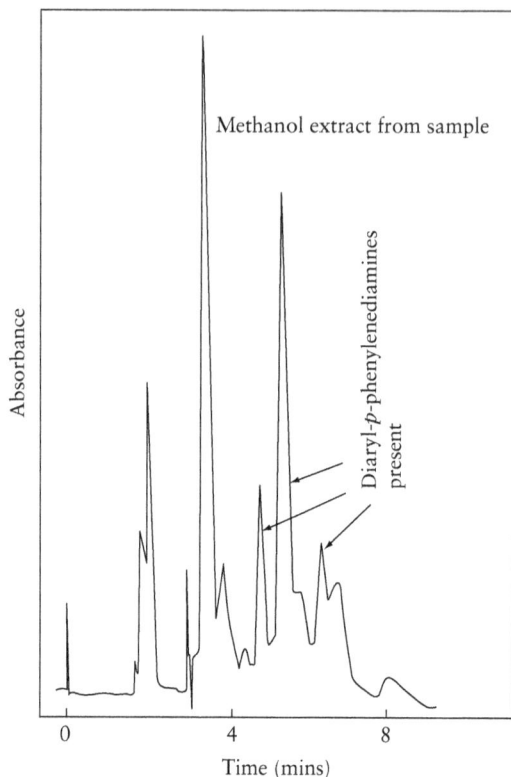

Figure 4.3: HPLC UV chromatogram of the methanol extract from a CR rubber compound. Reproduced with permission from M.J. Forrest in *Rubber Analysis – Polymers, Compounds and Products*, Rapra Review Report No.139, Smithers Rapra, Shawbury, UK, 2001, p.26. ©2001, Smithers Rapra [13].

Unfortunately, these waxes are added only at relatively low doses to the compounds (e.g., 2–5 phr) and so their detection by bulk analytical techniques [e.g., DSC, pyrolysis (Py) IR] is unlikely to succeed, and their relatively high MW and relatively poor solubility makes isolation by solvent extraction difficult.

TLC can be used for the analysis of *p*-phenylenediamines. The best results for the isolation and identification from extracts are obtained using isopropanol/chlorobenzene/ water/25% aqueous ammonia (52:33:10:5) as the developing solvent, and diazotised 4-amino-benzene sulfuric acid or 3,5-dibromo-*p*-benzoquinonechlorimine as the colour-forming reagent.

4.2.4 Degradation studies

Sections 4.2.1–4.2.3 have covered the analytical techniques and methods used to identify and quantify the different types of antidegradant used in rubber products.

A significant amount of investigative work has been carried out on the function of these additives in rubbers as well as the breakdown products and reaction products that can result. To illustrate the findings of this work, a selection of it will be included here. However, because the products of stabilisation and antidegradant activity are often substances with relatively low MW, they are of interest in other important areas of research and characterisation work. Hence, related information will also be found in the sections of this book that deal with those, for example:
  – Composition of rubber fume (Chapter 7)
  – Migration from rubber products into food (Chapter 8)
  – E&L studies on rubber products (Chapter 9)

The analytical techniques that can be employed to study the degradation of rubber and some examples of the literature that has published recently in this area are introduced below.

Although they cannot be used to identity the type of antioxidant in a rubber sample, it is possible to use the thermal analysis techniques TGA and DSC to obtain a measure of the thermal oxidative stability of the rubber compound that results from their use and, in the case of DSC, the energetics of their protective action. In the latter case, the studies that have been published include work on the use of p-phenylenediamines to protect polyisoprene rubber [14, 15] and that of Parra and Matos [16], who investigated the synergistic effects of a range of different antioxidants in NR.

In addition to being able to contribute towards research studies, this capability enables these techniques to be used in industry as quality-control tools and to assist in failure diagnosis where it is suspected that a sample does not possess sufficient protection because of the absence or an insufficient quantity of the antioxidant. One of the ways to use DSC for this type of work is to employ a relatively high experimental temperature (e.g., 100 °C), along with an oxidising atmosphere (e.g., air), and monitor the change in heat capacity as a means of indicating the onset of oxidation. When DSC is used for this purpose for plastics, the result obtained is referred to as the 'oxidation induction time', and there are standards that can be used to carry out the work (e.g., ISO 11357-6). With rubber compounds, because they tend to have more complex compositions and contain a higher proportion of volatile species, this type of investigative work can be more difficult because a change in heat capacity can also take place because of the volatilisation of substances or, if the sample is uncured, due to the volatiles released during curing. The use of sealed sample pans can help overcome these problems although, in the case of curing, sometimes the pressure that builds up within the pan is too great and it splits open. Another way in which DSC can be used to the study oxidation of rubbers is to use the fact that this type of event results in a change in heat capacity, with the glass transition temperature (T_g) occurring over a broader temperature range and at a higher temperature. This phenomenon is usually more obvious with rubbers that undergo some crosslinking during oxidation (e.g., diene-type rubbers).

The observation made above with regard to the disruptive effects of compounding ingredients volatilising during DSC is also applicable when TGA is used to obtain information on the stability of a rubber. In the case of the TGA experiment, it is a reduction in the temperature at which the rubber compound starts to pyrolyse that can be used. Because this event takes place at a relatively high temperature (e.g., >250 °C), a high experimental temperature in either an isothermal or temperature ramp mode has to be used, and so the influence of volatile ingredients often cannot be avoided. If all other factors are the same, the earlier and faster that the rubber starts to pyrolyse, the less thermally stable it is. It is important to point out, though, that this is a measure of non-oxidative thermal stability because an non-oxidising atmosphere (e.g., nitrogen) must be used to prevent the situation becoming even more complicated due to the oxidation of additives, particularly CB.

It is complications, such as those described above, with the use of TGA that means that DSC tends to be the more accepted technique for thermal stability work. However, researchers have shown how TGA can be used effectively in thermal stability studies, and this type of investigation can obviously have important contributions in the designing of a rubber compound to ensure that it has the appropriate stability for a particular end-use application. An example of this type of study has been performed by Denardin and co-workers [17], who used TGA with a ramped heating rate to show how the dynamic, non-isothermal Kissinger and Osawa methods can be used to determine the activation energies of the degradation processes that occur with CR rubber. Also, an extensive study of EPDM rubbers by Gamli and co-workers [18] involved using high-temperature (410 to 440 °C) isothermal TGA experiments and mathematical models based on proposed degradation mechanisms to establish the kinetic parameters of degradation.

Another thermal analysis technique that has a part to play in degradation and oxidation investigations and, hence provide information on the effectiveness of antioxidants, is DMA. In this work, use can be made of the ability of DMA to record comparative data on samples that have been through an ageing process, either an accelerated one in an oven or during use in service. When a rubber or plastic ages, free radicals are usually generated and, if not stopped by an antioxidant, these can lead to one of two effects:
-   Scission of polymer chains – resulting in a decrease in the modulus
-   Creation of crosslinks – resulting in an increase in the modulus

Often, these effects can both occur but usually one of them (the choice of which being dependant upon the type of rubber) will predominate. This means that the progress of ageing can be followed by DMA by monitoring the modulus of the rubber samples [19].

A number of researchers have used a combination of analytical techniques to study thermal stability. For example, Skachkova and co-workers [20] used isothermal TGA in conjunction with atomic force microscopy to study the thermo-oxidation of oil-extended EPDM/polypropylene thermoplastic rubbers. They found that the oil helped to stabilise the rubber by degrading and combining with the oxidised surface of the sample to form a partial barrier to oxygen. Pruneda and co-workers [21] used a number of techniques, including TGA, DSC and scanning electron microscopy (SEM), to obtain the activation energies associated with the degradation processes of NBR/ polyvinyl chloride blends that contain the antidegradant TMQ.

Ogawa and co-workers [22] used PyGC–MS to identify the volatile substances produced by NBR sheets that had been aged in an ozone-rich environment. The data they obtained revealed three substances related to the antioxidant in the NBR rubber. By focusing on the quinoline-type antioxidant itself, they also showed that when the concentration of this additive was reduced to $\leq$50% of its starting value by a ozone-ageing regimen, the mechanical strength of the 0.1-mm surface layer of the rubber became critically low.

Hakkarainen and co-workers [23] evaluated the behaviour of NBR during long-term thermal ageing by following the migration of low-MW additives and the thermal degradation products of additives. Volatile substances were detected and identified using headspace GC–MS analysis and the oxidation of the butoxyethyl phosphate plasticiser was followed by Fourier–Transform infrared (FTIR) spectroscopy.

PyGC–MS was mentioned in Section 3.2 as a means of identifying the polymers present in rubber samples. It can also be used to identify antioxidants and other additives and to obtain information on the changes that can occur to these ingredients when a rubber ages in service or under accelerated conditions. An example of this application is provided by Yang and co-workers [24]. They used a flash evaporation PyGC–MS technique to analyse the volatile additives and macromolecular structure of a number of polymeric materials, including a NBR in aged and unaged forms. The research team found that during the thermal oxidation of NBR seals, additional crosslinks were formed, but there were no additional volatile degradation products. The principal change that they detected was a decrease in the level of volatile additives, particularly paraffins, the antioxidant, and hindered phenol compounds, which was accompanied by the hardening of the rubber and a reduced level of protection from further oxidation. Depth profiling showed that different additives had different migration speeds within the rubber and, from the levels of additives that remained, it was possible to predict the degradation status of the rubber. Overall, the research team considered that PyGC–MS could provide extensive information on how polymers degrade and so could make a valuable contribution towards research into degradation mechanisms.

Tawfic and Hussein [25] prepared an adduct amine-epoxy (Am-Ep) by reacting together N-methyl-p-phenylenediamine and epoxidised oil. They then added this adduct in differing amounts to EPDM and NBR rubbers and assessed its influence on thermal stability, blend compatibility and mechanical properties. The analytical techniques used in the work included SEM and DSC and the results obtained revealed that the optimum addition level of the Am-Ep adduct was 7.5 phr.

4.3 Determination of cure system additives and their breakdown products

This section looks at the analysis of rubbers to obtain information on their cure systems. There are two principal types of cure system used in rubber products: sulfur or peroxide. Where either of the systems could have been used (e.g., EPDM rubbers) it is usually possible to differentiate between them by carrying out a sulfur determination. Although additives such as CB will contribute to the overall sulfur level in a rubber compound, a low result (e.g., <0.3% w/w) is usually indicative of a peroxide-cure system. Techniques to quantify the amount of sulfur in a rubber are discussed in Sections 2.2 and 2.3. In common with most other aspects of rubber analysis, there can be complications, and it is possible for technologists to add elemental sulfur into a rubber as a peroxide cure co-agent. Therefore, a relatively high sulfur level cannot always be taken to indicate a sulfur-type cure system and further work (e.g., GC–MS) needs to be carried out. The identification of peroxide co-agents such as triallyl cyanurate (TAC) are covered in Section 4.4.4.

The fact that some sulfur-cure system accelerators [e.g., tetramethylthiuram disulfide (TMTD)] produce nitrosamines during the curing process has given cause for concern for a number of years. Because of the potential health and safety concerns, a number of studies have been carried out on these species. Analysis of rubber compounds for nitrosamines usually takes the form of extraction followed by GC analysis using a specific detector (e.g., electrolytic conductivity or thermal electron analyser). A review of the subject of nitrosamines in rubber is provided by Willoughby and Scott [26].

In addition to the information presented in this chapter, there is also information that is applicable and relevant to the determination of curing system additives in Section 4.5 and, as mentioned in Section 4.1, in many other chapters of this book that deal with particular themes such as reverse engineering (Chapter 5) and curing studies (Chapter 6).

4.3.1 Analysis of uncured samples

In the case of failure and deformulation work it is usually only cured rubber which is available for analysis because the sample will usually have originated from the a final product. However, there are examples, and quality-control work is one, where the uncured sample will be available. This will provide the opportunity to look for un-reacted accelerators and curing agents such as peroxides. Particular care has to be taken because these chemicals are, by their nature, thermally labile and reactive. It is not possible, therefore, to use hot solvent extraction (e.g., reflux or Soxhlet extraction) or high-temperature analysis techniques, such as GC–MS, because these will bring about the same types of reactions that occur during curing. Rather, cold extraction is used along with an ambient temperature separation and identification technique such as TLC or HPLC. It should also be borne in mind that, even at ambient temperature, the choice of extraction solvent can be important. The use of acetone can cause thiuram disulfide accelerators (e.g., TMTD) to decompose to dithiocarbamates. For this reason, the use of solvents such as methanol, 2-propanol or DCM is preferred.

Gross and Strauss [27] described the use of HPLC for the analysis of accelerators. There are many reviews concerning the use of TLC, including those by McSweeny [28] and Hummel and Scholl [29].

In recent years, advances in technology have reduced the price and, hence, increased the accessibility of LC–MS instruments and these are often used along with GC–MS instruments to identify cure systems ingredients and their breakdown products and reaction products. Unlike GC–MS, which can use the published 70-eV electron-impact mass spectra libraries, there are no commercial libraries available for LC–MS and so in-house libraries need to be generated under specific conditions of use. Once these are available, however, these instruments offer the analyst an increased range of options over traditional HPLC systems (Section 2.5.5).

4.3.2 Analysis of cured samples

The analysis of cured samples is the more usual state of affairs. It is necessary to have a certain amount of knowledge on the types of species that are formed during vulcanisation reactions if success in identifying a cure system is to be forthcoming.

The most useful generally available technique for this type of work is GC–MS. There are two possible modes of sample introduction: solution injection and headspace (dynamic or static mode). In the former, a useful method of sample preparation involves cryogenic grinding of 0.3 g of the cured sample, followed by extraction using 2 ml of a non-selective solvent (e.g., diethyl ether or acetone) in an ultrasonic bath for 30 min.

The resulting solvent extract is then analysed, for example, under the following conditions:
- Instrument: Hewlett Packard 6890/ 5973 GC–MSD
- Carrier: helium at 2.0 ml/min constant flow
- Injection mode: 1 μl splitless (0.75 min; 50 ml/min)
- Injection temperature: 320 °C
- Column type: RTX5-MS 30 m × 0.25 mm, 0.25 μm film thickness
- Column temperature: 40 °C for 5 min, 20 °C/min up to 300 °C held for 12 min
- MS settings: 20 to 620 Da scanned every 0.33 s

The solvent extraction method enables the large majority of the diagnostic species of interest to be detected. Other non-cure system species such as antidegradants and process aids will also be identified. The main drawback with this approach is that volatile diagnostic compounds such as low-MW amines (e.g., dimethylamine) can be masked by the large solvent peak at the start of the chromatogram.

The headspace techniques provide a means of identifying the largest range of cure system breakdown products, particularly in conjunction with cryogenic cooling of the GC oven to improve resolution of the early eluters. In this approach, a relatively small amount of sample (e.g., 3 mg) is heated in the oven of a dynamic desorption unit and the volatiles produced are collected in a cryogenically cooled Tenax™-filled trap. The experimental conditions are, typically, as follows:

- Dynamic headspace:
  - Instrument: Perkin Elmer ATD 400
  - Desorption temperature: 150 °C for 10 min
  - Desorption flow: helium at 20 ml/min
  - Trap collection temperature: −30 °C
  - Trap injection temperature: 250 °C
  - Trap outlet split flow: 9 ml/min
  - Trap inlet split flow: off
  - Split: 10:1
- GC–MS – The conditions used are the same as for the solvent extraction method with the exception of the following:
  - GC column temperature: −30 to 50 °C at 5 °C/min, following by 20 °C/min to 300 °C and then held at 300 °C for 12 min
  - Carrier: helium at 10.5 psi

As discussed in the case of antioxidants (Section 4.2.2), the newer GC–MS-type instruments offer great potential in the area of additive analysis. Forrest and co- workers [11] used GC×GC–ToF–MS to determine the additives, including the curatives, in an acrylic rubber sample. Data were obtained from an acetone extract of the cured acrylic rubber sample using the same sample preparation conditions and instrument conditions as those shown in Section 4.2.2 for the analysis of the butyl rubber extract.

Figure 4.4 shows the 3D GC×GC–ToF–MS TIC chromatogram for the cured acrylic rubber extract. Figure 4.5 shows the TIC chromatogram recorded on the same extract with a conventional GC–MS instrument operated under the same conditions shown in Section 4.2.2.

By analysing and interpreting the data produced by these two techniques it is possible to obtain information on the cure system that was used to cure the acrylic rubber. This was found to include the amine N,N′-di-o-tolyl guanidine.

For either analytical approach, identification of the diagnostic cure system breakdown species is by reference to mass spectral libraries, such as those produced by the National Institute of Standards and Technology, the Royal Society of Chemistry, Wiley/NBS [30] and the USA Environmental Protection Agency/National Institutes of Health. The mass spectra of the majority of species of interest are in

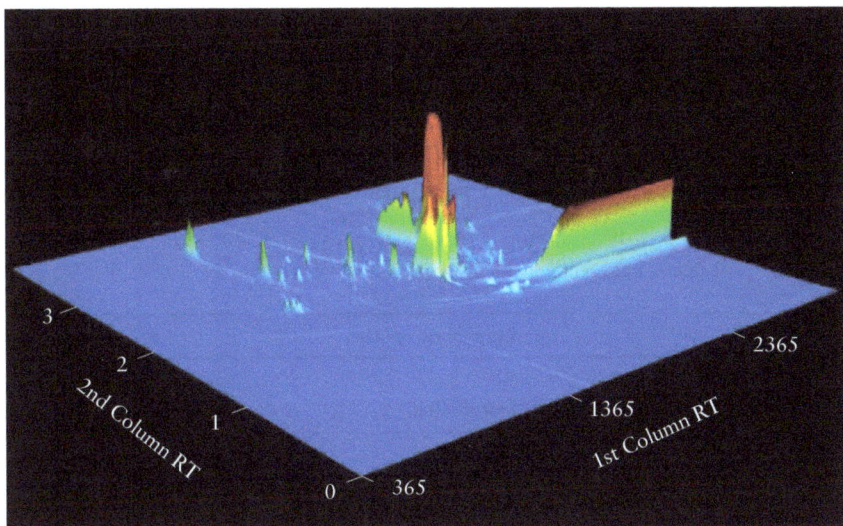

Figure 4.4: 3D GC×GC–ToF–MS TIC chromatogram for the acrylic rubber acetone extract. Reproduced with permission from M.J. Forrest, S. Holding and D. Howells, *Polymer Testing*, 2006, 25, 63. ©2006, Elsevier [11] and M.J. Forrest, S. Holding and D. Howells in *Proceedings of the High Performance and Specialty Elastomers*, 20–21st April, Geneva, Switzerland, Rapra Technology Ltd, Shawbury, UK, 2005, Paper No.2. ©2005, Rapra Technology Ltd [12].

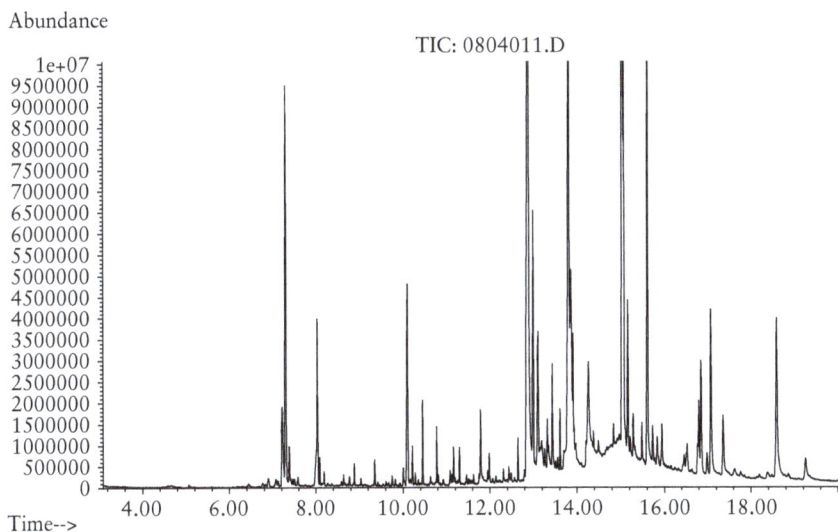

Figure 4.5: Conventional GC–MS TIC chromatogram for the acrylic rubber acetone extract. Reproduced with permission from M.J. Forrest, S. Holding and D. Howells, *Polymer Testing*, 2006, **25**, 63. ©2006, Elsevier [11] and M.J. Forrest, S. Holding and D. Howells in *Proceedings of the High Performance and Specialty Elastomers*, 20–21st April, Geneva, Switzerland, Rapra Technology Ltd, Shawbury, UK, 2005, Paper No.2. ©2005, Rapra Technology Ltd [12].

these libraries and so, providing that in-house knowledge to reconstruct cure systems is available, successful attempts can be made to interpret the results.

One problem associated with this task is that most accelerators and curing agents, such as peroxides, break down to give more than one product and a number of these are common to more than one cure system species. This is demonstrated in Table 4.4 and these types of diagnostic data are included in a book by Willoughby [31].

Table 4.4: Diagnostic species associated with different cure systems.

Sulfur cure system	Diagnostic species
2-(4-Morpholinothio)benzothiazole	BT
	Morpholine
Sulfrasan R	Morpholine
MBT	BT
2,2′-Dithio*bis*(benzothiazole)	BT
TMTD	Dimethylamine
	Tetramethylthiourea
	Dimethylformamide
Zinc dimethyldithiocarbamate	Dimethylformamide
	Dimethylamine
Peroxide curative	**Diagnostic species**
Perkadox 14/40	Tertiary butanol
	1,3-Di-(1,1-dimethylmethanol) benzene methanol
Tert-butyl cumyl peroxide	1,1-Dimethyl benzene methanol
	Acetophenone
	Tertiary butanol
Dicumyl peroxide	1,1-Dimethyl benzene methanol
	Acetophenone

BT: Benzothiazole
MBT: 2-Mercaptobenzothiazole
Reproduced with permission from M.J. Forrest in *Rubber Analysis – Polymers, Compounds and Products*, Rapra Review Report No.139, Smithers Rapra, Shawbury, UK, 2001, p.28. ©2001, Smithers Rapra [32]

Another example of GC–MS being used to identify cure system breakdown products and, hence, provide possible identities to the compounds that have been used in the original compound is provided by Sidwell [33] in the results obtained from the analysis of water that had contacted a NR compound (designated NR2). The formulation of this NR2 compound is shown in Table 4.5 and samples for analysis were obtained by curing for 6.25 min at 155 °C.

Table 4.5: Formulation of NR compound – NR2.

Ingredient	phr
NR (SMR-CV60)	100
ZnO (cure activator)	5
Stearic acid (cure activator)	1
4,4'-Thio-*bis*-(3-methyl-6-*tert*-butyl phenol) (antioxidant)	1
6PPD (antiozonant)	1
Hydrocarbon wax (Okerin 1944)	2
High abrasion furnace CB N330 (reinforcing filler)	35
Naphthenic oil (petroleum-based oil lubricant)	5
Sulfur (curative)	1
CBS	2.5

CBS: N-cyclohexyl-2-benzothiazole sulfenamide (cure accelerator) phr: parts per hundred of rubber
Reproduced with permission from J. Sidwell in *Proceedings of the Food Contact Polymers Conference*, 21–22nd April, Brussels, Belgium, Smithers Rapra, Shawbury, UK, 2009, Paper No.7. ©2009, Smithers Rapra [33]

The GC–MS TIC chromatogram that was obtained on a DCM-partitioned, distilled water sample that had contacted a cured sample of NR2 for 24 h at 40 °C is shown in Figure 4.6. The peaks of the three substances highlighted in this chromatogram (i.e., cyclohexanamine, BT and mercaptobenzothiazole) are the breakdown/reaction products of the CBS accelerator in the compound and can be regarded as diagnostic species, useful for the identification of this compound during reverse engineering work. However, care has to be taken during such evaluations because two of these species (BT and mercaptobenzothiazole) can also be formed by other accelerators (Sections 5.3.3 and 9.2.3.3). It is, therefore, important to build up an in-house database of such breakdown products and to consult other sources (e.g., the database generated during the FSA research project A03038) (Section 8.1).

In addition to the three major breakdown substances shown in Figure 4.6, a number of other products that were also related to the CBS accelerator were detected in the distilled water by GC–MS and these are all shown in Figure 4.7.

In the same study, Sidwell also reported on the use of headspace GC–MS for the detection of low-MW monomers, breakdown products of accelerators and reaction byproducts of the cure system in rubber samples. A sulfur-cured food-grade NBR sample was prepared and the volatiles that were released at 150 °C were examined by headspace GC–MS. Among the substances detected were:
- Acrylonitrile (ACN) (residual monomer)
- Dimethylamine [breakdown product of the accelerator tetramethylthiuram monosulfide (TMTM)]
- Carbon disulfide (reaction product of the sulfur cure system)

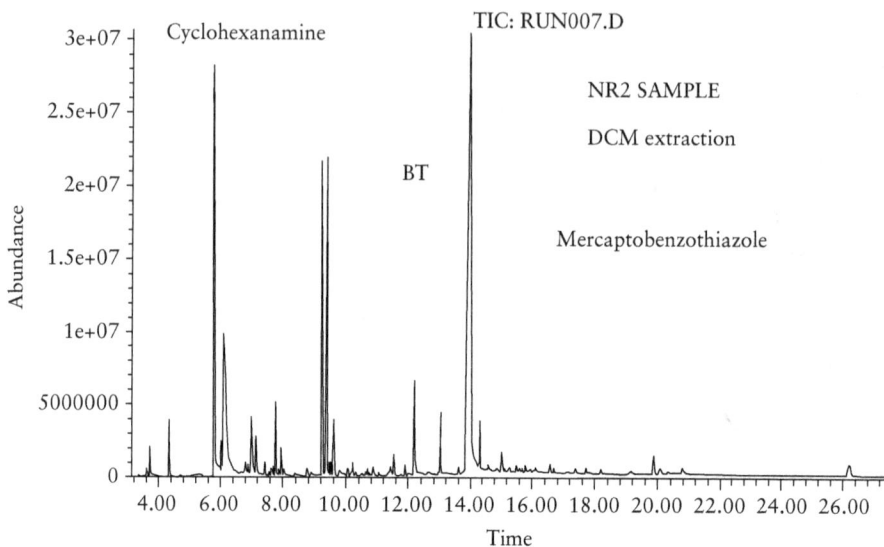

Figure 4.6: Water extractables obtained at 24 h at 40 °C from compound NR2 (partitioned into DCM and examined by GC–MS) [33]. Reproduced with permission from M.J. Forrest in *Food Contact Rubbers 2 – Products, Migration and Regulation*, Rapra Review Report No.182, Smithers Rapra, Shawbury, UK, 2006, p.28. ©2006, Smithers Rapra [34].

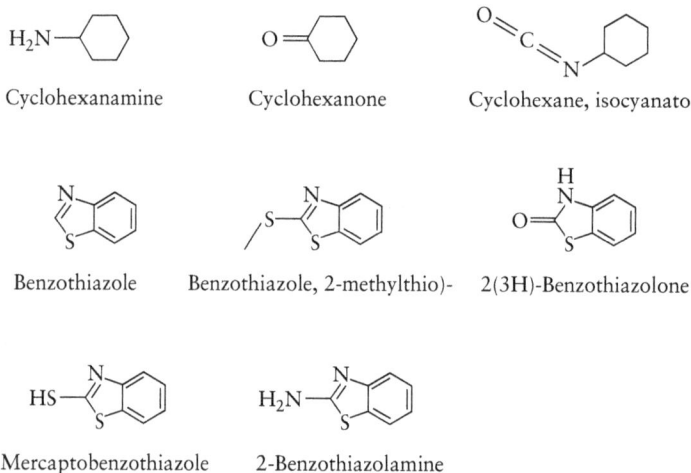

Figure 4.7: Breakdown products of CBS detected by GC–MS in distilled water that had contacted NR2 for 24 h at 40 °C. Reproduced with permission from M.J. Forrest in *Food Contact Rubbers 2 – Products, Migration and Regulation*, Rapra Review Report No.182, Smithers Rapra, Shawbury, UK, 2006, p.29. ©2006, Smithers Rapra [35].

The presence of these substances eluting early on in headspace GC–MS chromatogram is shown in Figure 4.8.

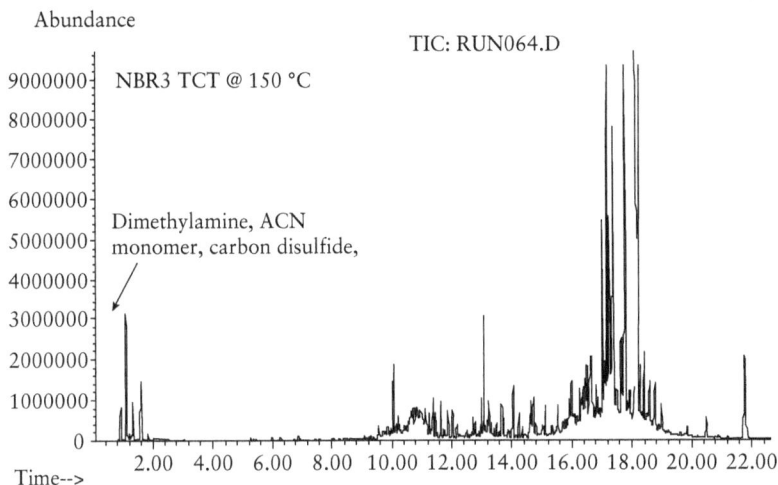

Figure 4.8: Headspace GC–MS TIC chromatogram of a cured NBR sample (TCT: thermal desorption cold trap injector). Reproduced with permission from J. Sidwell in *Proceedings of the Food Contact Polymers Conference*, 21–22nd April, Brussels, Belgium, Smithers Rapra, Shawbury, UK, 2009, Paper No.7. ©2009, Smithers Rapra [33].

LC–MS systems also assist in the analysis of vulcanised rubber to detect cure system species because the types of species being investigated will be amenable to reverse-phase HPLC. Examples of how LC–MS, in conjunction with GC–MS, has been applied to food-contact rubbers (e.g., EPDM) as part of an overall analysis of the low-MW species present are provided in Section 8.4.2.2. Another example is provided in Section 9.4, where LC–MS, again used in conjunction with GC–MS, is employed to identify the substances, including cure system species, that can be extracted from an NBR compound.

It is also possible to use PyGC to obtain information on the type of cure (e.g., sulfur or sulfurless) that has been used to cure rubbers. For example, a study carried out in China [36] showed how PyGC can be used to distinguish between the different cure systems employed to vulcanise NR. The method that was developed was based on the approximate linear relationship between sulfur content and the relative areas of characteristics Py products in the chromatogram. A standard deviation of 1.02 and a coefficient of variation of 2.96% was claimed for the method.

4.3.3 Cure state and cure-state studies

The sections above have described the various analytical approaches that can be used to identify and quantify the curatives and cure system components present in rubber samples. Due to the large amount of information that has been published, and its fundamental importance to rubber technology, rather than produce a sub-section within this section, analogous to that on ageing studies that appears at the end of the antidegradant and stabiliser section (i.e., Section 4.2.3), the subject of cure state has been given its own section in this book: Chapter 6.

4.4 Determination of miscellaneous additives

4.4.1 Blowing agents

There are two main types of blowing agent used with polymer systems physical and chemical, and these are defined below:

1. Chemical type – chemical compounds (principally organic) that undergo chemical decomposition at the processing temperature to form a gaseous species (e.g., nitrogen) which will create the cellular structure. Examples include azo compounds, nitroso compounds, sulfonyl hydrazide compounds and sodium bicarbonate.
2. Physical type – low boiling point organic compounds (e.g., DCM or pentane) which volatilise at the processing temperature to create the cellular structure.

In the case of rubbers, by far the most widely used types are chemical blowing agents, the majority of which decompose during curing to produce nitrogen. The reactivity of these compounds can make analysis difficult even if the rubber sample is uncured (e.g., they will decompose in the injection port of a GC–MS instrument), and the approach that is often employed to identify these additives is an indirect one *via* the analysis of the decomposition products. A GC–MS instrument is the best choice for this, used in the headspace mode, and an in-house library of typical breakdown fragments obtained by the analysis of standard rubber compounds is required.

Quantification of blowing agents in an uncured product is very difficult because their polarity and reactivity (e.g., hydrogen-bonding reactions and thermal instability) can cause problems in their isolation from the compound and subsequent analysis. In terms of the reverse engineering of commercial samples, this is unlikely to be required because the product will invariably be in the blown state. For laboratories that carry out quality control, other properties such as density and expansion ratios can be used to determine the level of a blowing agent in a compound.

If it is suspected that a physical-type blowing agent has been used to produce a foam, then it is also possible to use headspace GC–MS to obtain information on it. Even in the blown state, the high sensitivity of the technique enables an assignment of the decomposition gas to be made from the small amount of the residual blowing agent that will be present. It is uncommon for these types of blowing agents to be used in rubber compounds because the curing/processing temperatures of standard rubbers are usually too high to give controlled blowing. They tend to be used in amorphous plastics (e.g., polystyrene) which have a relatively low T_g ($\approx$100 °C). Given this application, there is some scope, therefore, for their use in thermoplastic rubbers.

Because the breakdown of a blowing agent (either physical or chemical type) is an energetic event, it is possible to use DSC to study their performance within a rubber compound. In the case of chemical types, an exotherm with result in the specific heat trace and, with the physical types, an endotherm because they remove heat from the sample to breakdown. By the use of reference samples, it is possible to use DSC to establish the:

- Type of blowing agent
- Amount of a particular blowing agent

It is important when using DSC for this kind of work that, as with curing studies (Chapter 6), sealed sample pans are used to prevent the gas that is generated escaping and influencing the specific heat data. In addition, a major complication that occurs with rubber samples that is rarely encountered with plastics is that, in order to achieve a satisfactory product, the blowing reaction occurs within the same temperature range as the curing reaction. The data obtained are, therefore, a composite of these two processes and empirical steps have to be taken to try and separate out the two events. One possible route is to analyse control samples that are equivalent in all respects except that they do not contain any blowing agent. The obvious potential flaw in this approach is that the blowing agent may influence the chemistry of the cure as it breaks down (e.g., by changing the pH or undergoing specific reactions) and so care has to be taken to obtain meaningful data.

4.4.2 Flame retardants

There are two main categories of flame retardant: organic and inorganic. The type of organic flame retardant present in a rubber compound can be determined in an analogous way to the plasticiser-type additives (Section 3.4) because this type of flame retardant is usually similar in MW and polarity to a number of these substances and, hence, they have similar properties (e.g., extractability and amenability to analysis by IR).

Once the type of organic flame retardant has been determined by FTIR spectroscopy, an accurate quantification can be obtained by carrying out an elemental analysis for a characteristic element (e.g., halogen or phosphorus) on the un-extracted sample. The extract value itself will rarely provide an accurate quantification because other species, particularly any plasticiser/oil, will contribute to it as well. In some instances (e.g., organophosphates), it may be possible to use an HPLC of the extract to quantify the flame retardant. GC-based techniques can also be used in some cases, but bromide types present problems for the GC-based techniques because of their relatively high MW (>900).

Care must be taken over the quantification of inorganic flame retardants because some of them (e.g., antimony trioxide) can react with any organic flame retardant present, or breakdown to produce volatile products (e.g., hydrated alumina), under quantitative ashing conditions, and also during analysis by thermal techniques such as TGA. A good initial approach involves carrying out a semi-quantitative elemental analysis on the rubber sample by X-ray fluorescence spectroscopy to see which type(s) of flame retardant are present. Accurate quantifications can then be obtained by precise elemental determinations.

Some inorganic flame retardants in a rubber sample can be quantified by making use of their specific characteristic properties once their presence has been confirmed. For example, if it is known that hydrated alumina is present in a sample, it is possible to quantify it reasonably well using TGA. This can be done by determining the early weight loss event that results from the loss of the water of hydration from the hydrated alumina and using the fact that this additive loses close to 35% of its original weight when it is heated to 550 °C.

It is possible to use other techniques to study flame retardants. Yang and co-workers [37] used a diverse combination (e.g., X-ray diffraction, SEM, TGA and oxygen index measurements) to study the morphological, mechanical, thermal and flammability properties of montmorillonite (MMT)-filled silicone rubber samples containing the synergistic flame retardants magnesium hydroxide and red phosphorous. The results showed that it required only 1% of the MMT nanomaterial to increase the decomposition temperature of the silicone rubber above the value of the control sample and that this high thermal stability was matched by good flame-retardant properties.

Ismawi and co-workers [38] evaluated the effects of various flame-retardant additives on the flammability and other properties of NR. The flame retardants that were used in the work, alone or in combination with one another, included the following:
- Decabromodiphenyl oxide
- Antimony trioxide
- Aluminium trihydroxide
- Organoclay
- Zinc hydroxystannate
- Chlorinated paraffin

The properties of the resulting NR samples were investigated by vertical burning tests, SEM, TGA, smoke density measurements, determination of the limiting oxygen index value, and cure characterisation work.

4.4.3 Process aids

4.4.3.1 Plasticisers and oils

The addition of hydrocarbon oil or a synthetic plasticiser (e.g., dioctyl phthalate) into a compound will improve its processability, in addition to changing a wide range of physical properties. The techniques used to identify and quantify these types of process aid are described in Section 3.4.

4.4.3.2 Resins

High-styrene resins are thermoplastic polymers of styrene and butadiene, and they are used to modify the hardness of rubber compounds. They are used in preference to high levels of CB because they do not increase the processing viscosity to the same extent. The high level of styrene ($\leq 85\%$) makes them easy to detect by looking for their characteristic T_g by DMTA or thermal mechanical analysis. The DMTA trace of a SBR containing a high-styrene resin is shown in Figure 4.9. The T_g for the rubber (at about -20 °C) and the T_g for the resin (at ≈ 50 °C) are clearly visible in the data. PyGC–MS is also an option and can be

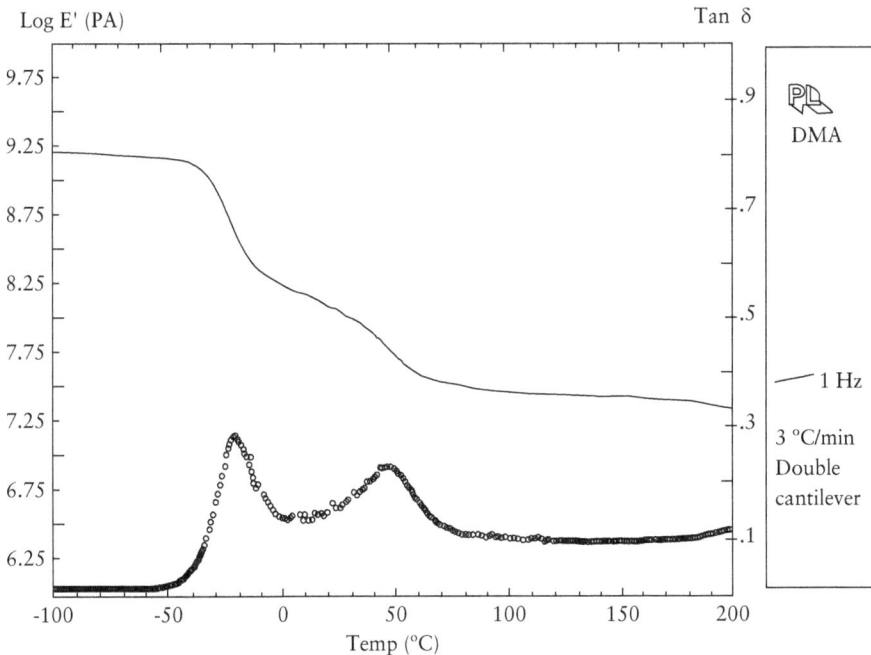

Figure 4.9: DMTA trace of a SBR containing a high-styrene resin. Reproduced with permission from M.J. Forrest in *Rubber Analysis – Polymers, Compounds and Products*, Rapra Review Report No.139, Smithers Rapra, Shawbury, UK, 2001, p.30. ©2001, Smithers Rapra [39].

employed in way that is analogous to the identification of polymers (Section 4.2) by looking for characteristic monomer species and other Py products in the pyrogram. This approach can obviously be compromised if the high-styrene resin is in a SBR rubber compound.

Coumarone resins are manufactured by the polymerisation of styrene, coumarone and indene, and are used as tackifiers and plasticisers. Varying the polymerisation conditions leads to a range of resins having melting points in the range 65–110 °C. If sufficient resin is present in a sample, its melting point can be detected by DSC. PyGC–MS can also be used to detect the presence of these resins in a compound.

Thermosetting alkyl phenol formaldehyde resins can be used instead of the high- styrene type to give excellent flow characteristics in moulding and extrusion. Non- reactive phenolic resins can be used as tackifiers instead of coumarone resins. With regard to analysis, phenolic fragments can be detected in both solvent extracts and pyrograms by GC–MS.

4.4.3.3 Pine tar and factice

Pine tar can be used as a processing aid in rubber compounds and an analysis of the solvent extract of a rubber sample by GC–MS will reveal diagnostic species (e.g., pinene) if this additive is present.

Factice, of which there are both 'white' and 'brown' versions depending upon their colour and manufacturing history, is used in rubber compounds and is particularly good at controlling the amount of die swell that occurs during extrusion. It is produced by crosslinking vegetable oil with sulfur and so a relatively-high sulfur content will give a good indication as to whether it is present or not in a rubber compound. Whether any sulfur has been contributed by a sulfur-based cure system will obviously have to be considered.

The presence of both pine tar oil and factice in a sample can be detected by the use of nuclear magnetic resonance due to its relative sensitivity and diagnostic capabilities, whereas FTIR may have problems in providing a definitive result.

4.4.4 Peroxide co-agents

Peroxide co-agents such as TAC and triallyl isocyanurate can be detected in solvent extracts of either the unvulcanised or vulcanised rubber by GC–MS because a sufficient amount of these additives remains in the 'unbound' state (i.e., not part of the crosslinked matrix) in the case of vulcanised samples. It is also possible to detect these species by headspace GC–MS because they are reasonably volatile and thermally stable.

4.4.5 Stearic acid

Stearic acid is an important ingredient in rubbers, being used in the majority of sulfur-cured compounds due to its role as a co-agent, along with zinc oxide, in the cure system. The quantitative analysis of stearic acid and its associated fatty acids (the industrial chemical usually also contains palmitic acid and myristic acid) can, therefore, be important from a quality-control viewpoint and Watanabe and co-workers [40] investigated two analytical methodologies for this purpose. In place of the conventional liquid-phase extraction procedure, the research team determined the fatty acids in an SBR sample directly by thermal desorption (TD) GC–MS and thermally-assisted hydrolysis and methylation (THM) GC–MS. The results showed that the precision of the analytical data was only fair (i.e., relative standard deviation (RSD) of 7.8%) for the TD GC–MS method due to the interaction between the polar fatty acids and the basic sites in the chromatographic system. The problem with interactions was overcome with the THM GC–MS approach because the fatty acids were derivatised to the methyl esters using tetramethylammonium hydroxide and the average result of 0.62% (RSD of 3.2%) of total fatty acids in the SBR agreed well with the actual, added weight of fatty acids of 0.64%.

4.5 General analysis work for the identification and quantification of additives

Sections 4.2–4.4 have addressed the determination of different additives according to their specific classes (e.g., UV stabiliser, antioxidant, antiozonant). This section includes the techniques and methods used by industrial scientists and researchers to provide more general information, including information on the:
- Additives present in a rubber for the purposes of reverse engineering.
- Function, behaviour and performance of additives to assist with degradation studies and curing studies.

The examples given below are taken from studies published within the last 15 years and have been selected because they are considered to provide a good overview to the industrially relevant work that is possible using the technology that is presently available.

A research team from Brazil [41] used a Py-based technique to characterise a rubber compound. In this case, FTIR was employed to investigate the gaseous Py products of two EPDM rubbers with a view to evaluating its potential to characterise the additives that were present within the samples. Both of the EPDM rubbers contained additives that would usually be expected to be employed in such a material and Py– FTIR characterisation data were obtained for the following ingredients:

- Paraffin oil
- Stearic acid
- TMQ
- TMTM
- Tetraethylthiuram disulfide (TETD)
- MBT

The work undertaken by this research team also included a comparative study involving the characterisation of uncompounded EPDM, unvulcanised EPDM and vulcanised EPDM. The Py–FTIR results obtained showed that the vulcanisation process did not interfere with the data and that it was possible to identify the functional groups of the additives in both cases and, thus, eliminated the need for a solvent extraction process to isolate them. As well as being shown as a fast and cost-effective analytical method for the determination of additives, another benefit was that it was sensitive enough to be capable of detecting sulfur-containing additives down to 1.26% (1.4 phr) in both unvulcanised and vulcanised EPDM. However, one limitation was its inability to differentiate between TMTM and TETD because of overlapping within the FTIR spectral information.

Il'yasov and co-workers [42] investigated the high-temperature volatility of 39 ingredients that can be used in different tyre rubber formulations (e.g., tread, sidewall). The research team used a gravimetric method that was capable of the quantitative analysis of the ingredients when they were heated under different regimens. The ingredients that were studied in this way included fillers, curing agents, resins, plasticisers, antioxidants, activators and modifiers, and the research team thought that the results could be used to develop a procedure for determining the weight loss of rubber mixes during their preparation and processing so that the formulation could be optimised.

FTIR, in its various forms, is an extremely versatile technique and its use to identify polymers is covered in some detail in Section 3.2. Tikhomirov and Kimstach [43] showed how it can also be used for the quantitative and qualitative analysis of polymers and their additives for quality control and other purposes. Their overview of the subject provides examples of the application of a number of FTIR-based techniques, including:
- Attenuated total reflection spectroscopy
- Diffuse-reflection FTIR
- GC–FTIR
- Near IR

Kuzdzal [44, 45] described the advantages of using LC–MS-ion-trap-ToF–MS for the identification and structural analysis of polymer additives, such as plasticisers and viscosity modifiers. The application of the technique to different scenarios for the purposes of quality control, quality assurance of raw materials, migration testing, or chemical modification of additives due to degradation or oxidation was presented. Two of the scenarios included were the important areas of the potential interaction and migration of species between packaging and drug products as well as food-contact materials and food products.

As Chapter 2 demonstrates, many analytical methods are available for the study and characterisation of rubbers and their compounds and products, including a number that are interfaced with MS, for example, GC–MS, LC–MS, Py–MS, and TGA–MS. These methods work well but some, particularly if initial preparation steps such as solvent extraction are taken into account, can be relatively time-consuming and so not always suited to a high-throughput industrial laboratory. The search for fast and accurate analytical methods is, therefore, always of interest, and Trimpin and co- workers [46] have discussed the use of three such methods. All three of these rapid methods are based on MS and are:

- Multi-sample matrix-assisted laser desorption/ionisation (MALDI) MS – for the analysis of low-MW polymers.
- Atmospheric pressure (AP)-solids analysis probe MS – for the analysis of additives.
- Atmospheric pressure Py MS – for the determination of polymer type.

The last two methods are stated as being able to provide information on a sample regardless of its composition or the MW of the polymer.

Hakkarainen [47] published a review of the use of the solid-phase microextraction technique in different polymer-related applications. Within the review, the analysis of various polymer additives was discussed, for example, plasticisers, flame retardants and the extraction of polymer additives from biological fluids. The analysis of polymer degradation products was also covered, with reference to thermo-oxidation, photooxidation and thermal oxidation products and the degradation products of degradable polymers. In addition, the analysis of residual monomers and residual solvents was included, as was the analysis of migrants from food-contact materials, pharmaceutical packaging and medicinal materials.

Buchberger and Stiftinger [48] reviewed the methods available for the analysis of the stabilisers present in polymeric materials. In their review they made the following points:

- The degradation products of stabilisers should be identified to understand the reactions that are occurring.
- Adequate analytical methods are required to ensure the optimisation of the performance of a polymer in service and ensuring that its manufacturing costs are well controlled.
- LC–MS with atmospheric pressure ionisation (API) has become the state-of-the-art for the identification of the additives and components within polymeric materials.

- The API techniques that are available include: electrospray ionisation, AP photoionisation and API.
- API has the advantages of low detection limits and wide applicability in terms of different additive structures.
- As an alternative to combinations of chromatography and MS (e.g., LC–MS), direct mass spectrometric techniques for solid samples are emerging because they offer advantages in terms of speed (e.g., for screening work) and avoidance of time-consuming preparation work.
- In addition to chromatography, capillary electrophoresis has demonstrated potential for the separation of polymer stabilisers and the characterisation of polymers.

A review provided by Geissler [49] described the use of PyGC–MS to quantitatively analyse the additives in polymers.

Aminlashgari and Hakkarainen [50] reviewed the emerging mass spectrometric techniques available for the analysis of polymers and polymer additives. The emerging techniques that are covered in this review include:

- New developments in laser desorption ionisation techniques (e.g., solvent-free MALDI and surface-assisted laser desorption ionisation).
- Developments in secondary-ion mass spectrometry (SIMS) such as gentle-SIMS and cluster SIMS.
- Desorption electrospray ionisation MS and direct analysis in real time MS for the analysis of solid samples.
- Ion-mobility spectrometry mass spectrometric analysis for evaluation of complex structures.

Although the potential of these analytical techniques still has to be fully explored, the authors stated that they will strengthen the position of MS as an irreplaceable tool for polymer characterisation.

Sanches and co-workers [51] used TGA–FTIR to analyse both unvulcanised and vulcanised EPDM rubbers with the objective of identifying some of the additives that are frequently used in these compounds. The data generated on standard compounds showed that the technique was capable of detecting the following additives at a level of 0.7 phr (0.63%):

- TMTM
- MBT

However, the technique was found to have some limitations because it was not capable of detecting a number of other additives, including:

- TMQ
- Paraffin oil
- Stearic acid

Despite these limitations, the authors believed that the technique had value because it was particularly suited for the detection of sulfur-containing additives, and the fact that no initial solvent extraction procedure was required reduced the time and effort needed to obtain compositional information on rubber compounds.

References

1. M.J. Forrest in *Rubber Analysis – Polymers, Compounds and Products*, Rapra Review Report No.139, Smithers Rapra, Shawbury, UK, 2001, p.55

2. R.P. Lattimer in *Proceedings of the ACS 132nd Fall Meeting*, 6–9th October, Cleveland, Ohio, USA, American Chemical Society Rubber Division, Washington, DC, USA, 1987, Paper No.86.

3. T.R. Crompton in *Determination of Additives in Polymers and Rubbers*, Rapra Technology, Shawbury, UK, 2007.

4. J.C.J Bart, *Polymer Degradation and Stability*, 2003, **82**, 2, 197.

5. J.A. Sidwell in *High Performance Liquid Chromatography – Analytical Applications in the Rubber and Plastics Industries*, Members Report Number 49, Rapra Technology Ltd, Shawbury, UK, 1980.

6. A.B. Sullivan, G.H. Kuhls and R.H. Campbell, *Rubber Age*, 1976, **108**, 3, 41.

7. W.C. Warner, *Journal of Chromatography*, 1969, **44**, 315.

8. J.G. Kriener, *Rubber Chemistry and Technology*, 1971, **44**, 381.

9. K. Nagasawa and K. Ohta, *Rubber Chemistry and Technology*, 1969, **42**, 625.

10. H.B.S. Conacher and B.D. Page, *Journal of Chromatographic Science*, 1979, **17**, 188.

11. M.J. Forrest, S.R. Holding and D. Howells, *Polymer Testing*, 2006, **25**, 63.

12. M.J. Forrest, S. Holding and D. Howells in *Proceedings of the High Performance and Specialty Elastomers*, 20–21st April, Geneva, Switzerland, Rapra Technology Ltd, Shawbury, UK, 2005, Paper No.2.

13. M.J. Forrest in *Rubber Analysis – Polymers, Compounds and Products*, Rapra Review Report No.139, Smithers Rapra, Shawbury, UK, 2001, p.26.

14. E. Klein, Z. Cibulkova and V.A. Lukes, *Polymer Degradation and Stabilisation*, 2005, 88, 3, 548.

15. Z. Cibulkova, P. Simon, P. Lehocky and J. Balko, *Polymer Degradation and Stabilisation*, 2005, **87**, 3, 479.

16. D.F. Parra and J.R. Matos, *Journal of Thermochemistry and Calorimetry*, 2002, **67**, 2, 287.

17. E.L.G. Denardin, D. Samos, P.R. Janissek and G.P. Souza, *Rubber Chemistry and Technology*, 2001, **74**, 4, 622.

18. C.D, Gamlin, N.K. Dutta and N.R. Choudhury, *Polymer Degradation and Stabilisation*, 2003, **80**, 3, 525.

19. M.J. Forrest in *Principles and Applications of Thermal Analysis*, Ed., P. Gabbott, Blackwell Publishing Ltd, Oxford, UK, 2008, p.225.

20. V.K. Skachkova, N.A. Erina, L.M. Chepal and E.V. Prut, *Polymer Science Series A*, 2003, **45**, 12, 1220.

21. F. Pruneda, J.J. Sunol, F. Andreu-Mateu and X. Colom, *Journal of Thermal Analytical Colorimetry*, 2005, **80**, 1, 187.

22. T. Ogawa, T. Yamagata, T. Hara, S. Osawa, Y. Yoshida and M. Kanazawa, *Kobunshi Ronbunshu*, 2003, **60**, 2, 64.

23. M. Hakkarainen, A-C Albertsson and S. Karlsson, *International Journal of Polymer Analysis and Characterisation*, 2003, **8**, 4, 279.

24. R. Yang, Z. Jiaohong and Y. Liu, *Polymer Degradation and Stability*, 2013, **98**, 12, 2466.
25. M. L. Tawfic and A.I. Hussein, *Kautschuk, Gummi, Kunststoffe*, 2015, **68**, 9, 30.
26. B.G. Willoughby and K.W. Scott in *Nitrosamines in Rubber*, Rapra Technology Ltd, Shawbury, UK, 1997.
27. D. Gross and K. Strauss, *Kautschuk, Gummi, Kunststoffe*, 1979, **32**, 1, 18.
28. G.P. McSweeny, *Journal of the Institute of the Rubber Institute*, 1970, **4**, 243.
29. D.O. Hummel and F.K. Scholl in *Atlas of Polymer and Plastics Analysis, Volume 3 – Additives and Processing Aids: Spectra and Methods of Identification*, 2nd Revised Edition, Verlag Chemie GmbH, Weinheim, Germany, 1981.
30. F. McLafferty and D.B. Stauffer in *Wiley/NBS Registry of Mass Spectral Data*, John Wiley and Sons, New York, NY, USA, 1989.
31. B. Willoughby in *Rubber Fume: Ingredients/Emissions Relationships*, Rapra Technology Ltd, Shawbury, UK, 1994.
32. M.J. Forrest in *Rubber Analysis – Polymers, Compounds and Products*, Rapra Review Report No.139, Smithers Rapra, Shawbury, UK, 2001, p.28.
33. J.A Sidwell in *Proceedings of the Food Contact Polymers Conference*, 21–22nd April, Brussels, Belgium, Smithers Rapra, Shawbury, UK, 2009, Paper No.7.
34. M.J. Forrest in *Food Contact Rubbers 2 – Products, Migration and Regulation*, Rapra Review Report No.182, Smithers Rapra, Shawbury, UK, 2006, p.28.
35. M.J. Forrest in *Food Contact Rubbers 2 – Products, Migration and Regulation*, Rapra Review Report No.182, Smithers Rapra, Shawbury, UK, 2006, p.29.
36. Z. Naidong and Z. Qingshan, *China Rubber Industry*, 1991, **38**, 1, 33.
37. L. Yang, Y. Hu, H. Lu and L. Song, *Journal of Applied Polymer Science*, 2006, **99**, 6, 3275.
38. D.H.A. Ismawi, J.F. Harper and A. Ansarifar, *Journal of Rubber Research*, 2008, **11**, 4, 223.
39. M.J. Forrest in *Rubber Analysis – Polymers, Compounds and Products*, Rapra Review Report No.139, Smithers Rapra, Shawbury, UK, 2001, p.30.
40. A. Watanabe, C. Watanabe, R. Freeman, M. Nakajima, N. Teramae and H. Ohtani, *Rubber Chemistry and Technology*, 2014, **87**, 3, 516.
41. N.B. Sanches, S.N. Cassu, M.F. Diniz and R. Dutra de Cassia Lazzarini, *Polimeros: Ciencia e Tecnologia*, 2014, **24**, 3, 269.
42. R.S. Il'yasov, E.G. Mokhnatkina, E.E. Potapov and E.V. Sakharova, *International Polymer Science and Technology*, 2009, **36**, 11, T/29.
43. S.V. Tikhomirov and T.B. Kimstach, *International Polymer Science and Technology*, 2008, **35**, 8, T/59.
44. S. Kuzdzal, *Rubber World*, 2010, **241**, 4, 19.
45. S. Kuzdzal, *Adhesives and Sealants Industry*, 2010, **17**, 3, 21.
46. S. Trimpin, K. Wijerathne and C.N. McEwen, *Analytica Chimica Acta*, 2009, **654**, 1, 20.
47. M. Hakkarainen in *Advances in Polymer Science*, Eds., A. Alberson and M. Hakkarainen, Springer-Verlag, Berlin, Germany, 2008, **11**, 23.
48. W. Buchberger and M. Stiftinger, *Advances in Polymer Science*, 2012, **248**, 39.
49. M. Geissler in *Proceedings of the 7th European Conference on Additives and Colors*, 16–17th March, Bonn, Germany, Society of Plastics Engineers, Antwerp, Belgium, 2011, Paper No.3.
50. N. Aminlashgari and M. Hakkarainen, *Advances in Polymer Science*, 2012, **248**, 1.
51. N.B. Sanches, S.N. Cassu and R.L. Dutra, *Polimeros*, 2015, **25**, 3, 247.

5 Reverse engineering and product deformulation

5.1 Introduction

One of the most frequent tasks asked of a rubber analyst is to deformulate a rubber compound in order to obtain as much information of the types of ingredients that have been used in its manufacture and the level of these ingredients. This work can be regarded as being analogous to being asked to piece together a jigsaw and is a satisfying and challenging task to undertake, particularly if the rubber analyst has sufficient knowledge of rubber technology to compliment his/her skills as an investigative analyst.

In order to provide the greatest opportunity for success in this area it is important that in addition to the skills and knowledge mentioned above that the rubber analyst has access, either solely in-house, or from a combination of in-house capability and that of a trusted service provided, of a full range of analytical techniques. These should include as a minimum:

- Fourier–Transform infrared (FTIR) spectroscopy for providing information on the generic type of base polymer(s), bulk additives (plasticiser/oil and inorganic fillers), and providing an initial insight into the complexity of the sample.
- Nuclear magnetic resonance (NMR) to provide information on the polymer microstructure (e.g., ratio of comonomers, cure site monomers) and, in the case of polymer blends, approximate blend ratio.
- Thermal techniques that include thermogravimetric analysis (TGA) for bulk composition analysis and dynamic mechanical analysis for determination of glass transition temperature and generation of modulus *versus* temperature plots.
- Chromatographic techniques that are capable of identifying and quantifying substances possessing a range of molecular weights (MW) and degrees of thermal stability, for example, gas chromatography (GC)–mass spectrometry (MS) and liquid chromatography (LC)–MS.
- An elemental technique that can be used in a semi-quantitative, scanning mode, such as X-ray fluorescence spectroscopy (XRF), to assist in identifying the type(s) or inorganic additives and fillers present.

The combined capability of the techniques stated above will provide a significant amount of information on a sample, but they may have to be complimented by some of the other specific or specialist techniques that are described in Chapter 2 depending upon the particular rubber compound under investigation and how the deformulation work progresses. Typical examples of these techniques include X-ray diffraction to determine the specific type of a particular inorganic filler (e.g., a silicate), or a quantitative elemental technique [e.g., inductively coupled plasma (ICP)] to quantify a metal or

https://doi.org/10.1515/9783110640281-005

a hetero-element as a means to quantify a polymer in a blend or the level of a specific additive. It is also often possible to obtain additional compositional information by complementing the chemical analysis work with a physical test. For example, combining the bulk compositional data obtained from TGA with a hardness determination can help in deciding the approximate type of carbon black (CB) used by referring to the compounding guidelines of rubber technologists.

There can be a number of reasons why a reverse engineering programme is embarked upon, including:

- To ensure the correct formulation was used initially because a product has failed in service.
- To assist in a research and development programme by obtaining information on products that are already in the market place.
- To investigate the regulatory compliance of part of the formulation (e.g., cure system ingredients).*

*Particular care needs to be taken in this case due to the limitations discussed below.

It is important to understand the limitations associated with these types of investigative study in order to ensure that unrealistic goals are not laid down at the outset. In the first two cases listed above, experience has shown that it is not possible from analytical work alone to determine with 100% accuracy the complete composition of the vast majority of rubber compounds. Although the degree of success depends upon a number of factors (see below), as a rough estimate, it is usually possible to obtain between 70 and 85% of the compositional information. These comments apply even if a substantial budget, extended time-scale and wide range of modern analytical instrumentation are available to a knowledgeable and experienced analyst. They also apply to both unvulcanised and vulcanised samples.

There are a number of reasons why there are practical restrictions on the degree of success, and some of these are listed below:

- Presence of thermally-labile substances (e.g., cure system compounds) that will breakdown during vulcanisation and often during analysis (e.g., during a solvent extraction step or when analysed using instruments such as GC–MS).
- Presence of substances that have no single diagnostic species, a good example being oligomeric antioxidants.
- Presence of additives that are volatile, or have a volatile fraction, and so are lost to a degree during vulcanisation or analysis.
- Variations in the manufacturing specifications of some ingredients (e.g., the amount of bromine and chlorine in halobutyl rubbers due to variable amounts of cure site monomer being used in different grades).
- Complications arising from the presence of two or more ingredients of the same 'type' being used, for example, process oils, CB, inorganic fillers, and a blend of two grades of polymer of the same type [e.g., styrene-butadiene rubber (SBR)].

- The fact that only the chemical identify of substances can be determined not the specific manufacturers grade that has been used.
- Once vulcanised, it is not possible to determine the MW characteristics of the original polymer(s) and, even in the unvulcanised state, complete isolation of the polymer(s) from the other ingredients in a rubber compound may be difficult due to strong polymer–filler interactions.

It is these types of challenges that can cause problems if reverse engineering work is used to try to assess compliance with regulations and regulatory guidelines even in a relatively minor way (e.g., cure system ingredients) and mean that it is not possible to use reverse engineering to show complete compliance with, say, a 'positive list' of ingredients in a food-contact regulation (e.g., BfR Recommendation XXI) because substances that were not on the list could have been missed.

Specific applications of the mainstream spectroscopic, chromatographic, thermal and elemental techniques to partial or extensive reverse engineering-type work is covered in other chapters of this book, particularly Chapters 3 and 4. Taken together, these show that by using combinations of these techniques a significant amount of qualitative and quantitative information can be obtained on the bulk components (i.e., plasticiser/oil, polymer and filler) and specific classes of additives (e.g., antidegradants and curatives).

As mentioned above, many examples of the specific aspects of reverse engineering work are covered in other chapters in this book and so, rather than repeat some of them in this chapter, we will instead review the capability of one of the more advanced versions of the mainstream techniques of which the author has had personal experience (Section 5.2). It is for this reason, as well as the fact that it is widely applicable to rubber compounds and products and offers advantages in this area, that this technique has been chosen and not because it is necessarily any better than the many others that are available and which are overviewed in Chapter 2.

This chapter also uses an in-depth case study to provide a practical illustration of an approach to reverse engineering that can be taken with a complex, real-life sample (Section 5.3). In addition, some published examples of reverse engineering work that has been carried out on rubbers and related polymer systems are provided in Section 5.4. For additional assistance on how to undertake these types of exercises, reference books that have been published on this subject include those of Gupta and co-workers [1] and Scheirs [2].

With regard to guidance as to what additives and ingredients are being used in the rubber industry, there are a number of books that provide extensive coverage of the range of additives and chemicals that can be compounded into rubber products [3, 4].

5.2 Use of GC×GC–ToF–MS for reverse engineering

Despite its high-resolution, GC–MS can encounter problems in trying to resolve all of the components present in rubber extracts. This is particularly true if the sample contains a relatively large amount of a multi-component species, such as a mineral oil. Inevitably, there is a considerable amount of co-retention in the middle part of the GC chromatogram and the ability to identify components within this region can be severely compromised, thereby limiting the success of reverse engineering work.

The GC×GC–time-of-flight (ToF)–MS technique offers the rubber analyst a greater potential capability for reverse engineering work and the background to how it delivers this is covered in Section 2.5.3. One of the principal features of the technique is that two types of GC column are used and this enables separations to be performed according to both volatility and polarity. Powerful deconvolution software also aids the processing and interpretation of the data.

The advantages of using GC×GC–ToF–MS in reverse engineering work are highlighted in this chapter by using the results obtained on five food-grade, high-performance rubber samples [5]. Firstly, the reverse engineering data that can be generated by the GC×GC–ToF–MS are illustrated using a hydrogenated-nitrile rubber (NBR) compound as an example. The compound was a peroxide-cured, hydrogenated NBR, designated 349N, which had the formulation shown in Table 5.1.

Table 5.1: Formulation of peroxide-cured hydrogenated NBR – 349N.

Ingredient	phr
Zetpol® 2000L (hydrogenated NBR)	100
High-abrasion furnace N330 (CB filler)	35
ZnO	3
Perkadox® 14/40 [1,3-*bis*(*tert*-butyl-peroxy-isopropyl)benzene]	6
Antioxidant 2246 [2,2′-methylene-*bis*(4-methyl-6-*tert*-butylphenol]	1

ZnO: Zinc oxide
Reproduced with permission from M.J. Forrest, S. Holding and D. Howells in *Proceedings of the High Performance and Specialty Elastomers*, 20–21st April, Geneva, Switzerland, Rapra Technology Ltd, Shawbury, UK, 2005, Paper No.2. ©2005, Rapra Technology Ltd [6]

A test sheet was prepared from the hydrogenated NBR 349N compound by curing a portion for 22 min at 165 °C, conditions which gave a degree of cure of ≈90%, as determined by the moving die rheometer (MDR), of the theoretical maximum.

A solvent extract was obtained for the analysis by finely cutting up 0.3 g of the cured test sheet, placing it into a vial along with 2 ml of acetone, and then subjecting it to ultrasonic agitation for 30 min. At the end of this period, the extract was removed and placed into a GC sample vial. A 20-ppm eicosane 'internal standard' was included in the acetone used for the extraction. Acetone was the preferred choice of solvent because it is a good, non-selective solvent for the removal of low-MW compounds from a wide range of rubber types. It is also a good solvent for GC–MS-type work due to its relatively high volatility and reasonable stability (i.e., it does not contain stabilisers that may interfere with analyte peaks).

The acetone extract of the hydrogenated NBR 349N compound was analysed by GC×GC–ToF–MS using the following experimental conditions:

- Instrument: Agilent 6890 gas chromatograph with LECO Pegasus III GC×GC–ToF–MS.
- Injection: PTV injection, 10 °C above primary oven temperature; 1 µl.
- Primary column: J&W scientific HP-5MS (30 m × 0.250 mm with film thickness of 0.25 µm).
- Secondary column: SGE BPX-50 (1.8 m × 0.100 mm with film thickness of 0.10 µm).
- Carrier gas: helium at 1.0 ml/min with constant flow.
- Primary oven programme: 40 °C for 10 min, then 10 °C/min to 320 °C and hold for 15 min.
- Secondary oven programme: 50 °C for 10 min, then 10 °C/min to 330 °C and hold for 15 min.
- Modulator offset: 30 °C.
- Modulator frequency: 4 s.
- Hot time: 0.30 s.
- Mass spectrometer: scanning 30 to 650 Da at 76 spectra/s.

The data generated by the GC×GC–ToF–MS analysis of the acetone extract are shown in Figures 5.1–5.3.

For comparison, the acetone extract from hydrogenated NBR 349N was also analysed by conventional GC–MS, using the conditions below, and the results obtained are shown in Figure 5.4:

- Instrument: Agilent 6890/5973 GC–MS.
- Injection: 1 µl splitless at 310 °C.
- Carrier: helium at 1.0 ml/min.
- Column: Restek RTX-5 amine (30 m × 0.32 mm with film of 1.0 µm).
- Column oven: 40 °C for 5 min, then 20 °C/min increase up to 300 °C held for 15 min (30-min run time).
- MS: scanning 35 to 650 Da every 0.25 s.

The conventional GC–MS chromatogram (Figure 5.4) confirmed that this technique would struggle to separate a large number of substances when they elute closely together.

The solvent extracts from a further four food-grade, high-performance rubbers were also analysed using both GC×GC–ToF–MS and conventional GC–MS. The formulations of these four compounds are shown in Tables 5.2–5.5.

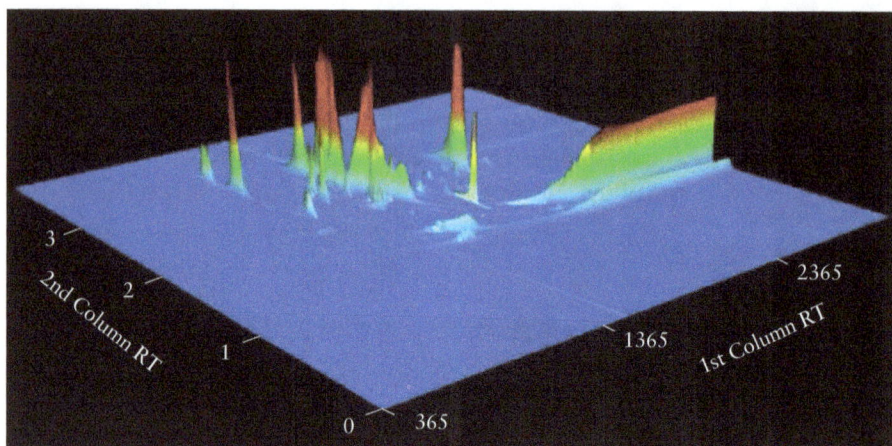

Figure 5.1: Three-dimensional (3D) GC×GC–ToF–MS total ion current (TIC) chromatogram of hydrogenated NBR 349N acetone extract (RT: retention time). Reproduced with permission from M.J. Forrest, S. Holding and D. Howells in *Proceedings of the High Performance and Specialty Elastomers*, 20–21st April, Geneva, Switzerland, Rapra Technology Ltd, Shawbury, UK, 2005, Paper No.2. ©2005, Rapra Technology Ltd [6].

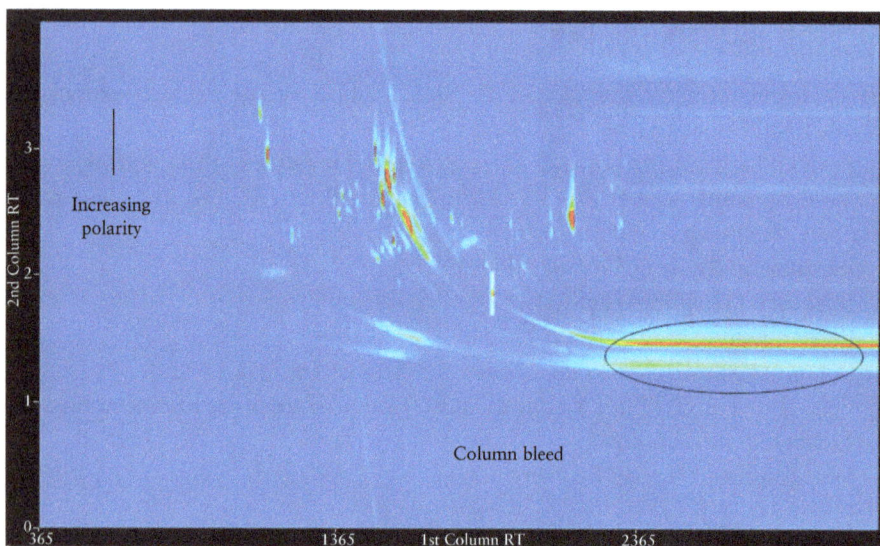

Figure 5.2: Plan view of GC×GC–ToF–MS TIC chromatogram of hydrogenated NBR 349N acetone extract – species due to 'column bleed' and a polarity marker are shown. Reproduced with permission from M.J. Forrest, S. Holding and D. Howells in *Proceedings of the High Performance and Specialty Elastomers*, 20–21st April, Geneva, Switzerland, Rapra Technology Ltd, Shawbury, UK, 2005, Paper No.2. ©2005, Rapra Technology Ltd [6].

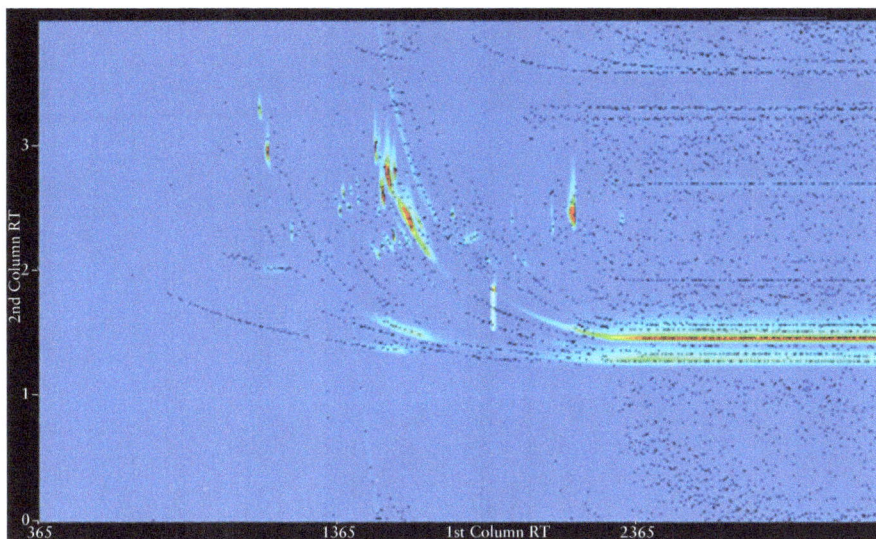

Figure 5.3: Plot view of GC×GC–ToF–MS TIC chromatogram of hydrogenated NBR 349N acetone extract with peak markers. Reproduced with permission from M.J. Forrest, S. Holding and D. Howells in *Proceedings of the High Performance and Specialty Elastomers*, 20–21st April, Geneva, Switzerland, Rapra Technology Ltd, Shawbury, UK, 2005, Paper No.2. ©2005, Rapra Technology Ltd [6].

The cure temperatures and times used to produce test sheets from these four rubber compounds are given in Table 5.6.

The five rubber compounds shown above represent a range of high-performance elastomers that could be used for food-contact type applications. They display a range of complexity, with the hydrogenated NBR 349N having the greatest number of constituents within it. In all five cases, the greater diagnostic power of the GC×GC–ToF–MS technique resulted in a larger number of species being detected in the acetone extract. Hence, more compositional information was provided compared with conventional GC–MS, which is invaluable in reverse engineering work where the degree of success is dependent on a good set of diagnostic compounds.

The difference in performance between GC×GC–ToF–MS and conventional GC–MS is demonstrated in Table 5.7, where a comparison of the approximate number of diagnostic substances found in the acetone extract of the five rubber compounds is shown. Diagnostic substances can be regarded as those that are the most useful in revealing information on the original formulation of a rubber. They can be defined as:
-   Compounds that are the known breakdown products of certain classes of additive (e.g., curatives and antidegradants).
-   Original ingredients in their unchanged form.

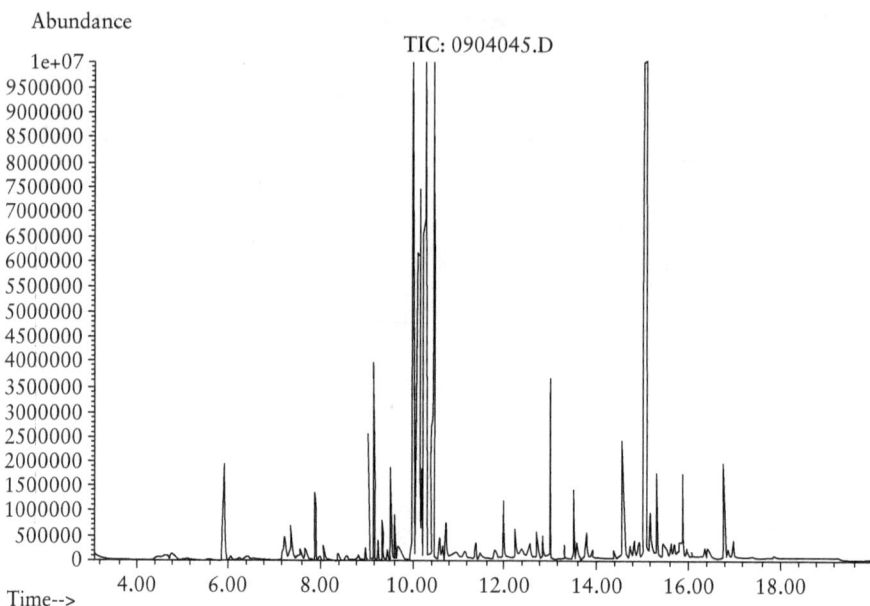

Figure 5.4: Conventional GC–MS TIC chromatogram of a hydrogenated NBR 349N acetone extract. Reproduced with permission from M.J. Forrest, S. Holding and D. Howells in *Proceedings of the High Performance and Specialty Elastomers*, 20–21 April, Geneva, Switzerland, Rapra Technology Ltd, Shawbury, UK, 2005, Paper No.2. ©2005, Rapra Technology Ltd [6].

Table 5.2: Formulation of fluorocarbon rubber – 49V.

Ingredient	phr
Viton™ GBL 200 (peroxide-crosslinkable fluorocarbon rubber tetrapolymer)	100
ZnO	3
Medium thermal N990 (CB filler)	30
Triallyl cyanurate	3
Luperco 101XL [2,5-dimethyl-2,5-(di-*tert*-butylperoxy)hexane]	4

Reproduced with permission from M.J. Forrest, S. Holding and D. Howells in *Proceedings of the High Performance and Specialty Elastomers*, 20–21st April, Geneva, Switzerland, Rapra Technology Ltd, Shawbury, UK, 2005, Paper No.2. ©2005, Rapra Technology Ltd [6]

This definition of diagnostic substances does not include non-specific compounds (e.g., general aliphatic and aromatic hydrocarbons) and other common compounds, such as siloxanes, which can originate from a number of sources, including processing materials (e.g., moulding-release agents).

The data in Table 5.7 show that, in most cases, the GC×GC–ToF–MS system has identified significantly more diagnostic substances than conventional GC–MS and,

Table 5.3: Formulation of acrylic rubber – Vamac® 8.

Ingredient	phr
Vamac G [ethylene-methyl acrylate (acrylic) rubber]	100
Stearic acid	1.5
FEF N550 (CB filler)	50
Diak™ No.1 (hexamethylenediamine carbamate)	1.5
DOTG	4
Antioxidant 2246	1

FEF: Fast extrusion furnace
DOTG: N,N′-di-*ortho*-tolyl guanidine
Reproduced with permission from M.J. Forrest, S. Holding and D. Howells in *Proceedings of the High Performance and Specialty Elastomers*, 20–21st April, Geneva, Switzerland, Rapra Technology Ltd, Shawbury, UK, 2005, Paper No.2. ©2005, Rapra Technology Ltd [6]

Table 5.4: Formulation of hydrin rubber – 20D.

Ingredient	phr
Hydrin 200 [epichlorohydrin-ethylene oxide copolymer (hydrin) rubber]	100
FEF N550 (CB filler)	50
Maglite® DE (magnesium oxide)	7
ZnO	3
Diak™ No.1	1.5
Antioxidant 2-mercaptobenzimidazole	2

Reproduced with permission from M.J. Forrest, S. Holding and D. Howells in *Proceedings of the High Performance and Specialty Elastomers*, 20–21st April, Geneva, Switzerland, Rapra Technology Ltd, Shawbury, UK, 2005, Paper No.2. ©2005, Rapra Technology Ltd [6]

Table 5.5: Formulation of butyl rubber 50B.

Ingredient	phr
Butyl 268 [a isobutylene-isoprene (butyl) rubber]	100
ZnO	3
Stearic acid	1
Vistanex™ LM-MS (low-molecular-weight butyl rubber process aid)	10
Intermediate super-abrasion furnace N220 (CB filler)	50
Sulfur	1
Rhenogran® MPTD 70 (dimethyldiphenylthiuram disulfide)	1.43
DOTG	0.3
Nocrac AW (6-ethoxy-1,2-dihydro-2,2,4-trimethylquinoline)	1

Reproduced with permission from M.J. Forrest, S. Holding and D. Howells in *Proceedings of the High Performance and Specialty Elastomers*, 20–21st April, Geneva, Switzerland, Rapra Technology Ltd, Shawbury, UK, 2005, Paper No.2. ©2005, Rapra Technology Ltd [6]

Table 5.6: Curing conditions for the four rubber compounds
(Tables 5.2–5.5).

Rubber compound	Cure conditions*
Hydrin rubber – 20D	31 min at 170 °C
Fluorocarbon rubber – 49V	11 min at 170 °C
Butyl rubber – 50B	21 min at 155 °C
Acrylic rubber – Vamac® 8	25 min at 170 °C

* These curing conditions gave a degree of cure of around 90%, as
determined by MDR, of the theoretical maximum
Reproduced with permission from M.J. Forrest, S. Holding and D. Howells
in *Proceedings of the High Performance and Specialty Elastomers*, 20–21st
April, Geneva, Switzerland, Rapra Technology Ltd, Shawbury, UK, 2005,
Paper No.2. ©2005, Rapra Technology Ltd [6]

Table 5.7: Comparison of the number of diagnostic substances found by
GC×GC–ToF–MS and conventional GC–MS.

Rubber compound	GC×GC–ToF–MS	Conventional GC–MS
NBR – 349N	>35	>30
Fluorocarbon – 49V	>20	>10
Acrylic – Vamac® 8	>30	>20
Hydrin – 20D	>10	>10
Butyl – 50B	>15	>5

Reproduced with permission from M.J. Forrest, S. Holding and D. Howells
in *Proceedings of the High Performance and Specialty Elastomers*, 20–21st
April, Geneva, Switzerland, Rapra Technology Ltd, Shawbury, UK, 2005,
Paper No.2. ©2005, Rapra Technology Ltd [6]

because the better resolution results in higher-quality mass spectra, a greater proportion of these will be positively identified by either automatic searching using a
mass spectral database or by manual interpretation.

To sum up the results presented above, the GC×GC–ToF–MS technique offers
two principal advantages over conventional GC–MS that can make it potentially
more effective for reverse engineering-type work. These advantages are:
1. A greater number of diagnostic compounds from sources such as the cure and
   antidegradant system can be identified.
2. The greater sensitivity of the technique, coupled with its greater resolving
   power and the deconvolution software offers a greater chance that small
   amounts of original additives (e.g., accelerators) can be detected.

With regard to comparing the ability of the two techniques to positively identify all
the low-MW compounds (i.e., including the diagnostic compounds mentioned

Table 5.8: Comparison of the total number of substances detected and the number positively identified for GC×GC–ToF–MS and conventional GC–MS.

Compound	GC×GC–ToF–MS	Conventional GC–MS
NBR – 349N	102 (49)	66 (34)
Fluorocarbon – 49V	184 (66)	36 (14)
Acrylic – Vamac® 8	217 (138)	55 (26)
Hydrin – 20D	46 (23)	37 (17)
Butyl – 50B	11 (70)	27 (12)

( ): Number of substances positively identified using the mass spectral database
Reproduced with permission from M.J. Forrest, S. Holding and D. Howells in *Proceedings of the High Performance and Specialty Elastomers*, 20–21st April, Geneva, Switzerland, Rapra Technology Ltd, Shawbury, UK, 2005, Paper No.2. ©2005, Rapra Technology Ltd [6]

above) in the five rubber extracts, Table 5.8 gives the total number of compounds detected by each of the techniques and, of these, how many have been positively identified from their mass spectra.

Care has to be taken in compiling the type of data shown in Table 5.8 because a significant degree of subjective interpretation can be required to carry out the matching process. However, it can be seen that the GC×GC–ToF–MS technique has produced a more extensive list of identified compounds and so, in addition to its value in reverse engineering work, offers greater advantages in the following type of studies:

– Those that involve profiling rubber compounds to assess the low-MW compounds that have the potential to migrate into food and pharmaceutical products (Chapters 8 and 9).
– Analysis of food and drug products which have contacted rubber components to determine if any species have migrated into them (Chapters 8 and 9).

The data in Table 5.8 also show that the proportion of compounds for which no assignment, or only a tentative assignment, can be found is approximately the same for the two techniques.

Finally, any improvement in the ability to detect and identify low-MW components in rubber extracts is welcomed by rubber analysts, and sophisticated chromatographic techniques, such as GC×GC–ToF–MS, have the potential capability to demonstrate positive technical advantages over conventional GC–MS systems.

5.3 Case study – deformulation of a rubber sample

5.3.1 Introduction

In order to illustrate how the various analytical techniques and methodologies that have been presented and discussed in Chapters 2–4 can be combined to deformulate a rubber compound, a reasonably complex example, designated 'unknown', will be worked through in this chapter. It should be remembered that there is considerable latitude available to the analyst in the approach that he/she can take in this type of work and that one single, 'correct' way does not exist.

The example chosen is a highly flame-retardant rubber compound used to produce a foam-insulating product for use in the cable industry. The overall objective that was set at the outset of this analysis work was to obtain information on the following:

– Polymer content and type(s)
– Flame-retardant content and type(s)
– Filler content and type(s)
– CB content and identification of cure system and antidegradants
– Plasticiser content and type(s)

The methodology and approach used to reverse engineer this 'unknown' sample are described in Sections 5.3.2 and 5.3.4 and the results obtained are presented in Sections 5.3.3, 5.3.5 and 5.3.6.

This case study shows that the work programme was broken down into two stages, with an initial suite of tests being performed on the unknown sample to obtain as much compositional information as possible in a time and cost-effective way. Once the results from this initial programme had been evaluated, additional tests were performed to fill in gaps in the information or clarify points that had been raised.

5.3.2 Initial suite of tests

5.3.2.1 Quantitative solvent extraction
One part of the sample was milled on a two-roll laboratory mill and then extracted with methanol for 16 h. At the end of the extraction period the extract was dried at 105 °C for 30 min and then weighed. This procedure was carried out in duplicate and both the extract and extracted portion retained for further work.

5.3.2.2 Ash content
The ash content of the sample was determined using the method described in International Organization for Standardizations' ISO 247:2006. This procedure was carried out in duplicate.

5.3.2.3 Thermogravimetric analysis
One part of the sample (≈10 mg) was analysed by TGA using a specific programme. Initially, the sample was heated from ambient temperature to 550 °C in a nitrogen

atmosphere. When a constant weight had been obtained, the temperature was reduced to 300 °C, the atmosphere changed to air and then the temperature raised to 950 °C. Weight loss and derivative weight loss curves were plotted throughout. The heating rate at each stage was 20 °C/min and the flow rate of nitrogen and air 50 ml/min.

5.3.2.4 Fourier–transform infrared spectroscopy
The following analytical procedures were carried out using FTIR:
- A transmission infrared (IR) spectrum was recorded of the sample extract prepared in Section 5.3.2.1.
- A pyrolysis IR spectrum was recorded of the methanol-extracted portion of the sample prepared in Section 5.3.2.1.
- A transmission IR spectrum was recorded of a liquid paraffin mull of the ash prepared in Section 5.3.2.2.

5.3.2.5 X-ray fluorescence spectroscopy
A semi-quantitative elemental scan was carried out by XRF on the following:
- Sample in the 'as-received' state
- Methanol-extracted portion of the sample
- Sample ash

5.3.2.6 Gas chromatography–mass spectrometry
One part of the sample was qualitatively extracted with chloroform for 16 h and then the concentrated extract analysed by GC–MS under the following conditions:
- Injection: 1 µl at a split of 30:1.
- Injection temperature: 320 °C.
- GC column: SGE BPX5 (25 m × 0.32 mm with a film of 0.2 µm).
- GC column temperature: 40 °C for 2 mins, followed by 20 °C/min to 320 °C.
- Mass spectrometer: scanning 25 to 400 Da every 2 s.

A GC–MS experiment was also conducted whereby the sample was heated to 180 °C for 10 min in a tube furnace and the volatile species trapped onto an adsorbent packed with Tenax™. The latter was then transferred to an automated thermal desorption unit and desorbed for 10 min at 300 °C. The desorbed species were trapped in a cold trap set at –30 °C and, once the desorption was complete, injected into the GC–MS by rapid heating of the trap to 300 °C. The compounds injected into the GC–MS were then analysed under the same conditions as those shown above.

5.3.3 Results from the initial suite of tests

The methanol-extract value of the sample was 34.73% and the FTIR spectrum of the extract was a composite, with evidence for the presence of following two flame retardants:
- Pentabromodiphenylether
- A phosphate ester

The pyrolysate FTIR spectrum for the methanol-extracted portion of the sample (Figure 5.5) gave data that were consistent for a blend of the following two polymers:
- NBR
- Polyvinyl chloride (PVC)

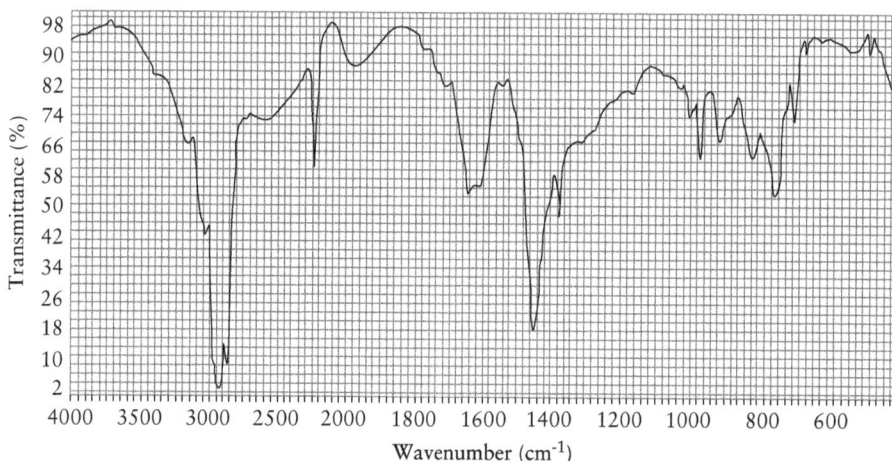

Figure 5.5: FTIR spectrum of the methanol extract of an unknown sample. Reproduced with permission from M.J. Forrest in *Rubber Analysis – Polymers, Compounds and Products*, Rapra Review Report No.139, Smithers Rapra, Shawbury, UK, 2001, p.34. ©2001, Smithers Rapra [7].

The ash content for the sample was found to be 18.28% and the FTIR spectrum of the ash was a composite with evidence for the presence of the following:
- Aluminium oxide (from the flame-retardant hydrated alumina)
- Metallic oxides

The semi-quantitative XRF data obtained on the three portions of the unknown sample are shown in Table 5.9.

The elemental data in Table 5.9 confirmed the qualitative FTIR assignments and in addition, showed that two metallic oxides were present in the sample:

- Antimony trioxide (a flame retardant)
- ZnO (part of the cure system for the NBR)

Table 5.9: Semi-quantitative XRF data on the three portions of the unknown sample.

Element	Sample in the 'as received' state (%)	Methanol extracted portion of the sample (%)	Sample ash (%)
Silicon	0.25	0.10	0.10
Titanium	<0.01	<0.01	<0.01
Aluminium	23.5	30.0	77.0
Iron	<0.01	<0.01	<0.01
Calcium	0.04	0.06	0.15
Magnesium	0.45	1.10	0.35
Potassium	0.05	<0.01	<0.01
Sodium	<0.03	<0.03	0.95
Phosphorus	0.25	<0.02	0.65
Zinc	3.5	2.7	6.1
Chlorine	23.5	23.5	0.2
Bromine	70.5	2.4	5.9
Antimony	6.5	12.0	5.0
Sulfur	2.2	2.5	0.45

Reproduced with permission from M.J. Forrest in *Rubber Analysis – Polymers, Compounds and Products*, Rapra Review Report No.139, Smithers Rapra, Shawbury, UK, 2001, p.35. ©2001, Smithers Rapra [8]

The TGA trace of the sample was unusually complex for a rubber product due to the multiple weight loss events of the PVC (dehydrochlorination, main chain breakdown and additional carbonaceous residue – the NBR also yields some) and the loss of water of the hydrated alumina. These events meant that it was not possible to obtain accurate quantifications of plasticiser, polymer, CB and inorganics and so it was only possible to obtain the data shown in Table 5.10.

Table 5.10: Data obtained from the TGA trace of the unknown sample.

Identification	Proportion of the unknown sample (w/w)
Plasticiser, organic flame retardants, pyrolysis fraction of PVC and NBR, and released water of hydration	72%
Total carbonaceous material (from PVC, NBR and carbon black pigment)	7%
Inorganic residue (inorganic additives)	21%

The GC–MS TIC chromatogram of the chloroform extract from the unknown sample is shown in Figure 5.6 and the qualitative assignments for the species peaks present in this chromatogram were as follows:

- Trichloromethane
- Formamide
- N,N-dimethylformamide
- Cyclohexanethiol
- Phenol
- N,N-dimethylurea
- 3-(Bromomethyl)-heptane
- Benzothiazole (BT)
- Urea
- Phthalimide
- 4-(1,1,3,3-Tetramethylbutyl)-phenol
- Aliphatic hydrocarbons
- Aliphatic carboxylic acids
- Di-(2-ethylhexyl)phthalate
- 2-Ethylhexyl diphenyl phosphate

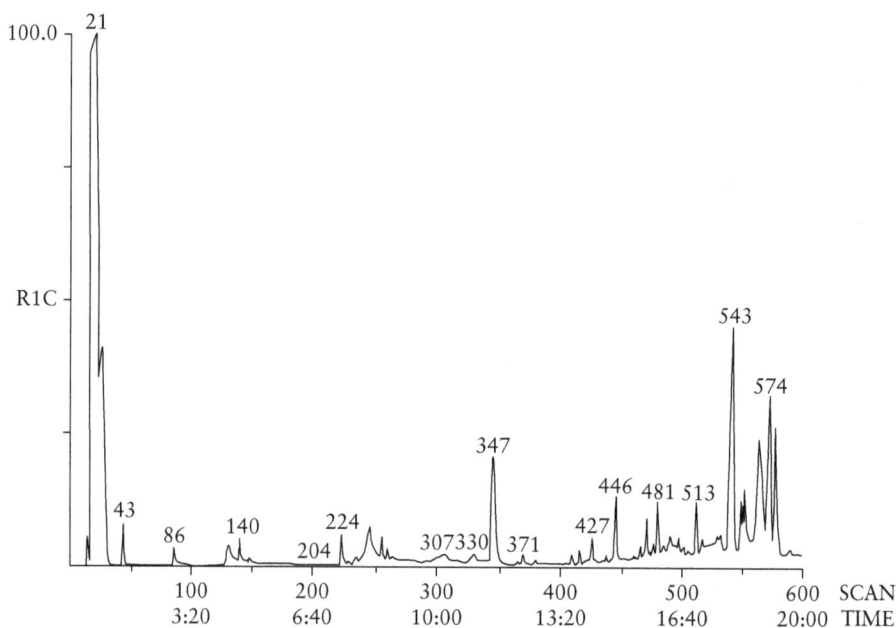

Figure 5.6: GC–MS TIC chromatogram of the chloroform extract from the unknown sample (RIC: reconstructed ion current). Reproduced with permission from M.J. Forrest in *Rubber Analysis – Polymers, Compounds and Products*, Rapra Review Report No.139, Smithers Rapra, Shawbury, UK, 2001, p.36. ©2001, Smithers Rapra [9].

The GC–MS TIC chromatogram of the volatile species from the unknown sample that were trapped onto Tenax™ is shown in Figure 5.7 and the qualitative assignments for the species peaks present in this chromatogram were as follows:
- Carbon disulfide
- Benzene
- Toluene
- Aliphatic hydrocarbons
- N,N-dimethylformamide
- Cyclohexanone
- Cyclohexanethiol
- Phenol
- Acetophenone
- 3-(Bromomethyl)-heptane
- BT
- 2-Methylpropyl butanoate
- 4-(1,1,3,3-Tetramethylbutyl)-phenol
- N-phenylbenzeneamine
- Phthalimide

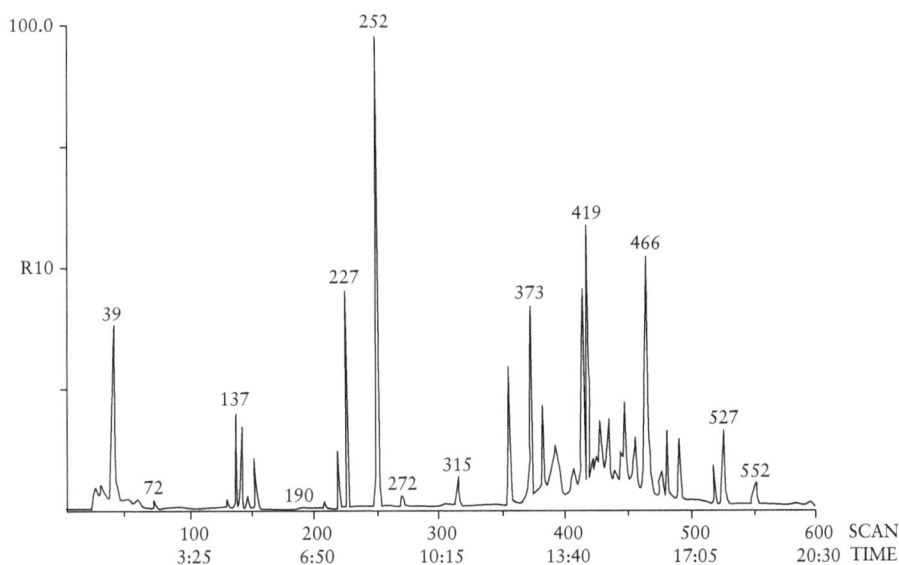

Figure 5.7: GC–MS TIC chromatogram of the volatile species from the unknown sample that was trapped onto Tenax™. Reproduced with permission from M.J. Forrest in *Rubber Analysis – Polymers, Compounds and Products*, Rapra Review Report No.139, Smithers Rapra, Shawbury, UK, 2001, p.36. ©2001, Smithers Rapra [9].

The two sets of GC–MS data confirmed the presence of a phosphate in the unknown sample, and identified it as 2-ethylhexyl diphenyl phosphate, and identified the presence of a phthalate, namely di-ethylhexylphthalate. In addition, the identification of urea in the chloroform extract could mean that a flame retardant such as guanyl urea was present in the unknown sample.

The GC–MS data also provided information on the cure and antidegradant systems, and these are shown in Table 5.11.

Table 5.11: Cure system and antidegradant system information obtained on the unknown sample by GC–MS.

Identified species	Possible origin
Cyclohexanethiol	CBS accelerator
BT	MBT/MBTS (also the CBS) accelerators
N,N-dimethylformamide	TMTD accelerator
Phthalimide	Santoguard PVI
N-phenylbenzene amine	Possibly an acetone/diphenyl amine antioxidant

CBS: N-cyclohexyl-2-benzothiazole sulfenamide
MBT: 2-Mercaptobenzothiazole MBTS: 2,2′-Dithiobis(benzothiazole) PVI: Prevulcanisation inhibitor TMTD: Tetramethylthiuram disulfide

5.3.4 Additional tests

Having determined from the results obtained from the initial suite of tests that the sample was based on a blend of polymers and contained a number of flame retardants (both organic and inorganic), a number of tests could be carried out to obtain quantification data on the constituents identified in Section 5.3.3.

These additional tests were also used to look for substances that may, because of logical assumptions based on the initial findings, be present in the unknown sample, but which had not been apparent in the results (e.g., other flame retardants).

5.3.4.1 Quantitative elemental determinations
A number of tests were conducted on the unknown sample in order to obtain quantitative elemental information and these are listed below:
a. The following elements were quantified by ICP in the unknown sample in the as-received state and the ash from the unknown sample:
   - Antimony
   - Zinc
   - Aluminium

- Phosphorus
- Boron

b. The combined chlorine and bromine content of the unknown sample as-received state and in the methanol-extracted portion were determined using the oxygen flask combustion technique with titration as the final stage. The method used was that described in British Standard BS 7164-22.2:1992.

c. The bromine content of the as-received sample was determined using the oxygen flask combustion technique, but with ion chromatography as the final step to separate the bromine- and chlorine-containing species.

d. The sulfur content of the methanol extracted portion of the sample was also determined using the oxygen flask combustion technique.

e. The nitrogen content of the sample in the as-received state was determined using a PerkinElmer 2400 CHN analyser.

5.3.4.2 Solid-state nuclear magnetic resonance

The methanol-extracted portion of the unknown sample was analysed by cross-polarised solid-state NMR in order to provide additional data on the polymers present within it, and to ensure that the FTIR work had not missed anything.

5.3.5 Results from the additional tests

The chlorine content of the extracted portion of the unknown sample was 13.73%. This chlorine will have originated from the PVC in the sample and could be used to calculate the amount present in the original sample by adjusting for the level of methanol extract and then using the fact that PVC contains 56.8% chlorine. The calculation that was performed and the result obtained is shown below:

$$\frac{(100-\text{extract value}) \times 13.73}{56.8} = 15.8\% \text{ PVC in the unknown sample} \qquad (5.1)$$

Assuming the acrylonitrile (ACN) content of the NBR in the PVC/NBR blend was 36% (i.e., a medium ACN content NBR), the amount of NBR in the unknown sample could be calculated using the same approach as that for PVC content. The nitrogen content of the extracted portion of the sample was 3.39% and the percentage of nitrogen in 36% NBR was 9.5%:

$$\frac{(100-\text{extract value}) \times 3.39}{9.5} = 23.3\% \text{ NBR in the unknown sample} \qquad (5.2)$$

The ICP quantifications for the other elements are shown in Table 5.12.

The data in Table 5.12 show that the flame-retardant zinc borate was not present in the unknown sample. However, it confirmed that the flame retardants hydrated

Table 5.12: ICP quantification results for other elements in the unknown sample.

Element	Sample in the 'as received' state	Sample ash
Antimony	1.38%	2.12%
Zinc	0.41%	2.35%
Boron	<0.05%	<0.05%
Aluminium	7.03%	37.8%
Phosphorous	0.22%	1.36%

Reproduced with permission from M.J. Forrest in *Rubber Analysis – Polymers, Compounds and Products*, Rapra Review Report No.139, Smithers Rapra, Shawbury, UK, 2001, p.38. ©2001, Smithers Rapra [10]

alumina and antimony trioxide were present, at 20 and 2%, respectively, and that ZnO was present at 2%.

The bromine content of the as-received sample was 8.63%. This will have originated from the brominated flame retardant, pentabromodiphenylether, and showed that the sample contained ≈13% of this compound by weight.

With regard to the NMR work that was carried out, the solid-state NMR spectrum for the methanol extracted portion of the unknown sample (Figure 5.8) showed peaks at shift values of 45.6 and 57 ppm, which were due to PVC, and shifts at 33 and 130 ppm, which were due to NBR. No other polymers were apparent in the unknown sample from this data and so it corroborated the assignments made by pyrolysis FTIR (Section 5.3.2.4).

5.3.6 Summary of the information obtained on the unknown sample

By combining all of the information that had been obtained on the unknown sample using the initial suite of tests (Sections 5.3.2 and 5.3.3) and the additional tests (Sections 5.3.4 and 5.3.5), it was possible to present an approximation of its composition as it stood at this stage and this is shown in Table 5.13. For the reasons that have been discussed during this chapter and in Chapters 3 and 4, the quantifications in Table 5.7 are only approximate and it is not possible to provide quantifications for a number of the compound ingredients (e.g., accelerators).

As with all investigations of this type, the characterisation work on the unknown sample could continue, and one of the next steps could be to carry out further work by solvent extraction LC–MS to obtain quantifications for 2-ethylhexyl diphenyl phosphate and di-(2-ethylhexyl)phthalate. The nature and direction of any additional work would depend upon the original objectives of the reverse engineering programme.

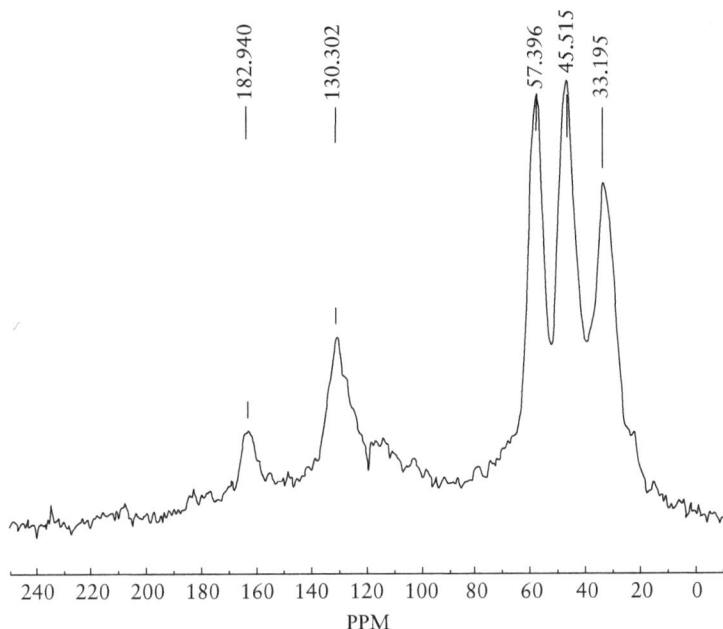

Figure 5.8: Solid-state NMR spectrum for the methanol extracted portion of the unknown sample. Reproduced with permission from M.J. Forrest in *Rubber Analysis – Polymers, Compounds and Products*, Rapra Review Report No.139, Smithers Rapra, Shawbury, UK, 2001, p.39. ©2001, Smithers Rapra [11].

Table 5.13: Summary of the compositional data obtained on the 'unknown' rubber sample.

Polymers present	NBR and PVC in a blend having the approximate ratio: NBR 55:PVC 45
	Total polymer content: 39%
Organic flame retardants*	Pentabromodiphenyl ether (13%)
	2-Ethylhexyl diphenyl phosphate
Plasticiser	Di-(2-ethylhexyl)phthalate
Carbon black content	2%
Total inorganic content	24%
Inorganic compounds	ZnO (2%) Hydrated alumina (20%) Antimony trioxide (2%)
Antidegradants	Diphenylamine/acetone condensation product
Cure system	Sulfur-based system
Accelerators	CBS TMTD MBT/MBTS
Other cure system species	PVI

* In addition to the flame retardants given above, guanyl urea could be in the sample (see GC– MS data in Section 5.3.3). If present, this type of additive can be present in rubber compounds at around the 3% level

Reproduced with permission from M.J. Forrest in *Rubber Analysis – Polymers, Compounds and Products*, Rapra Review Report No.139, Smithers Rapra, Shawbury, UK, 2001, p.38. ©2001, Smithers Rapra [10]

It is worth reiterating, though, that the complexity of rubber compounds precludes a complete characterisation by chemical analytical techniques alone. It is always necessary for the analytical data to be supplemented by input from a rubber technologist in order for a satisfactory copy of the compound to be obtained. In order to achieve this, it is usually necessary to use the analytical data as a basis for a number of test mixes to ensure that all the relevant properties (e.g., cure behaviour, processibility, durability) and, in this case, flame retardancy, are met.

5.4 Published reverse engineering studies

Because of the deficiencies that still exist regarding the amount of information that can be recovered from a rubber compound by reverse engineering work (Section 5.1), research is still being undertaken to improve the accuracy that can be obtained from these investigations. For example, Datta and co-workers [12] described how they carried out reverse engineering work on four rubber compounds of known composition in order to investigate the effectiveness of a new analytical methodology. This methodology, which used a range of analytical techniques and involved fractional mass transfer from the acetone extract to the TGA data, produced bulk compositional data that were very consistent with the formulations and were thought to offer advantages in areas such as the analysis of waste tyre rubbers of unknown composition.

A recent example of how a range of analytical techniques can be combined in a reverse engineering study was provided by Harada [13] and, in this study, the author also covered the analysis of vulcanised rubber to obtain information on crosslink structure and the dispersion of fillers.

Another study describing an extensive analytical programme carried out to deformulate rubber compounds was published by Coz and co-workers [14]. In this work, reverse engineering work was carried out on four rubber compounds that were used to manufacture the following automotive products:
- Radial passenger tyre tread
- Radiator hose
- Oil pan seal
- Engine gasket

A computer programme was used to reconstruct the formulations from all the analytical data and these were found to be very similar to the actual recipes.

Datta and co-workers [15] carried out a quantitative analysis on three batches of end-of-life ground tyre rubber using a non-pyrolytic FTIR method and derivative TGA. Two batches were based on truck tyres and the other batch was based on passenger tyres. The primary objective of the work was to establish an analytical method to characterise the composition of the rubber blend in waste tyres to assist with efforts to recycle them. A secondary objective was to evaluate the validity of

the method's independence of the particle size of the ground rubber. The group used attenuated total reflectance (ATR)–FTIR to analyse the rubber samples. The data generated revealed the presence of a blend of NR, SBR and polybutadiene (BR) in all three batches by use of the following characteristic IR absorption bands:
- NR – 1,375 cm^{-1}
- SBR – 699 cm^{-1}
- BR – 738 cm^{-1}

The same group of workers had derived a general IR blend parameter in a previous study [16] and this was used, in conjunction with the derivative TGA information, to quantify the three rubbers in the samples. The blend parameter had been calculated by using the exact height for the ATR–FTIR absorption bands of NR (1,375 cm^{-1}) and SBR (699 cm^{-1}) obtained by modifying the spectra using a novel numerical algorithmic method of baseline creation on, and subsequent subtraction from, the original spectrum.

Wesdemiotis [17] described a multi-dimensional MS method for the analysis of polymers. The technique interfaces a suitable ionisation technique and mass analysis with fragmentation by tandem mass spectrometry (MS2) and an orthogonal on-line separation method. Examples of techniques that can be used for the separation stage include LC for liquid-phase work or ion-mobility spectrometry (IMS), in which separation takes place in the solution state or post-ionisation in the gas phase. The mass analysis step in the process provides elemental composition information, whereas the MS2 step exploits differences in the bond stabilities of a polymer to provide connectivity and sequence information. The LC work can achieve separations using differences in polarity, end-group functionality, or hydrodynamic volume, whereas IMS adds selectivity according to macromolecular shape and architecture. Together these techniques can be combined to determine the information listed below on a wide range of polymeric materials, including homopolymers, copolymers, polymer blends and crosslinked structures:
- Constituents
- Structures
- End groups
- Sequences
- Architecture

A review of the analytical procedures and techniques that can be used to separate and analyse the constituents in rubber and plastic formulations has been published by Bart [18].

Bacher and co-workers [19] presented a multi-technique methodology to reverse engineer rubber compounds to identify the polymers and ingredients that are present in order to assist with duplication or enhancement work. The combined approach that was employed used the following techniques:

- TGA to determine the quantities of polymer, plasticiser and CB.
- FTIR to verify the identity of the polymer, plasticiser and some of the additives.
- ICP–optical emission spectrometry to determine the elementary filler composition.
- Range of 'wet chemistry' methods to determine certain additives and fillers.

Nimkar and co-workers [20] used an integrated multi-spectral range Fourier– Transform (FT) spectrometer to characterise a range of polymeric materials. The combination that was developed included IR, Raman spectroscopy and TGA to provide a versatile and cost-effective tool for laboratories engaged in reverse engineering work. Some attributes of the workstation included:
- Extended range ATR–FTIR that enabled IR spectra to be acquired below 100 cm^{-1}.
- FT–Raman accessory configured with a 1,064-nm laser that enabled spectra to be recorded from materials that fluoresce with visible excitation.
- TGA–IR that provided an easy way to identify the volatile components.

The research team hopes that their results will demonstrate the complementary nature of the various analytical techniques towards providing a complete solution to a challenging problem.

Bhatt and co-workers [21] carried out work to establish a correlation between the FTIR data and TGA data obtained on the composition of rubber blends used in tyre formulations. Determining this correlation then enabled the blend ratio of rubbers such as NR, SBR and BR in passenger tyres and off-the-road tyres to be calculated, which could be used to assist with important industrial activities, such as quality assurance work.

Nawale and co-workers [22] explored the feasibility of using a mathematical enhancement, such as the second-order derivative of ATR–FTIR spectra, for carrying out compositional analysis work on NR, polychloroprene and their blends in CB-filled vulcanisates. A germanium crystal was used to record the ATR spectra and the second-order derivative approach was found to be successful for the qualitative and quantitative analysis of the two rubbers and their blends. The research team also considered that this method and approach was applicable to other rubber blends, including those of NR/SBR and NR/BR.

Work carried out at PerkinElmer showed how the AutoStepwise TGA method could be used for the compositional analysis of tyre samples [23]. This technique is a subset of the controlled rate thermal analysis technique developed independently by Rouquerol [24] and Paulik and Paulik [25]. They used the following conditions and TGA programme for the analyses:
- Sample mass: 13 mg (approximate weight).
- Instrument: Pyris 1 TGA.
- Method: AutoStepwise.
- Heating rate: 20 °C/min from 30 to 900 °C.

- AutoStepwise entrance threshold: 4%/min.
- AutoStepwise exit threshold: 0.5%/min.
- Pan open: aluminium pan in a ceramic holder.
- Purge gas: nitrogen to 700 °C and air to 900 °C (both 30 ml/min).

The workers regarded the results as showing excellent resolution between all the weight loss events, including the difficult oil and polymer steps, and this enabled the following compositional data to be recorded on the tyre sample:
- Oil: 21.8 %
- Polymer: 43.9%
- CB: 32.2%
- Inert filler: 2.11%

The AutoStepwise approach was regarded by the PerkingElmer workers as having an advantage over the variable heating rate TGA method (i.e., the high-resolution TGA method). This advantage was due to it handling diffusion-controlled weight loss steps much better because the high-resolution method tends to move through the oil weight loss step much quicker and so may not provide the full evolution of oil before the polymer degradation step is encountered. Due to its speed and applicability for a range of different oil types, the AutoStepwise method was also thought to be a better approach, as would high-resolution TGA, for quantifying the oil-in-rubber samples than the following methodologies covered by Maurer [26]:
- Isothermal analysis at a temperature significantly below the temperature at which polymer degradation occurs.
- Use of a reduced pressure (or vacuum) to aid volatilisation of the oil.
- Quantitative extraction of the rubber with a solvent to remove the oil prior to TGA analysis.

References

1. S.D. Gupta, R. Mukhopadhyay, K.C. Baranwal and A.K. Bhowmick in *Reverse Engineering of Rubber Products*, CRC Press, Taylor and Francis Group, Boca Raton, FL, USA, 2013.
2. J. Scheirs in *Compositional and Failure Analysis of Polymers: A Practical Approach*, John Wiley & Sons, New York, NY, USA, 2000.
3. *The Complete Book on Rubber Chemicals*, NPCS Board of Consultants & Engineers, ASIA Pacific Business Press, Inc., Kamla Nagar, Delhi, India, 2009.
4. The Blue Book – Materials, Compounding Ingredients and Services for the Rubber Industry, Rubber World Magazine, Akron, OH, USA, 2016.
5. M.J. Forrest, S. Holding and D. Howells, *Polymer Testing*, 2006, **25**, 63.
6. M.J. Forrest, S. Holding and D. Howells in *Proceedings of the High Performance and Specialty Elastomers*, 20–21st April, Geneva, Switzerland, Rapra Technology Ltd, Shawbury, UK, 2005, Paper No.2.

7. M.J. Forrest in *Rubber Analysis – Polymers, Compounds and Products*, Rapra Review Report No.139, Smithers Rapra, Shawbury, UK, 2001, p.34.
8. M.J. Forrest in *Rubber Analysis – Polymers, Compounds and Products*, Rapra Review Report No.139, Smithers Rapra, Shawbury, UK, 2001, p.35.
9. M.J. Forrest in *Rubber Analysis – Polymers, Compounds and Products*, Rapra Review Report No.139, Smithers Rapra, Shawbury, UK, 2001, p.36.
10. M.J. Forrest in *Rubber Analysis – Polymers, Compounds and Products*, Rapra Review Report No.139, Smithers Rapra, Shawbury, UK, 2001, p.38.
11. M.J. Forrest in *Rubber Analysis – Polymers, Compounds and Products*, Rapra Review Report No.139, Smithers Rapra, Shawbury, UK, 2001, p.39.
12. S. Datta, R. Stocek, Ivo Kuritka and P. Saha, *Polymer Engineering and Science*, 2015, **55**, 6, 1450.
13. M. Harada, *International Polymer Science and Technology*, 2016, **43**, 2, T/45.
14. D. Coz. K. Baranwal and T.M Knowles in *Proceedings of the ACS Rubber Division Fall Meeting*, 8–11th October, Louisville, KY, USA, American Chemical Society Rubber Division, Washington, DC, USA, 1996, Paper No.68.
15. S. Datta, J. Antos and R. Stocek, *Polymer Testing*, 2017, **59**, 308.
16. S. Datta, J. Antos and R. Stocek, *Polymer Testing*, 2017, **57**, 192.
17. C. Wesdemiotis, *Angewandte Chemie, International Edition*, 2017, **56**, 6,1452.
18. J.C.J. Bart in *Proceedings of the 60th ANTEC Technical Conference*, 5–9th May, San Francisco, CL, USA, Society of Plastics Engineers, Brookfield, CT, USA, 2002, Paper 012.
19. B. Bacher, M. Walker and A. Riga, *Rubber World*, 2013, **247**, 4, 22.
20. A. Nimkar, T. Strother and S. Lowry in *Proceedings of the ANTEC Technical Conference*, Mumbai, India, 6–7th December, Society of Plastics Engineers, Brookfield, CT, USA, 2012, p.467.
21. J. Bhatt, B.K. Roy, A.K. Chandra, S.K. Mustafi and P.K. Mohame, *Rubber India*, 2003, **55**, 9, 7.
22. M. Nawale, J. Jarugala, S. Sahoo, K. Rajkumar, B. Abbavaram and R. Sadiku, R. Pritti and D.J. Maurya, *Polymer Testing*, 2017, **22**, 447.
23. *Compositional Analysis of Tire Elastomers using AutoStepwise TGA*, PerkinElmer, Waltham, MA, USA, 2009.
24. J. Rocquerol, *Bulletin of the Society de Chemie France*, 1964, p.31.
25. F. Paulik and J. Paulik, *Analytica Chemica Acta*, 1971, **56**, 2, 328.
26. J.J. Maurer in *Thermal Characterisation of Polymeric Materials*, Ed., E. Turi, Academic Press, New York, NY, USA, 1981, p.653.

6 Curing and cure state studies

6.1 Introduction

This chapter provides an overview of the various chemical and thermal techniques that have been used to characterise the curing behaviour and cure state of rubber compounds and rubber products. It is an essential element of quality control (QC) work to routinely monitor and assess these characteristics, but the need also arises for a number of other reasons. For example, during research and development (R&D) activities, and it often plays a crucial part in failure diagnosis investigations.

There are standard pieces of equipment within a rubber laboratory that are used to measure and characterise curing behaviour (e.g., rate of cure at a given temperature) and cure state (i.e., degree of cure) and these include the oscillating disc rheometer (ODR), the moving die rheometer (MDR), and the rubber process analyser (RPA). These are excellent QC and R&D instruments, but they have a practical constraint concerning the minimum amount of material that is required for an examination, and they also tend to be used on rubber samples that start off in the unvulcanised state. The information that can be obtained by these instruments can be complimented by the use of thermal analysis techniques [e.g., differential scanning calorimetry (DSC) and dynamic mechanical analysis (DMA)] and, in the case of liquid samples, specialist equipment [e.g., the RAPRA Vibrating Needle Curemeter (VNC)]. Thermal analysis techniques offer a number of advantages, including being able to record data on small samples (e.g., 5 mg in the case of DSC) and being able to analyse cured, partially cured, and uncured samples.

It is also possible to use physical testing techniques to establish the state of cure (i.e., crosslink density) of a rubber sample. Two classic approaches are equilibrium swelling in a solvent and extensional experiments involving a tensometer. The former method has the advantage in that it can be carried out on a sample having an irregular geometry; the latter requires standard test sheets and test pieces (e.g., dumbbells). The use of these techniques for this type of work has been discussed at length in other publications [1] and, although they can often contribute to analytical investigations, they will not be covered in detail in this chapter.

When a rubber cures, a wide range of volatile organic compounds are released as rubber vulcanisation fume. The composition of this fume has been studied (Chapter 7) to assess its affect on the health of employees in the rubber industry and general environment. This research work has established relationships between the fume and the compounding ingredients in a rubber compound. This has given the analysis of rubber volatiles a role in the determination of additives (Chapter 4) and reverse engineering work (Chapter 5).

Two illustrations of an ODR curve that has been recorded on a rubber compound are shown in Figures 6.1 and 6.2. Both versions of this curve have been

https://doi.org/10.1515/9783110640281-006

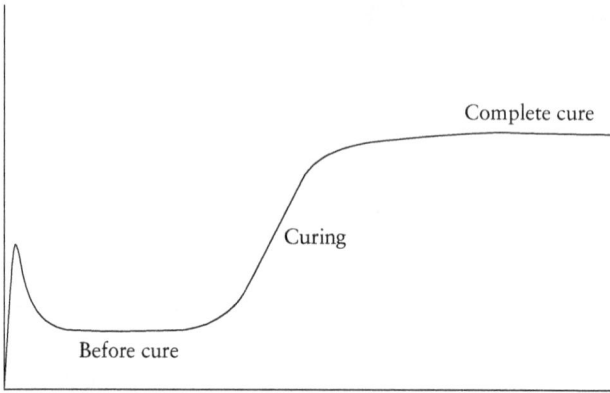

Figure 6.1: Typical ODR curve for a rubber compound. Reproduced with permission from A. Ciesielakiin *An Introduction to Rubber Technology*, Rapra Technology Ltd, Shawbury, UK, 1999, p.85. ©1999, Rapra Technology Ltd [2].

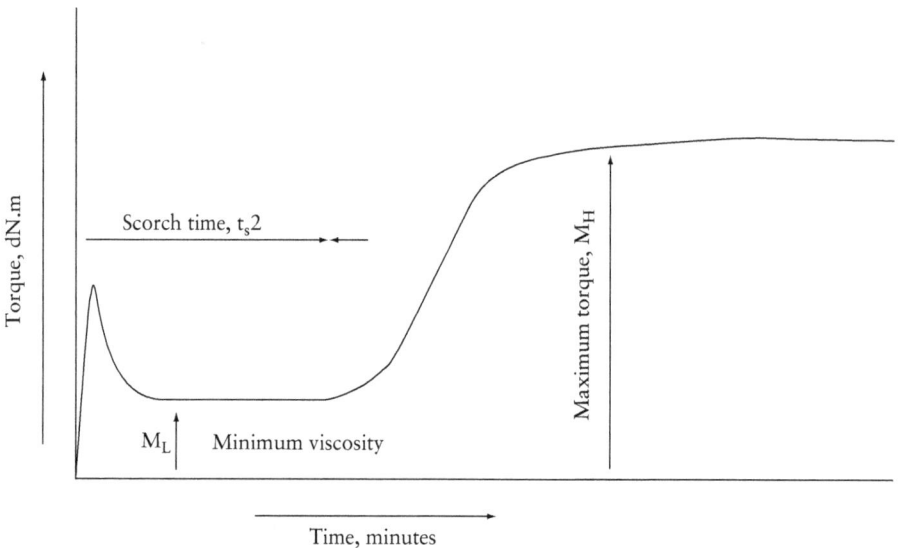

Figure 6.2: The trace in Figure 6.1 with important features highlighted. Reproduced with permission from A. Ciesielski in *An Introduction to Rubber Technology*, Rapra Technology Ltd, Shawbury, UK, 1999, p.85. ©1999, Rapra Technology Ltd [2].

obtained by plotting the change in torque as a function of time and demonstrate how the torque increases as the rubber cures and a three-dimensional crosslink network develops within the material. The optimum cure time for a particular rubber compound is usually regarded as the time to achieve 90% maximum torque at a given temperature and is designated 't90'. The version of the ODR trace shown in Figure 6.2 provides examples of the other properties that can be obtained from such a trace, such as the scorch time, minimum viscosity value (i.e., minimum torque value) and maximum torque value. Figures 6.1 and 6.2 show the torque value attaining a plateau once the rubber has become fully cured but, in some cases [e.g., with natural rubber (NR) compounds], the rubber undergoes 'reversion', and the value of the torque begins to decrease as the matrix undergoes thermal degradation.

R&D organisations such as Smithers Rapra have been very active in studying the cure behaviour of rubber compounds and other polymeric systems. A comprehensive review of the use of chemical and physical analytical techniques to assess cure has been published by Willoughby [3]. Additional information on the techniques and methodologies that can be used to study curing can be found in a number of other publications, such as those edited by Gabbott [4] and De and co-workers [5].

6.2 Thermal analytical techniques for curing and cure state studies

6.2.1 Differential scanning calorimetry

One analytical technique that is particularly useful for this type of work is DSC. A review of the use of DSC to study sulfur and peroxide vulcanisation curing processes in rubbers has been provided by Brazier [6].

DSC is a very effective tool for this application because the curing reaction of a rubber, in common with many other types of chemical reactions, is very exothermic. This exotherm, which results from a number of reactions that can take place during vulcanisation, is easily detected by a significant change in the specific heat plot of a sample.

A number of different methods for using DSC to characterise the cure of rubbers have been developed [7]:
- Borchardt and Daniels method [8].
- American Society for Testing and Materials' (ASTM) E698 method [9, 10].
- Isothermal method [11, 12].

The size and position (i.e., temperature range) of the curing exotherm obtained in a DSC analysis is dependent upon the:
- Type of polymer.
- Proportion of polymer in the sample.
- Type of cure system (e.g., peroxide or sulfur).
- Composition of the cure system (e.g., type of peroxide or type of accelerator(s) if the system is sulfur-based).

- Overall amount of curative(s).
- Level of individual curatives in the blend if a blend of curatives is present.

In addition to the factors stated above, which are sample-related, it has also been demonstrated using inter-laboratory trials that experimental conditions and instrumental effects can influence the cure data that are obtained [13, 14].

An example of a curing exotherm (from ≈90 to ≈240 °C) is shown in Figure 6.3. This is a DSC thermogram of a fluorocarbon rubber undergoing vulcanisation due to the presence of a peroxide-based cure system. It is apparent from this thermogram that values for the cure onset temperature and total area of the exotherm can be obtained which, by use of the sample weight, can be converted into J/g. The size of the sample is this case was 12.9 mg and the value for the curing exotherm was calculated to be 110 J/g.

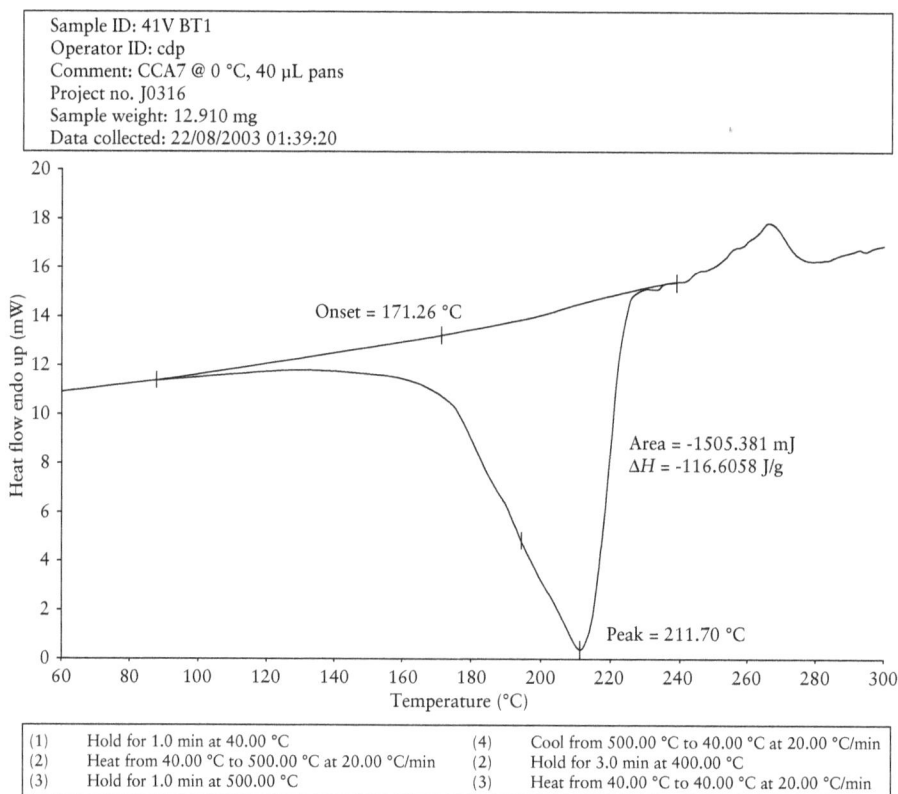

Sample ID: 41V BT1
Operator ID: cdp
Comment: CCA7 @ 0 °C, 40 µL pans
Project no. J0316
Sample weight: 12.910 mg
Data collected: 22/08/2003 01:39:20

Onset = 171.26 °C

Area = -1505.381 mJ
ΔH = -116.6058 J/g

Peak = 211.70 °C

(1)	Hold for 1.0 min at 40.00 °C	(4)	Cool from 500.00 °C to 40.00 °C at 20.00 °C/min
(2)	Heat from 40.00 °C to 500.00 °C at 20.00 °C/min	(2)	Hold for 3.0 min at 400.00 °C
(3)	Hold for 1.0 min at 500.00 °C	(3)	Heat from 40.00 °C to 40.00 °C at 20.00 °C/min

Figure 6.3: DSC trace showing the curing exotherm of a peroxide-cured fluorocarbon rubber. Reproduced with permission from M.J. Forrest in *Principles and Applications of Thermal Analysis*, Ed., P. Gabbot, Blackwell Publishing, Oxford, UK, 2008, p.242. ©2008, Blackwell Publishing [15].

In Figure 6.3, the exotherm is given a negative value but this is only due to the style of presentation that has been used – with exotherms being plotted in this case in a negative direction and endotherms in a positive one. The endotherm that occurs after the curing exotherm, with a peak at ≈270 °C, is due to the curing volatiles lost from the sample. This analysis was carried out using a sealed sample pan to try and contain these volatiles, but a large quantity were generated and the resulting pressure ruptured it. This situation could be addressed by reducing the size of the sample, but it can also occur if there is a relatively high proportion of additives in the rubber compound (e.g., plasticisers or oils) that are particularly volatile.

Two advantages of using DSC for curing investigations are that only small sample sizes are required (e.g., 5 mg) and no specific sample geometry is necessary. This type of investigation can be carried out for a number of purposes, for example:

– Quantification of cure system species
– Qualitative data on cure system type
– Characterisation of a cure (e.g., initiation temperature and kinetics)

As listed above, a number of factors, including a number of other components in a rubber compound and experimental conditions, will influence the DSC data and so this type of work, and other DSC applications, are usually carried out in a comparative way. For example, to use DSC to determine the amount of cure system in a rubber, because the exotherm obtained is very specific to the rubber matrix and the cure system used, standards need to be available if quantitative data are to be obtained on a test sample from the J/g exotherm value. Once these standards have been obtained and the reference data generated, this type of application obviously has uses from a QC point-of-view as well as other applications, such as failure diagnosis.

If an inconsistent cure state is thought to exist through the cross-section of a relatively large rubber product, by taking and analysing small samples by DSC it is possible to obtain a depth profile of this property. The ability of DSC to detect residual unreacted curatives can be used in a more general sense to investigate if a product has been cured appropriately as part of a QC process. In these examples, the exotherm generated can be relatively small and can occur in a similar region in the thermogram to where additives, such as plasticisers and oils, are volatilising and creating an endotherm. For these reasons, such exotherms can be very difficult to detect and it is, therefore, important to modify the experimental conditions to optimise the system in favour of their detection. For example, larger samples should be used in sealed sample pans to stop substances volatilising, although care has to be taken because rupturing of the sample pan can occur.

The examples given above illustrate how DSC can be used for practical purposes that benefit the day-to-day activities of industry and test houses. It can also be used in more fundamental R&D work to assist researchers who are interested in cure kinetics, as well as other important properties, such as optimising the degree of cure and the cure state that exists within different phases of polymer blends.

Information on cure kinetics can be of great value to R&D chemists trying to development new curing agents or curing systems. The achievement of an optimised state of cure within a rubber product is vital to it realising its full potential with respect to its properties and performance in service. Using DSC for this type of quantitative work, and others, can be challenging and a number of researchers and scientists have discussed its conduct and the possible problems that can be encountered [13, 14, 16].

One example of the curing kinetics studies that have been carried out is the one reported by Simon and Kucma [17]. They vulcanised rubber compounds using both isothermal and non-isothermal (i.e., temperature ramp) conditions and, among other features, described the temperature dependence of the cure induction period using an Arrhenius-type equation. Another example is the work carried out by Ou and co-workers [18], who used DSC and ODR to study the crosslinking reaction of silicone rubber. From the experimental data that were obtained, different kinetic analytical methods were used to determine the activation energy of silicone rubbers, such as the Kissinger, Ozawa, Friedman, Flynn–Wall–Ozawa, Kissinger–Akahira–Sunose and integral methods. The results obtained by their study showed how the composition of the silicone rubber, the analytical technique and kinetic analytical method affected the activation energy and its evolution during the crosslinking reaction.

Even though sulfur vulcanisation has been used by the rubber industry for over well over 100 years, it has often been refined and advanced in a mainly empirical way. Over the years, various studies have been conducted to advance the fundamental understanding of how elemental sulfur, the various classes of accelerators (e.g., thiurams) and different co-agents [e.g., zinc oxide (ZnO)] that can be used to vulcanise rubber react with one another and the rubbers that they are compounded into. One major contribution to this body of work is the extensive series of experiments conducted by researchers at Port Elizabeth University, who used DSC extensively and its contribution features in a number of publications [19, 20].

DSC can be used to study the cure state within different phases of polymer blends. For example, Wong-on and Wootthikanokkan [21] used DSC to determine the degree of cure that existed within the rubber phase of a plastic–rubber blend. To carry out the work, the two researchers prepared PVC/acrylic rubber blends using a twin-screw extruder and then cured samples for testing using compression moulding. The curing characteristics of blends consisting of two rubbers have also been investigated by DSC. For example, Baba and co-workers [22] used DSC to evaluate the crosslink density of rubber blends. Their work involved ethylene propylene diene monomer (EPDM) rubber and polybutadiene (BR) rubbers that had been crosslinked to varying degrees by either photooxidation or dicumyl peroxide.

To understand the influence that additives have on curing chemistry, Karbhari and Kabalnova [23] used DSC to investigate the curing kinetics of rubber-modified, carbon fibre (CF)-reinforced vinyl ester resins. Their work reported on how the quantity of CF, and the amount of sizing present on its surface, affected the cure rate, degree of cure, time to achieve the maximum cure state, and cure rate constant.

It is well known that the crosslinking of polymer chains increases the glass transition temperature (T_g) of the polymer [24]. For example, Zeng and Ko [25] reported that crosslinking *cis*-1,4-BR to a higher degree increased the T_g of the rubber from −103 to −96 °C. Cook and co-workers [26] studied the T_g of various rubbers with different crosslinking densities using both DSC and nuclear magnetic resonance (NMR), and found that the T_g *versus* crosslinking data followed a linear regression. In addition, they found that the NR and BR vulcanisates displayed a similar T_g *versus* crosslinking density plot, but that the increase in T_g of the styrene-butadiene rubber (SBR) vulcanisate with crosslink density was much greater than that of NR or BR.

In addition to being used to study the vulcanisation of solid rubbers, DSC can also be employed to characterise the cure of rubber latex. Peres and Lopes [27] used DSC to investigate the efficiency of the vulcanisation of NR latex using two types of cure system – a conventional sulfur cure system and an efficient [i.e., efficient vulcanisation (EV)] cure system. The DSC data were complemented with tensile strength (TS) tests and optical analysis. The results showed that the conventional sulfur cure system provided a higher curing rate, better processing safety (i.e., longer scorch time) and superior physical properties.

The cure of rubber-modified epoxy resin systems has also been investigated by Restrepo-Zapata and co-workers [28]. The curing reaction of an aliphatic epoxy resin/ EPDM rubber system was modelled from DSC data using a methodology proposed by Hernandez-Ortiz and Osswald. The kinetics of the reaction were represented by a Kamal–Sourour model, with and without diffusion reaction control, and was extracted using a non-linear regression method coupled with heat and mass balance equations.

6.2.2 Dynamic mechanical analysis

Another technique which can be used to investigate the degree of cure of a rubber product is DMA. It suffers from two drawbacks when compared with the DSC technique: a certain sample size and geometry (normally 2 cm × 1 cm × 1 mm) is usually required and it is often less sensitive, relying on the effect that crosslink density has on the T_g of the rubber compound (as shown by the tan δ plot) or its modulus. Although the sample dimension described above is often regarded as the 'standard' one, modern instruments have become more accommodating as new sampling devices have come onto the market. For example, it is now possible to obtain comparative DMA data on smaller samples and even those with a non-specific geometry (e.g., in the form of a powder).

Use of the tan δ plot in the DMA thermogram to obtain an indication of cure state is possible because the reduction in storage (i.e., elastic) modulus as the rubber passes through its T_g is reduced successively as the crosslink density is increased. The degree of reduction that is observed is not uniform and is dependent upon the type of rubber (i.e., polymer) that the compound is based upon. The approximate percentage reduction in the height of the tan δ peak moving from an uncured state to the optimum cured state for three types of rubber is shown below [29]:

- Nitrile rubber (NBR) – 15%
- Polyisoprene – 76%
- BR – 64%

Having established the nature of the relationship shown above by the use of standard compounds, it can then be used to provide an assessment of the degree of cure of a test sample. Although this task could be carried out using other tests (e.g., hardness determination, solvent swelling), DMA offers the advantages of a relatively quick diagnosis on, potentially, a relatively small portion of sample.

With respect to using the change in the modulus to reflect the cure state of a rubber compound, the effectiveness of this approach can be influenced by the levels of the principal additives (e.g., fillers and plasticisers) that are present. This is because high levels of these additives will have a significant influence on the modulus of the rubber compound and reduce the polymer fraction of the sample; the faction altered by crosslinking. What is found in practice is that for a highly filled rubber compound, the difference in modulus between it being fully cured and uncured is less than for the unfilled version. This illustrates that these types of studies and a lot of others in this chapter are best carried out in a comparative way on samples that have been fully characterised.

It is possible in some cases when using the typical temperature ramp (e.g., 3 °C/min) to observe a rubber sample that is a poor state of cure initially suffering a decreasein modulus once it has passed its T_g due to softening and then, as the temperature continues to rise, to gain modulus as the crosslinking reaction takes place. This effect is more likely to be observed when the poorly cured rubber sample contains curatives that have relatively low temperatures of initiation (e.g., certain peroxides) and is lightly modified by fillers.

As shown in Figures 6.1 and 6.2, the ODR and MDR can be used to plot graphs of torque *versus* time at a constant temperature to show how a cure is progressing and, based on the results obtained, the optimum cure time is usually regarded as the time required for the torque to reach 90% of its maximum achievable value, and is designated 't90'. Khimi and Pickering [30] assessed the use of DMA in the determination of t90. The DMA experiments were carried out using shear mode isothermal tests to measure the changes in material properties caused by vulcanisation. The results obtained revealed that the shear storage modulus (G'), shear loss modulus (G") and tan δ all reflected the vulcanisation process, but the tan δ plot gave the best representative level of vulcanisation. In fact, the tan δ plot could be used to obtain a value for t90, and this value was in the good agreement with the value of t90 obtained using a MDR.

Formela and co-workers [31] used DMA to contribute to a programme of work designed to characterise ground-tyre rubber samples that had been reclaimed using a twin-screw extruder and then re-vulcanised using different types of standard rubber industry vulcanisation accelerators [e.g., 2-mercaptobenzothiazole and tetra-methylthiuram disulfide (TMTD)] formulated into conventional sulfur and EV cure systems. The results obtained showed that the static and dynamic mechanical prop-erties of the re-vulcanised rubber depended strongly on which type of cure system had been used. The highest crosslink densities were obtained with the samples that had been cured using a TMTD-based EV system, whereas the best processing and mechanical properties resulted from the use of a N-*tert*-butyl-2-benzothiazole sulfenamide (TBBS)/N-cyclohexyl-2-benzothiazole sulfenamide (CBS) accelerated conventional sulfur cure system.

The dynamic mechanical properties of electron beam-irradiated NBR vulcani-sates containing varying levels of sulfur have been studied by DMA [32]. DMA was performed from −80 to +80 °C, at frequencies that varied from 0.32 to 32 Hz, and at strains of between 0.001 and 10%. The results showed that the irradiation of the samples caused significant changes in the tan δ peak temperature and storage mod-ulus. The vulcanisates containing high amounts of sulfur formed intense crosslink networks and crosslink rearrangements, which were supported by an increase in storage modulus and the shift in tan δ towards higher temperatures compared with the control. The research team also found that there was an increase in the tan δ peak height due to chain scission and subsequent plasticisation.

6.2.3 Other thermal techniques

It is possible to use thermal mechanical analysis (TMA), either in penetrometry or ex-pansion mode, to evaluate the state of cure of finished rubber products, and its use for this purpose has been discussed by a number of authors [33–36]. In the expansion approach, the sample is heated from ambient to a high temperature (e.g., 200 °C) and the small dimensional change that occurs is monitored. This dimensional change in-creases as the cure state of the sample increases. For example, a degree of cure of 50% (based on a rheometer reading) might give a change of ≈20 μm, which increases to ≈40 μm for a fully cured sample. Providing that reference data for a particular system are available, the technique can be used to investigate the state of cure of a test sample.

The use of TMA *via* the penetrometry mode for calculating crosslink density was advanced by Prime [36], who demonstrated a correlation between Young's modulus and the solvent swelling ratio of a sample. This correlation then enabled the crosslink density 'e' of a sample to be obtained by, firstly, calculation of its elas-tic modulus 'Em' from the TMA penetrometer measurements using the equation de-veloped by Gent [37], and then by use of the following equation [38]:

$$e = Em/3RT \tag{6.1}$$

where 'T' is 239 K.

R (universal gas constant) = 8.314×10^7 ergs/mol/K

Provder and co-workers [39] used diffusing mechanical analysis, thermal analysis, and DMA to study the cure of a polydimethylsiloxane (PDMS) top coat/epoxy coating for primer marine applications. The data obtained by this combination of analytical techniques provided the research team with information on a wide range of properties, including curing, drying and relaxation behaviour, film-forming parameters, coating thickness, thermal transitions, stress–strain properties, elastic modulus, substrate effects and dynamic mechanical properties.

Zhao and co-workers [40] investigated the use of a number of novel compounds for the high-temperature curing of brominated butyl rubber. In addition to recording thermogravimetric analysis data on the compounds, their curing behaviour and physical properties were evaluated and the results showed that N,N′-phenylenedimaleimide, N,N′-(4,4′-methylenediphenyl)dimaleimide and diammonium phosphate were suitable compounds for curing brominated butyl rubber at high temperatures.

Maistros and co-workers [41] used dynamic dielectric analysis (DDA) to monitor the isothermal cure in rubber-modified epoxy systems. Their results showed that in the frequency range 0.1 to 20 kHz, DDA was capable of showing the onset of gelation in the modified resins *via* a sudden increase in the relaxation time. The separation of phases within the material was detected by an increase in relative permittivity due to interfacial polarisation. Overall, the results obtained by DDA were found by the workers to be in good agreement with those obtained by turbidity measurements and solubility determination.

6.3 Use of chromatographic and spectroscopic techniques for curing and cure state studies

Andersson and co-workers [42] used pyrolysis (Py) gas chromatography (GC) to study the sulfur bridges in filled NR vulcanisates. The gas chromatograph was fitted with a sulfur-selective flame photometric detector and the main Py products detected were carbon disulfide and thiophenes. Two NR formulations were analysed, one cured with sulfur and CBS, and the other with the sulfur donor compound TMTD. The yields of the Py products were found to be different for the two rubber compounds and to vary with the cure time in both cases.

It is possible to use spectroscopic techniques to study cure state and crosslinked structures within rubbers as demonstrated by the work reported by Ito and Sawanobori [43]. The two researchers used NMR to determine the crosslink density of both solid and foamed sulfur-cured EPDM compounds. With the solid EPDM samples, the results showed a good correlation between the spin-relaxation time (T2) at high temperatures (e.g., 120 °C) and the crosslink density obtained by equilibrium swelling and extensional experiments. They also found that the T2 of the EPDM foams was not affected by their cellular structure and, hence, it was possible to use

NMR to obtain the crosslink density of these products, something that is not possible by equilibrium swelling or extensional experiments.

Brown and co-workers [44] demonstrated how Fourier–Transform NMR can be used to estimate the crosslink densities of unfilled and carbon black (CB)-filled rubber blends. The different rubbers that were included in the study included NR, BR, NBR and EPDM. Using the Fourier–Transform approach was shown to improve the resolution of NMR data and enabled work to be performed on filled compounds that contained filler additives at realistic, commercial levels.

Pazur and Walker [45] discussed the issues surrounding the use of NMR spin echo measurements for assessing the state of cure of crosslinked rubber. They suggested that a technique known as the 'state of cure number' could be used for the analysis of the data and that this overcame the problems (e.g., variable test results) found with spin echo measurements. The state of cure number correlated to the crosslink density in a rubber sample and the published work showed that it could be applied to the measurement of the state of cure of specialty rubbers for automotive sealing applications and optimisation of the injection moulding curing conditions of a developmental EPDM part.

A research team at Shandong University [46] used *in situ* Fourier–Transform infrared (FTIR) spectroscopy to study the curing kinetics of a two-part, room temperature-cured, silicone rubber. The two-part rubber, which had excellent low-temperature properties, was prepared by mixing a random silicone copolymer (PDMS-*co*-diethylsiloxane), synthesised using anionic ring-opening polymerisation, with a hydrosilicone. The cure kinetics that resulted from the *in situ* FTIR analysis were used to forecast the tack-free time of the silicone rubber at different curing temperatures, and these were found to agree with the measured tack-free time of test samples.

6.4 Rheometric studies

The data shown in Figure 6.4 illustrate how rheometer curves (e.g., ODR curves) can change as the formulation of a rubber compound is varied, and how the information obtained (e.g., scorch time, cure time and rate of cure) can be compared directly. The curves shown in Figure 6.4 are for a series of NR compounds that contain a sulfur cure system accelerated using five types of sulfenamide accelerator, each at a concentration of 0.6 phr. The other formulation details are:

- NR: 100
- ZnO: 5.0
- Stearic acid: 2.0
- Sulfur: 2.5

Mention has been made in Section 6.1 of the use of RPAs as a QC tool in the rubber process laboratory. ASTM D6204 Part C describes a test method for studying cure

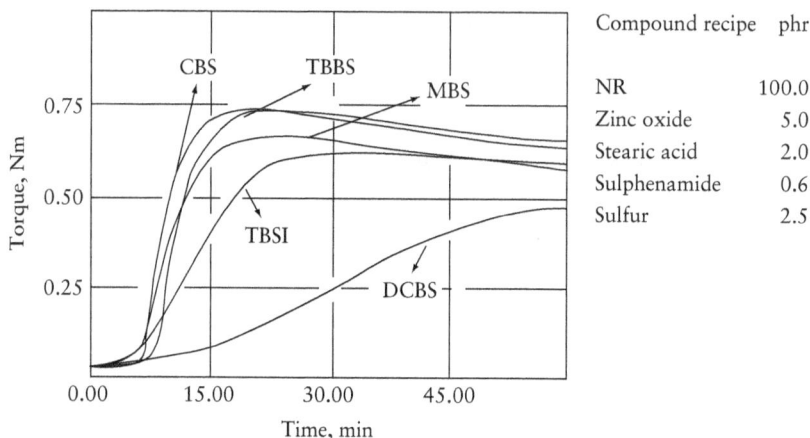

Figure 6.4: Overlaid rheometer curves showing how varying the type of sulfenamide accelerator within a NR compound can alter its curing characteristics (DCBS: N,N-dicyclohexyl-2-benzothiazyl sulfenamide; MBS: 2-(4-Morpholinothio)benzothiazole and TBSI: N-*tert*-butyl-2-benzothiazole sulfenimide). Reproduced with permission from R.N. Datta and F.A.A. Ingham in *Rubber Technologist's Handbook*, Eds., J.R. White and S.K. De, Rapra Technology Ltd, Shawbury, UK, 2001, p.175. ©2001, Rapra Technology Ltd [47].

kinetics and conducting variable temperature analysis (VTA) using the RPA manufactured by Alpha Technologies. Two articles by Dick and Norton [48, 49] provide reviews of how reaction cure kinetics data and VTA data, generated using the Alpha Technologies RPA, can be used to investigate rubber curing problems and improve QC in manufacturing operations. To illustrate this situation, several laboratory designs of experimental studies were carried out on selected rubber compounds to compare the effectiveness of the cure kinetic and VTA approaches. These model compounds were based on sulfur-cured SBR and NR compounds and peroxide/promoter-cured EPDM, chlorinated polyethylene, polychloroprene and fluorinated compounds. In addition, in an article written in 2015, Dick and Xue [50] provided an overview of the improvement of the capabilities of the RPA made by Alpha Technologies to undertake a range of analyses, including the analysis of scorch and curing profiles, and the prediction of the performance of cured rubber products by measuring their dynamic properties. The report also describes how improvements in the range and precision of strain and frequency sweeps resulted in improvements in the predictions that can be made using the RPA. The examples provided included the better prediction of the state-of-the-mix through ASTM D6204 Part B (high strain) and better predictions of the degree of silanisation through improved, wider-range strain sweeps for the Payne effect.

Rosa and Vergnaud [51] developed a new method for using an MDR. By increasing the temperature of the instrument with the square root of time, the kinetic parameters of the curing reaction were determined. These included the activation energy of the cure, the pre-exponential factor, and the order of the overall reaction. Other information that was obtained using this method included the determination of the temperature profiles developed through the thickness of the rubber sample and profiles of the state of cure.

Torque curves were used by Marzocca and Mansilla [52] to characterise the vulcanisation of SBR compounds prepared using different cure systems based on sulfur and TBBS. The time to achieve maximum torque was evaluated and the density and type of elastically active crosslinks were examined by swelling measurements.

Madhuri Nanda and Deba Kumar Tripathy [53] carried out a comparative study involving four cure systems for chlorosulfonated polyethylene rubber (CSM). The systems investigated were based on sulfur, peroxide (dicumyl peroxide), metal oxide (lead oxide) and epoxy resin, and their curing characteristics were determined using a cure rheometer. The highest values of maximum rheometric torque and scorch safety were found with the peroxide cure system, whereas the sulfur system gave superior mechanical and rheological properties. Analyses of the cured CSM rubbers by FTIR using the attenuated total reflectance technique enabled the researchers to propose possible crosslink structures for the different types of cure system.

A research team in China carried out a rheological study on the cure kinetics of a two-component addition-cured silicone rubber [54] and monitored the effect of the reaction temperature. Their research showed that the Kamal–Sourour (autocatalyst) model fitted the cure kinetics of the two-part silicone rubber better than the Kissinger model. A second-order reaction was derived for the time period up the point in the reaction where the crosslinked network started to restrict the diffusion of reaction components; after this point the order of reaction was >2 because of this restriction. The reaction rate constants at different cure temperatures and the activation energy were also calculated.

Willoughby [55] examined the rheological outputs that can be produced by a cure rheometer using the Cole–Cole plot, a procedure that is usually applied to a reaction system. The Cole–Cole plot was applied to rheological data obtained during the curing of a NR sample at ≥160 °C and above, conditions which ensured that both curing and reversion behaviour could be covered. This approach was considered by Willoughby to be capable of identifying the competition between intermolecular and intramolecular sulfurisation, and between crosslink formation and main chain scission, and so to have the potential to expand the capabilities of cure rheology.

Ghoreishy and Asghari [56] used a rheometer, a RPA, and a multi-recorder to study the cure kinetics of an SBR compound under both non-isothermal and isothermal conditions. Changes to the degree of curing were followed using kinetic parameters, direct determination of the temperature of the compound in the centre of the mould, and time–temperature profiles. The results obtained showed that curing behaviour under both non-isothermal and isothermal conditions could be explained using an autocatalytic model.

An oscillating disc rheometer was used by da Costa and co-workers [57] to study the vulcanisation of novel silica-filled NR vulcanisates at different curing temperatures. Overall, the objective of the study was to evaluate the suitability of caster oil fatty acid to function as an activator in the NR compounds, and the levels of stearic acid, castor oil and polyethylene glycol were varied while the amounts of silica filler and other ingredients were kept constant.

6.5 Finite element analysis and mathematical modelling

Another approach used to study state of cure is finite element analysis (FEA). During the curing of actual rubber articles, as opposed to standard 2 mm-thick test sheets, there is a temperature gradient (i.e., non-isothermal and non-uniform cure conditions) within the product. Therefore, the physical properties of the product may vary from those obtained from the test sheet. FEA was used by Gregory and Muhr [58] and they reviewed the theoretical framework for calculating isothermal cures that were representative of particular temperature histories. The theory was implemented for specific cases by solving the transient thermal problems using FEA together with user- defined variables representing the 'state of cure' and the 'representative temperature' of an equivalent isothermal cure. The predictions that were obtained were compared with experimental results and, although the overall agreement was good, there were limitations and these were discussed in their article.

A team of researchers from the Qingdao University of Science and Technology used FEA to study the curing process in rubber [59]. They used the technique to solve the Maxwell equation and heat transfer equation, employing ANSYS finite element software to calculate and analyse the process of conventional heating in which heat transfers from outside to the inside of a rubber sample and microwave heating (in which the reverse occurs). The results obtained showed that the time of microwave heating was shorter and efficiency higher. Another team from the South China University of Technology and Shanghai Tire & Rubber Group used thermo-electric couple cure temperature, DSC, and FEA to study the effect of the heat of reaction on the cure temperature of rubber blocks and tyres [60]. This approach enabled cure temperatures to be predicted and these were then compared with those determined experimentally.

A team from Jiangsu University of Science and Technology [61] presented a systematic model for the FEA of the rubber curing process. By monitoring how the thermal conductivity and specific heat capacity varied with temperature they were able to assess the accuracy of five models (N^{th}-order, Piloyan, Kamal–Ryan, Kamal–Sourour and Refei) in providing the optimum kinetics from the curing curves. By using a rubber piston washer as an example, and ABAQUS software, the workers were able to calculate the curing model and analysed the induction time, temperature and state of cure during the curing process. The findings of the study enabled them to demonstrate that FEA is suitable for providing cost-effective and practical predictions for the rubber curing process. ABAQUS software was also employed by a team at Qingdao Technological University, who used FEA to simulate and analyse the curing of radial tyres [62]. The output of their research included the temperature distributions for different curing stages and curing temperatures, and the degree of cure at various time points during the curing process.

Researchers drawn from Shandong University, Changchun Institute of Applied Chemistry and the Chinese Academy of Sciences developed a mathematical model for predicting the vulcanisation kinetics of silicone rubber [63]. The thermal effects of the curing process were computed using the increment method, and the vulcanisation process was described by FEA. The work by this team illustrated the importance of the exothermic curing reaction for ensuring that energy was used efficiently, and the optimum cure temperature for achieving an effective and uniform cure. The effect of increasing the forced convective heat transfer coefficient above a certain level was also investigated during the work programme. Further studies on the curing of silicone rubbers using FEA was reported by Jia and coworkers [64] and the team presented a methodology for the integrated processing–structure–property analysis of the vulcanisation of silicone rubber within a mould. This enabled the temporal evolution, the spatial distribution characteristics of the hot processing parameters, the crosslinking structure parameters, and the mechanical property parameters to be obtained by FEA. The benefits of this approach included optimisation of the curing conditions and ability to design the curing process to meet specific requirements.

Gough [65] used FEA and a model to simulate the curing behaviour of a laminated rubber bearing. He made a comparison between the temperatures in a real bearing with those of the finite element simulation, and the results obtained enabled guidelines to be composed that can be used to assist with the curing of thick rubber articles.

Karaagac and Deniz [66] described how Adaptive Neuro–Fuzzy Interference Analysis Systems (ANFIS) can be used to predict the cure of complex and thick rubber articles. These workers carried out an experimental study that involved 11 rubber compounds cured using 10 curing temperatures. The obtained data showed that ANFIS methods, when compared with conventional techniques, gave the lowest average errors for optimum cure times.

Vebber and Pereira [67] developed a predictive model for choosing cure agents for rubber blends. The objective was to enable curing agents to be chosen that would not show a preference for one or other of the rubbers in a blend (as can often happen due to differences in polarity and double bond concentration) and which can result in an over-curing of one phase and under-curing of the other. The successful method that was developed included a new tool based on a genetic optimisation algorithm for assessing the partitioning of cure agents in different rubbers and their blends. The quantitative data that resulted enabled a series of analyses of the solubility of the cure agents in the different phases, and was validated by correlation with the physico-chemical and mechanical properties of the prepared blends.

6.6 Other techniques and methodologies for studying curing and cure state

Dinzburg [68] described how a modified version of the Flory equation in combination with the Cluff–Gladding–Pariser method could be used to provide information about the cure state and structure of both filled and unfilled hydrogenated-NBR compounds. The results of the investigation showed that rubber–filler interactions play a significant part in the network structure. Dinzburg derived an empirical correlation to make it easier to find the molecular weight between crosslinks without the determination of the interaction parameter. The study also evaluated the influence of unsaturation in the polymer on the properties of filled and unfilled compounds.

Nichetti [69] described the use of mechanical property data and rheometric data to develop a curing kinetic model based on a constitutive equation to predict the mechanical properties of silica-filled SBR compounds during a curing cycle. The results obtained using a model for isothermal curing were compared with experimental data, and the application of the model to non-isothermal cures was considered.

A team in China used the solvent swelling approach to investigate the crosslink density of different types of rubbers. They published work on unfilled and CB-filled EPDM compounds cured using both the sulfur-donor compound TMTD and conventional sulfur cures [70] and on unfilled and CB-filled SBR rubbers that were cured using different cure systems [71]. Their investigations looked into a number of parameters, including the relationship between the number of moles of sulfur donor and the numbers of crosslinks formed, and how the swelling behaviour was influenced by the type of cure used.

Arrillage and co-workers [72] also used solvent swelling as a means of studying vulcanisation and the degree of cure of a rubber sample. They also attempted to combine the use of solvent swelling with DSC to measure the residual exothermicity, but found that the level of residual curatives was insufficient to provide a measurable exotherm. Swelling in cyclohexane was used to investigate the cure of a NR compound vulcanised using a sulfur-based cure and an EPDM compound that had been crosslinked using peroxides.

El Labban and co-workers [73] developed an experimental device capable of measuring and controlling the heat transfer within a mould cavity and moulded part with a view to optimising the curing cycles of thick-sectioned NR mouldings. An optimisation method based on the calculation of the vulcanisation field and its sensitivity was established. This enabled a performance index to be derived which qualified the process by considering a compromise between quality (assessed using the state of cure) and productivity (which was influenced by the cure time).

6.7 Comparative studies

Comparative studies have been published of the different techniques that can be used to investigate the cure state of rubbers. An example is that produced by Warley and Del Vecchio [74], in which they compared a number of physical tests (e.g., TS and modulus values) with DSC, solvent swelling, and the Mooney–Rivlin equilibrium modulus. Their results showed that the DSC method had advantages (e.g., small samples and relatively rapid), but was limited to materials with sufficiently high heats of vulcanisation, and the Mooney–Rivlin method proved useful when employing the small strain modulus equation. The two-solvent technique of the swelling method was good for determining the polymer–solvent interaction parameter, but gave poor resolution at high states of cure.

Jaunich and co-workers [75] have investigated the cure state of ethylene-vinyl acetate copolymers using rheology, DSC and FTIR to see if they were suitable alternatives to the time-consuming method of gel content determination. The results that the group obtained showed that all of these methods were capable of being used for this type of analysis and some of the alternatives gave very similar results to the gel content approach. Another aspect that they evaluated was the ability of methods to differentiate between samples that had been aged for a year prior to curing and those that had not been aged. They found that the alternative methods were able to detect differences in cure behaviour between the samples, whereas the determination of the gel content could not.

References

1. R Brown in *Physical Testing of Rubber*, Chapman and Hall, London, UK, 1996.
2. A. Ciesielski in *An Introduction to Rubber Technology*, Rapra Technology Ltd, Shawbury, UK, 1999, p.85.
3. B.G. Willoughby in *Cure Assessment by Physical and Chemical Techniques*, Rapra Review Report No.68, Rapra Technology, Shawbury, UK, 1993.
4. M.J. Forrest in *Application to Thermoplastics and Rubbers*, Ed., P Gabbott, Blackwell Publishing, Oxford, UK, 2008, Chapter 6.
5. *Thermal Analysis of Rubbers and Rubbery Materials*, Eds., P. De, N.R. Choudhury and N.K. Dutta, Smithers Rapra, Shawbury, UK, 2010.
6. D.W. Brazier, *Rubber Chemistry and Technology*, 1980, **53**, 457.
7. A.K. Naskar and P.P. De in *Thermal Analysis of Rubbers and Rubbery Materials*, Eds., P.P. De, N. Roy Choudhury and N.K. Dutta, iSmithers, Shawbury, UK, 2010, Chapter 3.
8. H.J. Borchardt and F. Daniels, *Journal of the American Chemical Society*, 1957, **79**, 1, 41.

9.   N.S. Schneider, J.F. Sprouse, G.L. Hagnauer and J.K. Gillham, *Polymer Engineering and Science*, 1979, **19**, 4, 304.
10. T. Ozawa, *Journal of Thermal Analysis*, 1970, **2**, 3, 301.
11. R.B. Prime in *Thermal Characterisation of Polymeric Materials*, Ed., E.A. Turi, Academic Press Inc, New York, NY, USA, 1981, p.431.
12. A.A. Duswalt, *Thermochimica Acta*, 1974, **8**, 1–2, 57.
13. M.J. Richardson, *Pure and Applied Chemistry*, 1992, **64**, 11, 1789.
14. S. Affolter and M. Schmid, *International Journal of Polymer Analysis and Charaterisation*, 2000, **6**, 35.
15. M.J. Forrest in *Principles and Applications of Thermal Analysis*, Ed., P. Gabbot, Blackwell Publishing, Oxford, UK, 2008, p.242.
16. M.P. Sepe in *Elastomer Technology Handbook*, CRC Press, Boca Raton, FL, USA, 1993, p.105.
17. P. Simon and A. Kucma, *Journal of Thermal Analytical Calorimetry*, 1999, **56**, 3, 1107.
18. H. Ou, M. Sahli, T. Barriere and J-C Gelin in *Proceedings of the MATEC Web of Conferences*, 2016, **80**, Paper No.16007.
19. M.H.S. Gradwell and D. Grooff, *Journal of Applied Polymer Science*, 2001, **80**, 12, 2292.
20. K.G. Hendrikse, W.J. McGill and J. Reedijik, *Journal of Applied Polymer Science*, 2000, **78**, 13, 2290.
21. J. Wong-on and J. Wootthikanokkan, *Journal of Applied Polymer Science*, 2003, **88**, 11, 2657.
22. M. Baba, S.C. George, J.L. Gardette and J. Lacoste, *Rubber Chemistry and Technology*, 2002, **75**, 1, 143.
23. V.M. Karbhari and L. Kabalnova, *Journal of Reinforced Plastic Composites*, 2001, **20**, 2, 90.
24. T.G. Fox and S. Loshaek, *Journal of Polymer Science*, 1955, **15**, 371.
25. X-R. Zeng and T-M. Ko, *Journal of Applied Polymer Science*, 1998, **67**, 13, 2131.
26. S. Cook, S. Groves and A.J. Tinker, *Journal of Rubber Research*, 2003, **6**, 3, 121.
27. A.C.C. Peres and L.M.A. Lopes, *Polimeros: Ciencia e Tecnologia*, 2006, **16**, 1, 61.
28. N.C. Restrepo-Zapata, A.T. Osswald and J. Pablo in *Proceedings of the ANTEC 69th SPE Annual Technical Conference*, 1–5th May, Boston, MA, USA, Society of Plastics Engineers, Brookfield, CT, USA, 2011, p.2264.
29. M.J. Forrest in *Rubber Analysis – Polymers, Compounds and Products*, Rapra Review Report No.139, Smithers Rapra, Shawbury, UK.
30. S.R. Khimi and K.L Pickering, *Journal of Applied Polymer Science*, 2014, **131**, 6, 7.
31. K. Formela, D. Wasowicz, M. Formela, A. Hejna and J. Haponiuk, *Iranian Polymer Journal*, 2015, **24**, 4, 289.
32. V. Vijayabasker and A.K. Bhowmick, *Journal of Materials Science*, 2005, **40**, 11, 2823.
33. E.M. Barrall, M.A. Flandera and J.A. Logan, *Thermochima Acta*, 1973, **5**, 4, 415.
34. P.M. James, E.M. Barrall, B. Dawson and J.A. Logan, *Journal of Macromolecular Science, Part A; Pure and Applied Chemistry*, 1974, **A8**, 1, 135.
35. E.M. Barral and M.A. Flandera, *Journal of Elastomers and Plastics*, 1974, **6**, 1, 16.
36. R.B. Prime, *Thermochimica Acta*, 1978, **26**, 1–3, 165.
37. A.N. Gent, *Transactions of the Institute of the Rubber Industry*, 1958, **34**, 1, 46.
38. R.S, Rajeev and P.P. De in *Thermal Analysis of Rubbers and Rubbery Materials*, Eds., P.P. De, N. Roy Choudhury and N.K. Dutta, iSmithers, Shawbury, UK, 2010, Chapter 5.
39. T. Provder, S. Malliprakash, S.H. Amin and A. Majid, *Macromolecular Symposia*, 2006, **242**, 279.
40. T-Q. Zhao, Q-C.Wang, Z-F. Zhou, M-X. Chen, R-J. Chen and S-M. Cai, *Luntai Gongye*, 2015, **35**, 7, 440.
41. G.M. Maistros, H. Block, C. Bucknall and I.K. Partridge in *Proceedings of the Polymer Blends Symposium*, 24–26th July, Cambridge, UK, 1990, p A7/1–4, 6125.

42. E.M. Andersson, I. Ericsson and L. Trojer, *Journal of Applied Polymer Science*, 1982, **27**, 7, 2527.
43. M. Ito and J. Swanobori, *Nippon Gomu Kyokaishi*, 2003, **76**, 3, 81.
44. P.S Brown, M.J.R. Loadman and A.J. Tinker in *Proceedings of the ACS Fall Meeting*, 8–11th October, Detroit, MI, USA, American Chemical Society Rubber Division, Washington, DC, USA, 1991, Paper No.36.
45. R.J. Pazur and F.J. Walker in *Proceedings of the 176th ACS Fall Technical Meeting*, 13–15th October, Pittsburgh, PA, USA, American Chemical Society Rubber Division, Washington, DC, USA, 2009, Paper No.45.
46. B. Li, Z. Zhang, D. Ma, Q. Zhai, S. Feng and J. Zhang, *Journal of Applied Polymer Science*, 2015, **132**, 41, 7.
47. R.N. Datta and F.A.A. Ingham in *Rubber Technologist's Handbook*, Eds.,J.R. White and S.K. De, Rapra Technology Ltd, Shawbury, UK, 2001, p.175.
48. J.S. Dick and E. Norton, *Kautschuk, Gummi, Kunststoffe*, 2014, **67**, 4, 12.
49. J.S. Dick and E. Norton in the *Proceedings of the 182nd ACS Fall Technical Meeting*, 9–11th October, Cincinnati, OH, USA, American Chemical Society Rubber Division, Washington, DC, USA, 2012, Paper No.90.
50. J.S. Dick and T. Xue, *Rubber Fibres Plastics International*, 2015, **10**, 3 186.
51. I.D. Rosca and J.M. Vergnaud, *Polymer*, 2003, **44**, 14, 4067.
52. A.J. Marzocca and M.A. Mansilla, *Journal of Applied Polymer Science*, 2007, **103**, 2, 1105.
53. M. Nanda and D.K. Tripathy, *Polymers and Polymer Composites*, 2010, **18**, 8, 417.
54. D-B. Guan, Z-Y. Cai, X-C. Liu, B. Lou, Y-L. Dou, D-H. Xu and W-G. Yao, *Chinese Journal of Polymer Science*, 2016, **34**, 10, 1290.
55. B.G. Willoughby, *Journal of Applied Polymer Science*, 2016, **133**, 41, 9.
56. M.H.R. Ghoreishy and G.H. Asghari, *Iranian Journal of Polymer Science*, 2007, **20**, 1, 11.
57. H.M. da Costa, T.A.S Abrantes, R.C.R. Nunes and L.L.Y. Viscont, *Polymer Testing*, 2003, **22**, 7, 769.
58. I.H. Gregory and A.H. Muhr in *Proceedings of the International Rubber Conference (IRC) 1997*, 6–9th October, Kuala Lumpur, Malaysia, 1997, p.858.
59. T. Li, G-Z. Yang, H-L. Chen and Q-L. Li, *China Rubber Industry*, 2013, **60**, 1, 42.
60. L-T. Kou, Q. Gu, J-P. Ding and Z-Y. Yao, *China Rubber Industry*, 2007, **54**, 10, 619.
61. J. Zhang, W-X. Tang, G-L. Wang and X. Zhang, *China Rubber Industry*, 2012, **59**, 7, 401.
62. H-J. Lu, T. Chen, Y-S. Wang and H. Ma, *Luntai Gongye*, 2009, **29**, 3, 172.
63. X. Wang, Y. Jia, L. Feng and L. An, *Macromolecular Theory and Simulations*, 2009, **18**, 4–5, 268.
64. Y. Jia, X. Wang, L. Feng and L. An, *European Polymer Journal*, 2009, **45**, 6, 1759.
65. J. Gough, *Malaysian Rubber Technology Developments*, 2007, **7**, 1, 4.
66. B. Karaagac and V. Deniz in the *Proceedings of the International Rubber Conference (IRC) 2010*, 17–19th November, Mumbai, India, 2010, Paper No.20.
67. C. Canete Vebber and C. Nunes Pereira, *Journal of Applied Polymer Science*, 2015, **132**, 18, 10.
68. B.N. Dinzburg, *Kautschuk, Gummi, Kunststoffe*, 1999, **52**, 6, 413.
69. D. Nichetti, *European Polymer Journal*, 2004, **40**, 10, 2401.
70. T. Songchao, L. Chong, L. Xuedai and W. Zhangyong, *China Synthetic Rubber Industry*, 2002, 25, 5, 289.
71. T. Songchao, L. Chong, L. Xuedai and W. Zhangyong, *China Synthetic Rubber Industry*, 2003, 26, 1, 21.
72. A. Arrillaga, Z. Kareaga, E. Retolaza and A.M. Zaldua, *Gummi Fasern Kunststoffe*, 2011, **64**, 11, 669.

73.   A. El Labban, P. Mousseau, R. Deterre and J.L Bailleul, *Rubber Chemistry and Technology*, 2010, **83**, 4, 331.
74.   R.L. Warley and R.J. Vecchio in *Proceedings of the ACS Spring Meeting*, 26–29th May, Akron, OH, USA, American Chemical Society, Washington, DC, USA, 1987, Paper No.29.
75.   M. Jaunich, M. Bohning, U. Braun, G. Teteris and W. Stark, *Polymer Testing*, 2016, **52**, 133.

7 Characterisation of rubber process fume

7.1 Introduction

The composition of rubber processing fumes is complex because it is generated by the activities of an industry, which in itself is very complex in a number of ways (e.g., in the ingredients that it uses, the products that it produces, and the processing methods that it employs to manufacture them). It is because of this complexity, and the pressing need to obtain accurate information on its composition due to increasing health and safety legislation, that its study has attracted so much attention from rubber chemists/analysts over the last 50 years. This chapter assists the rubber analyst who is tasked with studying rubber fumes by providing a background to their origin, nature and composition as well as reviewing the various approaches and analytical techniques that have been used to characterise them over the last 40 to 50 years.

Due to their chemical complexity, there are a large number of reactions and interactions that occur when rubber compounds are subjected to the high temperatures used during their processing and these can generate low-molecular weight (MW) compounds that contribute to rubber fumes [1]. Another principal variable that can influence the composition of rubber fumes is the way that rubber is processed, and the main processing stages for the majority of rubber components involve the following operations [2]:

- Mixing of the ingredients [3]
- Shaping operations (e.g., extrusion) [2]
- Vulcanisation to crosslink the rubber [4, 5]
- Post-vulcanisation processes (e.g., post-curing in an oven) [5]

Of the factors stated above, the processes where fumes are most likely to be generated are during the mixing, vulcanisation and post-curing stages.

The only international definition of rubber fumes believed to be available at this time is that produced by the International Organization for Standardization (ISO) during its work on standards for the study of rubber fumes. This ISO definition is: 'A variety of substances emitted from rubber compounds into a workplace atmosphere as a result of industrial processing, the composition of which depends on the formulation of the compounds concerned, the process technology in use and the associated process parameters'. In 2013, they published a standard, ISO/TS 17796:2013 [6], for the qualitative screening of rubber fumes in the workplace and storage environments using a thermal desorption (TD) gas chromatography (GC)–mass spectrometry (MS) method. ISO have also published a standard for the quantitative analysis of the substances in rubber process fumes [7]. This standard, which should be published within the next couple of years, will include the hazardous

https://doi.org/10.1515/ 9783110640281-007

substances that have been identified in rubber fumes and, for each one, provide examples of validated test methods.

Another definition of rubber fumes is provided by the UK Health and Safety Executive (HSE) in its EH40 publication and in its method for the determination of rubber dust and fume: Method for Determination of Hazardous Substance (MDHS) 47/2:1999 [8]. This HSE method has been used to determine the levels of rubber fume in the UK rubber industry for many years. The HSE definition of rubber fumes is: 'Rubber fumes is evolved in the mixing, milling and blending of natural rubber and rubber or synthetic elastomers, or of natural rubber and synthetic polymers combined with chemicals, and in the processes which convert the resultant blends into finished products or parts thereof, and including any inspection procedures where fumes continues to be evolved'.

From the comprehensive research work that has been carried out over the years, it is apparent that any analytical programme to characterise the chemical nature and composition of rubber fumes should embrace the following classes of species:
1. Gases.
2. Vapours.
3. Aerosols.
4. Special case compounds – N-nitrosamines, aromatic amines and polyaromatic hydrocarbons (PAH).

Considering all of these physical forms and classes of chemical compound ensures that a comprehensive analysis of rubber fumes is undertaken, with no species being overlooked or disregarded. A comprehensive review of the composition and nature of vulcanisation fumes in the rubber industry was published recently by Forrest [9]. Another comprehensive review of the composition of rubber fumes, based on the information provided by 95 publications, was prepared by ISO/TC 45 (rubber and rubber products) and published as an ISO Technical Report in February 2017 [10].

The information presented in Section 7.2 provides a summary of the nature and composition of rubber fumes with regard to the variety of chemical compounds that can be present and their origins, the different physical forms in which they can exist within the factory environment (i.e., gases, aerosols) and their origins. The principal factors that influence the composition of rubber fumes and the sophisticated analytical approaches that have been developed to capture and analyse them are introduced in Sections 7.3 and 7.4, respectively. Finally, in Section 7.5 an overview is presented of the characterisation data that has been published on rubber fumes, both as a result of laboratory studies and from carrying out monitoring work at commercial rubber processing sites.

7.2 Composition of rubber process fume

This section provides a brief overview of the chemical substances that can be found in rubber process fumes and where they have originated from within the rubber compound.

Experimental work that has been carried out, both in laboratory and factory environments, has shown that rubber fumes comprise a wide range of chemical species with respect to MW and compound type (i.e., polarity and structure). A number of studies have reported over 30 species in the fumes from a particular rubber compound (Section 7.5).

The substances that are present in rubber fumes can be from a wide range of different chemical classes and a number of scientists [11, 12] have identified a large range of chemical species, with the following types being represented:
- Aliphatic hydrocarbons – straight chain and cyclic
- Aromatic hydrocarbons
- Halogenated compounds
- Isothiocyanates
- Ketones
- Nitrosamines
- Thiazoles
- Aldehydes and alcohols
- Esters and ethers
- Amines
- Sulfur compounds

The relative contribution of these substances to any given sample of rubber fumes is dependent on the ingredients present in each rubber compound that is contributing to the rubber fumes at a particular moment in time. Also, these compounds are present in rubber fumes in a number of physical forms (i.e., gases, vapour and aerosols) and the relative levels of these different physical forms vary in different areas (e.g., mixing, moulding, storage) within a rubber factory.

The different physical forms of the compounds in rubber fumes leads to a way of referring to these species based on the ability of the human eye to detect them in the atmosphere: 'visible' or 'invisible' fractions. The definition of these two classes is provided in Section 7.4.1.

Research work has shown that there are three distinct sources of rubber fume compounds [13]:
- Volatile ingredients (e.g., antioxidants and plasticisers).
- Volatile impurities in ingredients (e.g., residual monomers and manufacturing impurities).
- Volatile reaction and breakdown products of chemical reactions that occur during processing (e.g., vulcanisation and antioxidant activity [14, 15]).

7.3 Principal factors affecting the composition of rubber fume in the workplace

As mentioned in Section 7.1, there are a number of factors that can alter the composition of rubber fumes and which need to be considered to understand their variable nature. This section provides an overview of the influence of two of the most important factors: rubber composition and processing temperature.

7.3.1 Influence of rubber compound formulation on composition of rubber fume

A number of authors have demonstrated that the chemical species in rubber fumes directly relate to the polymer and compounding ingredients present in a rubber compound.

With respect to the influence of the polymer in the rubber compound, Donskaya and co-workers [16] listed some of the volatile compounds that can be present in the gas phase of rubber fumes when different synthetic rubbers are in the compound:

- Styrene-butadiene rubber (SBR): styrene, butadiene, ethyl benzene.
- Polybutadiene: butadiene, 4-vinylcyclohexene, cyclododecatriene.
- Nitrile rubber (NBR): butadiene, acrylonitrile, butadiene trimers, 4-cyclohexenitrile.

An example of the published work that has demonstrated the influence of the compounding ingredients is that of Blanden and Isherwood [17]. They found that the substances collected were members of a number of different classes of compound (e.g., acid, basic, neutral) and some examples from four rubber compounds that were in the basic (i.e., alkaline) class are shown below:

- SBR: aniline, phthalimide, dicyclohexylamine, 2-(4-morpholinyl)benzothiazole, N-(1,3-dimethyl butyl)-N'-phenyl-p-phenylenediamine (6PPD).
- Ethylene propylene diene monomer (EPDM): morpholine, N-cyclohexyl formamide, N-butyl-1-butanamide, 2-cyclohexyl benzothiazole.
- Butyl rubber: benzidine, N-(2,2-dimethylpropyl)-N-methyl benzenamine, N-ethyl-2-benzothiazolamine, benzothiazole (BT), phenyl benzimidazole.
- NBR: aniline, dimethyl-2-butanamine, tetrabutyl urea.

Further information on the work carried out by Blanden and Isherwood is provided in Section 7.5.1; this section and Section 7.5.2 provide other examples that illustrate how the composition of a rubber compound has a profound influence on the substances that are present in its process fumes.

7.3.2 Influence of different processes and processing temperatures on the composition of rubber fumes

The type of rubber process that is being carried out in a rubber factory will have an influence on the rubber fumes that are generated. Taking into account both visible and invisible fractions of fume, Dost and co-workers [18] showed that in the general rubber goods (GRG) sector, the overall exposure to rubber fumes varies according to the process that is being carried out, as shown below:

moulding > extrusion > milling

The differences found for these three generic processes relate to factors such as the process temperature and the amount of the surface area of the rubber exposed to the atmosphere.

Regarding the influence of processing temperature, work by Willoughby [19] using the Rapra gas transfer mould (GTM)/GC–MS combination with an SBR compound illustrated how it can affect the types and levels of hydrocarbon species that are detectable in its fumes. For example, in the case of styrene, increasing the vulcanisation temperature from 145–165 °C and 180–240 °C increased the amount present in the fumes from 2–180 to 9–500 $\mu g/m^3$. For 4-vinylcyclohexane, none was found at the lower temperature, but 30–210 $\mu m/m^3$ was observed at the higher temperature. The data also showed that the situation was complicated, with the levels of some compounds reducing and others increasing as the vulcanisation temperature was increased.

Changing the moulding temperature can also affect the amount of a particular reaction product. For example, Maisey [20] reported the effect that changing the moulding temperature has on the amount of carbon disulfide (CS_2) released from a sulfur-cured EPDM rubber containing the accelerator tetramethylthiuram disulfide (TMTD). Increasing the temperature from 150 to 200 °C and then 250 °C increased the amount of CS_2 released from 1.03 μmol to 1.91 μmol and, finally, to 2.34 μmol. This work, which was carried out at Rapra, also showed that by modifying the cure system, CS_2 was not liberated.

The total amount of rubber fumes can be influenced by processing temperature as well, as demonstrated by Willoughby [21]. In his study, as expected, the total mass of fumes released from a rubber sample was found to increase with temperature, rising from 0.18% *w/w* at 160 °C to 0.75% *w/w* at 190 °C.

7.4 Capture and analyses of rubber process fumes

7.4.1 Different sampling techniques and analytical methods

A number of methods and techniques have been used and they have been designed bearing in mind the compositional complexity of rubber fumes, the objective being to achieve the most accurate characterisation possible.

In addition to the comments made regarding their composition in Sections 7.1–7.3, in general, rubber fumes can also be considered as being comprised of visible and invisible fractions [22]. Gases and vapours in the fumes make up what can be regarded as the invisible component of rubber fumes, with aerosols forming the visible component.

The visible fraction is believed to be produced at two stages in the production process: the mixing stage (due to the large mechanical stresses and local overheating within the rubber compound), and the vulcanisation stage (due to the condensation of vapours of high-boiling-point substances). In both of these cases, the aerosol particles contain numerous substances that are captured by the oil-like material that is forming the aerosol. The 1974/75 BRMA/Rapra environmental survey reported that the solvent-soluble portion of airborne particulates provides an effective measure of visible fumes from hot rubber. MDHS Method 47/2:1999 [8] is the standard method used in the UK and regards the cyclohexane-soluble fraction of the airborne contamination collected on a glass fibre filter as being the rubber fumes portion, with the insoluble fraction being regarded as the rubber dust fraction. The analytical methods used in a number of other countries are based on this MDHS method. However, difficulties can be encountered if cyclohexane-soluble organic dust and rubber chemicals, such as stearic acid, are present, resulting in overestimation of rubber fumes. Also, an analysis of the rubber dust and fumes data on the European Union Concerted Action programme database [23] has shown that using Teflon[TM] filters, rather than the glass fibre filters recommended by MDHS 47/2:1999, can influence the results.

To obtain analytical data on the invisible (i.e., gases and vapours) fraction of rubber fumes, two principal approaches have been used by analysts over the years. The first involves trapping the fume compounds (e.g., using adsorbents such Tenax[TM]) in a rubber factory and then using an analytical technique (e.g., GC–MS) in a laboratory to separate and identify the compounds trapped. This type of approach, which is covered in more detail in Section 7.5.1, is incorporated into an ISO standard published in 2013, ISO/TS 17796:2013 [6], which describes a method for the trapping and identification of volatile components of rubber fumes using active sampling on a poly(2,6-diphenylphenylene oxide)-type sorbent (i.e., Tenax[TM]) and subsequent analysis by TD GC–MS. This method can be used to screen emissions from the processing of rubber compounds in the workplace and in storage environments, but one potential limitation is that the data will be influenced by the specificity of the sorbent (Section 7.4.2).

Data generated on a number of rubber compounds using this ISO method have been published [24].

The second type of approach is laboratory-based and specialised equipment (e.g., the Rapra GTM) has been manufactured that closely mimics the curing processes used in industry (e.g., compression moulding). This equipment is interfaced with the analytical equipment (e.g., GC–MS) so that the fumes are transported directly into it, thereby minimising the losses that may take place if an intermediate 'trapping' stage is employed. More information is provided on this approach, and the results obtained as a result of it, in Section 7.5.2.

7.4.2 Influence of different sampling and analytical techniques on rubber fumes data

The complex chemical composition of rubber fumes means that ensuring that all of their components (i.e., gases, volatile compounds, aerosols) are trapped in an efficient manner to enable a full analysis to be performed is a challenging task. Accordingly, a number of scientists have carried out research to obtain information on how various 'sampling trains' of absorbent tubes, other sampling media, and sampling systems, can influence the data obtained. Work has also been carried out to investigate which analytical techniques and methodologies offer the most effective means of identifying and quantifying the compounds present in rubber fumes.

Giese and co-workers at the German Institute for Rubber Technology [25] described the development of a sampling technique, the selection and characterisation of adsorbents, and the development and testing of a sampling apparatus for the sampling and analysis of vulcanisation vapours and emissions. In this way, they developed and reported on a multi-faceted sampling and analysis methodology capable of collecting, and then identifying and quantifying, the various types of components present in rubber fumes. The overall procedure has been validated for ≈50 single chemical substances.

A generic illustration of the types of sampling analysis combinations used by Giese and co-workers and others in the occupational health and fire gas analytical fields is provided in Table 7.1.

The existence of so many types of substances can result in one species interfering with the collection and identification/quantification of one or more of the others. In order to check on the validity and accuracy of the 'Giese' multi-faceted system, both laboratory trials and workplace monitoring at tyre and GRG factories have taken place. A good level of reliability for the system, and accuracy of the results obtained using it, has been demonstrated by these trials and a wide diversity of compounds from the following substance groups were detected in rubber factories [25]:

- Highly volatile hydrocarbons
- Amines
- Aldehydes and ketones
- Chlorinated hydrocarbons
- Low-volatility vapours
- Aerosols

Table 7.1: Examples of sampling-analysis combinations for trapping various types of airborne substances.

Substance	Collection media	Analysis method
Gaseous compounds (e.g., carbon disulfide)	Gas sampling bag	Direct injection GC–FID or GC–MS
Volatile aliphatic and aromatic hydrocarbon compounds	Activated charcoal adsorbent sampling tube	Thermal desorption GC–FID or GC– MS
Amines and amides	Silica gel adsorbent sampling tube	Solvent desorption GC–nitrogen-phosphorus detector
Aldehydes and ketones	Modified silica gel sampling tube	Solvent desorption high-performance liquid chromatography–ultraviolet detection
General organic compounds, including those with lower volatility	Tenax/XAD-2/Porapak adsorbent sampling tubes	Thermal desorption GC–MS
Aerosol compounds	Glass fibre filter	Solvent extraction – FTIR analysis

FID: Flame ionisation detector
FTIR: Fourier–Transform infrared

A number of other studies have been published by the scientists at the German Institute for Rubber Technology. Some of these address more specific components that can be found in rubber fumes. For example, Kuhn-Stoffers and co-workers [26] described investigations into the use of activated carbon and silica gels (surface-modified with dinitrophenyl hydrazine) in terms of their suitability for sampling aldehydes and ketones in vulcanisation fumes. The study demonstrated that the derivitisation reaction was preceded by adsorption on the substrate material and that subsequent quantitative transformation to the hydrazone required different reaction times depending on the reactivity of the carbonyl components.

A study has been published which looks at the influence of plasticiser aerosols on the sampling and analysis of rubber fumes in the rubber industry [27]. A sampling system for fumes and aerosols from rubber products needs to be developed with special attention to the influence of plasticiser aerosols on the sampling process itself. In this study, the efficiency of sampling with a combination of selective adsorbents in the presence of plasticiser aerosols was evaluated using test gas mixtures and test aerosols in experiments that simulated the atmospheres that may be found in rubber factories. The analytical determination of individual substances and aerosols was carried out by GC and FTIR after desorption of the adsorbents with a solvent. Interferences caused by plasticiser aerosols were found by the scientists to occur only at high concentrations.

The collaborative study presented by Mai (LRCC in France) and Giese (DIK in Germany) [28] at the International Rubber Conference in 2013 included the recovery rate data for a number of the collection systems available for sampling rubber fumes. Some examples of those presented are shown below:

- Highly volatile aliphatic and aromatic compounds collected on activated charcoal: 90.2–100% recovery.
- Highly volatile sulfur compounds (e.g., CS_2) collected using a sampling bag: 80.0–97.5% recovery.
- Aerosols collected using a glass fibre filter: 94.0–98.9% recovery.

7.5 Analytical studies carried out on rubber fumes

This section provides an overview of the analytical results obtained on rubber fumes by the two principal two routes employed by research scientists, namely, studies conducted on the fumes present in rubber factories (Section 7.5.1) and those carried out on fumes generated in laboratories (Section 7.5.2).

7.5.1 Rubber fume data resulting from factory studies

The Nordic Curing Fumes Project [29] investigated the effect that curing fumes have on the environment. This project was set up by the Swedish Board for Environmental Protection, supported by Norsam (the trade association of the Nordic rubber industries), and looked into the:

- Composition of curing fumes.
- Environmental consequences of curing fumes.
- Possibilities to control the release of curing fumes from factories.

Overall, the results showed that amines and sulfur compounds were the main compounds in the vapour phase of rubber fumes from sulfur curing, and that amines, together with aliphatic and aromatic hydrocarbons, were the main components of the vapour phase of rubber fumes from peroxide curing. The authors also referred to the generation of N-nitrosamines from the nitrosation of secondary amines formed as a result of the action of certain rubber accelerators and the way that this could be avoided. Carbon black (CB) and aromatic process oils as sources of PAH in rubber fumes were also mentioned, and the authors stated that 14 PAH compounds had been found in the air in a Russian tyre works [30] and that these had the potential to react with nitrogen in the air to form nitro-derivatives [31] which are carcinogens and mutagens.

A group of industry specialists, on behalf of the EU-EXASRUB consortium, used the rubber dust and rubber fumes data from European Union Concerted Action to establish exposure trends in the EU rubber industry between the 1970s and 2003 [32]. One of the overall conclusions was that the average level of exposure to rubber dust and its cyclohexane soluble fraction (i.e., rubber fume) had declined, on average, by 4.3 and 3.3%, respectively, per year from the mid-1970s to 2003. This study also indicated that differences in sampling methodology, for example, the use of either glass fibre or Teflon™ filters, could influence the results obtained. The results of this study were broadly in line with those reported by other authors, such as Kromhout and co-workers [33].

Frolikova and co-workers [34, 35] reported on the levels of various substances in the air in different departments of a tyre factory. The quantity of an important PAH compound, benz(a)pyrene, was determined in the working environment in four areas of the factory – the CB storehouse, mixing department, mixing area, and during transportation and mixing. The concentrations found varied between 0 and 0.45 mg/m^3, with the highest levels detected in the CB storehouse. Also, levels of N-nitrosamines in vulcanisation fumes and in the vulcanisates after curing were determined, with the levels shown to vary between 0.1 and 9.1 µg/m^3 in fumes, and between 10 and 482 µg/kg in the vulcanisates. The total levels of N-nitrosamines present in the working atmosphere of different areas within the factory were established; levels between 2.1 µg/m^3 (buffing area), 11.65 µg/m^3 (vulcanisation area) and 94.27 µg/m^3 (finished product storehouse) were found.

Nudel'man reported on both the aerosol and gaseous components of the rubber fumes generated by a number of rubber compounds, including those based on NBR and NBR/polychloroprene (CR) blends [36]. Aerosols were found to be formed at two stages of production: mixing (due to large mechanical stresses and local overheating) and vulcanisation (due the condensation of vapours of high-boiling-point substances). In both cases the aerosol particles contained numerous substances captured by the drops of oil-like material forming the aerosol. Nudel'man listed the components that had been trapped and identified in an aerosol formed during the vulcanisation of a NBR and, although nine compounds were detected, 90% of the fumes comprised dibutyl phthalate.

Blanden and Isherwood [17] published data generated on rubber fumes collected at five unnamed sites in the UK engaged in different rubber manufacturing operations (e.g., car tyre curing, injection moulding and compression moulding). Particulate samples were collected on glass fibre filters and an electrostatic precipitator. Tenax™ absorption tubes fitted with glass fibre pre-filters were used to collect compounds of medium volatility. No attempt was made to collect gases or highly volatile compounds. Prior to analyses, each extract was separated into five fractions (basic, acidic, polar neutral, aromatic and aliphatic) and then the substances in each fraction identified qualitatively by GC–MS. In all cases, the fumes samples were complex mixtures of chemical substances with a number of compound classes represented:

- Aliphatic compounds
- Aromatic compounds
- Polar, neutral compounds
- Acidic compounds
- Basic compounds

Some of the compounds detected in the 'basic compound' class for three types of rubber have been presented in Section 7.3.1.

Cocheo and co-workers [37] reported the results of an industry study that involved the collection of samples in the vulcanisation areas of a shoe-sole factory and a tyre retreading operation. Samples were also taken in the extrusion areas of the retreading company and an insulated cable manufacturer. In all cases, the fumes samples were trapped on activated charcoal, desorbed in the laboratory with trichlorofluoromethane and analysed by GC–MS. A very large number of species were identified in the four areas, with some overlapping in the chromatograms, thereby resulting in a number of 'unidentified' peaks. An extensive table of assignments was provided for all of the individual compounds found in the four areas, together with an indication of their probable source within the rubber compound (i.e., the ingredient). The total number of compounds identified and quantified was 99, with some examples of the number in each class given below:

- Alkanes (16 compounds)
- Aliphatic hydrocarbons (32 compounds)
- Chlorinated compounds (4 compounds)
- Phenols (6 compounds)
- Esters (6 compounds)

The authors appreciated that charcoal is not a universal sorbent and some compounds would not have been trapped in detectable quantities. However, the study did produce results that were in general agreement with other studies that had been carried out.

In a Russian study [38], the gases produced during the manufacture of silicone rubber tubes for medical applications were found to include a range of siloxanes, ethyl formate, methanol, ethylene and o-dichlorobenzene, but at concentrations below the maximum permitted in the factory. In addition, ventilating discharges from the site were found to contain mainly formaldehyde.

In a study presented at the International Rubber Conference in 2013, Khalfoune [24] of the LRCCP Rubber and Plastics Research and Testing Laboratory described how ISO/TS 17796:2013 for the trapping and analysis of rubber fumes (Section 7.1) had been used to determine the substances present in the fumes generated by a number of rubber compounds of known composition. The presentation also included information on laboratory studies that had been carried out on standard compounds, and this work enabled a link to be made between the volatile species in rubber fumes and specific compounding ingredients in the rubber. The results presented by Khalfoune for a peroxide-cured EPDM included the following ingredient/substance relationships:

- Ethylene norbornene from the Vistalon EPDM rubber.
- Six organic compounds from the Peroximon F 40 peroxide curative.
- Ethanol from the Silane ST69 coupling agent.
- Ethyl palmitate from the stearic acid.

The information presented by Khalfoune was consistent with the work carried out by other scientists, such as Willoughby [13, 39] (Section 7.5.2).

7.5.2 Rubber fume data resulting from laboratory studies

The composition of fumes evolved from rubber compounds when they are heated to processing temperatures using laboratory-based systems has been investigated by a number of scientists. Using these types of experiments, a considerable amount of fundamental scientific work was carried out on the nature and composition of rubber vulcanisation fumes by Rapra Technology. These studies, which began in the 1970s and carried on through into the late-1990s, were led by Bryan Willoughby [40–42]. The results obtained, and the understanding of the chemical relationships that resulted, enabled the composition of rubber fumes from a given rubber compound to be predicted with a reasonable degree of accuracy.

Rapra carried out a series of over 40 vulcanisation experiments in their GTM using industrially relevant formulations produced from a 'pool' of 75 ingredients. The species that were given off in the fumes were studied by interfacing the GTM with GC–MS. This approach ensured that representative fumes samples were generated and then transferred into the GC–MS system quickly with negligible loss. More than 150 chemical species were detected and important ingredient/process conditions/emission relationships established, for example:

- A list of the formulations which yielded a single, specific compound and the common ingredients in such formulations.
- A list of the formulations which contained a specific ingredient, and the common compounds emitted from these formulations.
- The overall effect of the formulation, curing temperature and temperature of the rubber at the time of analyses.

The complexity of rubber compounds meant that the relationship between the ingredients and the emissions obtained was not necessarily simple. For example, an individual component of rubber fumes may have more than one source in a formulation. This work at Rapra did reveal certain key trends however. For example, three distinct sources of volatile emissions were recognised (Section 7.2). This information was provided by an in-depth study published by Willoughby [13] that contained sections on the origin of specific individual compounds, such as CS_2, and a table that listed the potential origins of over 40 volatiles species that could be detected in rubber fumes. The results of the Rapra Vulcanisation Project and description of the predictive rubber fumes software that resulted from it are described in a book authored by Willoughby [39].

In addition to these summary publications, a large number of other studies and publications have been published by Willoughby and other scientists at Rapra during the course of their work on rubber fumes. For example, in an article entitled *'Prediction of on-site performance for vulcanisation fume'*, Smith and Willoughby [43] described rubber fumes as comprising two distinct phases. These two phases, which have been referred to in Section 7.2 and 7.3.2, were:
- Visible portion – composed of aerosols formed by the condensation of hot vapours, which can be trapped onto filters providing that the sampling flow rate is high enough.
- Invisible portion – composed of a complex mixture of species that remain in the gaseous phase. The wide range of species present mean that either a range of adsorbents, or a range of liquid media for trapping by absorption, have to be used to ensure that the data subsequently obtained by analysis are fully representative.

Levin [11] and Asplund [12] both reported on a study carried out by the company Trelleborg Industri AB on the composition of curing fumes released from a range of rubber compounds under both industrial and laboratory conditions. All five compounds were CB-filled with four containing typical sulfur-curing systems and one compound with a peroxide-curing system. The laboratory-based phase of the work was carried out at Rapra Technology using their GTM/GC–MS combination. A combination of sampling media had to be used to capture fume samples in the factory environment due to the multi-component, complex nature of rubber fumes. The following sampling train was used in the study:

- Adsorbent tubes for sulfur compounds
- Tenax™ tubes for organic compounds
- Charcoal tubes for less volatile organic compounds
- Quartz wool filters for aerosols and particulates
- Impinger tubes for amines
- Continuous registering instrument for total hydrocarbon content (THC)

The analysis of the fumes that were captured by this sampling train from the rubber compounds resulted in the identification of 221 substances which, as introduced in Section 7.1, could be divided into a large range of chemical groups.

Becklin and co-workers [44, 45] described how vulcanisation fumes from a number of rubber compounds were collected directly from a laboratory-scale moulding press and introduced into a GC–MS system for identification and quantification of the organic compounds present. This system was similar to the one set up by Rapra Technology and enabled rapid screening of various rubber formulations [e.g., natural rubber (NR), EPDM, SBR, CR and NBR] for the identification and quantification of organic compounds present in their vulcanisation fumes. The effect of curing time and pressure were also investigated as well as recipe changes within a given compound series. The analytical results of over 35 organic compounds were presented and the pathways for the generation of some of these chemicals described.

Rozynov and co-workers [46] used a mini-curing system/GC–MS combination for characterising rubber vulcanisation fumes. The system enabled the rapid screening of various rubber formulations to identify and quantify the organic compounds present in their vulcanisation fumes. Some of the compounds that were detected in rubber fumes using this system were *bis*(2-ethylhexyl)phthalate, dimethylformamide, CS_2, acetophenone and isopropylbenzene.

Zietlow and Schuster [47] applied analytical methods based on dynamic headspace sampling in combination with GC–MS to the analysis of the components present in fumes generated during laboratory-scale vulcanisation experiments with a 'heatable reactor cell' and a compression mould that simulated workplace conditions. The results obtained from a small number of rubber compounds made it possible to identify, or classify, ≈300 individual compounds and, in common with other studies of this type, ingredient/emission relationships were apparent, enabling the assignment of a given compound to a specific raw material.

Lucas and co-workers [48] carried out laboratory investigations into the crosslinking efficiency and emissions behaviour of a number of peroxides used in EPDM-based compounds and compared the results obtained with those achieved with a N-*tert*-butyl-2-benzothiazole sulfonamide/tetrabenzylthiuram disulfide accelerated sulfur-cured EPDM. The choice of peroxide was shown to be important because using one with a high crosslinking efficiency meant that less had to be employed to achieve a successful cure with a rubber compound, and so the of level moulding fumes was lower. In addition to fogging tests for Deutsches Institut für Normung e.V., DIN 75201, the authors used thermodeposition GC–MS [as specified by the German Association of the Automotive Industry (VDA) 278 method] to identify the emissions produced by the rubber compounds, particularly the breakdown and reaction products of the peroxides. A number of principal cleavage products for five commercial peroxides were identified and their share of the total emissions estimated. The number of compounds and their respective levels in the emissions, along with their chemical nature and MW, varied considerably according to the type of peroxide used. However, the results obtained showed, in common with the work carried out by Rapra Technology, that it is possible to predict the range of breakdown products that will be produced by a peroxide of known chemical structure.

Nudel'man, in conjunction with Antonovski, [49, 50], also looked into the volatile decomposition products of three commercially available peroxides used to vulcanise rubber compounds. The investigation was performed by compounding the peroxides into an EPDM rubber formulation and then curing sheets of the rubber using a typical commercial moulding temperature (i.e., 150 °C). This moulding process was carried out between aluminium sheets that trapped all of the volatile compounds that were formed during the vulcanisation process within the rubber. Once the rubber sheets had cooled, samples were removed and placed into a TD unit that was connected to a GC–MS system. In addition to low-MW substances that were unrelated to the peroxide (e.g., oligomers from the rubber, antioxidants), around ≥10 compounds were found to have been formed from each of the peroxides investigated. The formation of cyclic compounds, such as tetrahydrofuran compounds, was thought to be due to the cyclisation of intermediate breakdown products.

A couple of studies have looked into the presence of specific species in rubber fumes. Chikishev and co-workers reported the results of a Russian study [51] that used experiments using MS to determine qualitatively the accelerator and antioxidant content of gases evolved from rubber compounds heated at 75 °C. Four compounds based on SKN-26 NBR with three types of crosslinking accelerator and two antioxidants were used as models. Over 15 types of compound were detected by their mass spectra and assigned, where possible, to the various antioxidants and accelerators used in the compounds.

In a study by Wommelsdorff and co-workers [52], trace amounts of carcinogenic PAH were found to be present in the fumes resulting from the vulcanisation of rubber products. Results of laboratory-based curing studies on 17 rubber compounds based on two rubber types (EPDM and SBR) and containing different filler and oil additives were presented. They showed that a reduction of the PAH content of vulcanisation fumes was achievable through the use of a mineral oil with low-PAH levels. Varying the compounding ingredients (e.g., use of silica as opposed to CB) also influenced PAH emissions.

190 Characterisation of rubber process fume

One of the contributory factors to rubber fumes is the low-MW compounds released from the base polymer at high temperature. Sakdapipanich and co-workers [53] and Hoven and co-workers [54] looked into the volatile components released from NR. These scientists from Mahidol University and Chulalongkorn University, respectively, used headspace GC–MS to identify the volatile compounds emitted by seven grades of raw NR when they were heated at 60 °C for 2 h. A large range of chemical compounds was detected and these were generated by a number of reactions, for example, esters by microbial esterification and sulfur-containing compounds by amino acid degradation. The types of volatile compounds detected were found to vary according to the type of NR. For example, the sulfur-containing compounds found in the skim crumb rubber, which are thought to be formed due to amino acid degradation, were not found in the other six types of NR. These results suggested that a wide range of compounds was present in the emissions and that the types and levels were dependent upon the grade of NR used in the rubber formulation.

Aarts and Davies [55] reviewed the published information on the presence of harmful components in fumes produced during the vulcanisation of rubber compounds and the conclusions that have been reached regarding the mechanisms associated with their formation. They also reported on the results that they produced from the analysis of fumes produced during the curing of four rubber compounds – two NR-based and two oil-extended-SBR based – in a cure simulation apparatus that enabled continuous sampling to take place. The total amount of condensate within the system and total volatiles produced by the four compounds were reported, with a high proportion of the condensate being due to the oil extender in the case of the SBR compounds. Major contributory components of both the condensate and volatiles were found to be the compounding ingredient 6PPD, and the breakdown products BT and 'parent' amines from the accelerators.

After reviewing various systems for analysing vulcanisation fumes, Schuster and co-workers [56] selected two reproducible and quantitative methods for the analysis of the vulcanisation fumes of rubber mixtures using laboratory curing equipment. Among the results presented were analytical results obtained on unvulcanised rubbers, with over 40 volatile compounds being detected in the vapours emitted by the NR, SMR CV 50, and over 45 from the SBR rubber, Buna 1712, at 180 °C. The objective was to show the contribution that base polymers make to vulcanisation fumes. The species detected in the fumes of other rubbers (e.g., EPDM and CR) were also discussed. Some of the data presented by Schuster were also published by Linde in a dissertation [57].

A collaborative presentation by Mai (LRCC in France) and Giese (DIK in Germany) [28] at the International Rubber Conference in 2013 in Paris described the results obtained by analysing rubber fumes in the laboratory using a dynamic headspace GC–MS method. To carry out the work, an EPDM compound was prepared and heated to 200 °C in a TD instrument to generate fumes. Examples of the reaction products generated by the accelerators in the compound and detected by the GC–MS are listed below:

- Benzylisothiocyanate and N-benzylidenebenzylamine from zinc diethyldithiocarbamate. BT and methylthiobenzothiazole from 2-mercaptobenzothiazole.
- Tetramethylthiourea from the TMTD.

In order to obtain a link between the substances detected in rubber fumes and ingredients present in a compound, Mai and Giese used a methodology that involved the successive addition of ingredients to the base polymer and analysed the compounds produced using the TD GC–MS system. This study also included data that had been obtained on rubber fumes produced by EPDM, NR and NBR compounds using the ISO standard for rubber fumes analysis, ISO/TS 17796:2013 [6] (Section 7.5.1). In addition, it included information on the sampling options for the collection and identification of rubber fumes in factories and the recovery rates for certain types of organic substances from different types of collection media e.g., (sorbents, filters and sampling bags) (Table 7.1).

7.6 Conclusions

Rubber fumes are an extremely complex mix of chemical substances (e.g., over 30 from a single rubber compound) that have a wide range of possible sources and origins. In addition, the physical form (i.e., gas, vapour or aerosol) that these species exist in the factory environment can be influenced by the type of process carried out and the process temperature. The type of process and process temperature can also influence the reactions that take place in a rubber compound and, hence, the types of species that can be emitted and the proportion of the rubber compound which falls within the volatile fraction and so has the potential to be released as process fumes.

There are three distinct sources of species in rubber fumes:

- Volatile ingredients (e.g., antioxidants and plasticisers).
- Volatile impurities in ingredients (e.g., residual monomers and manufacturing impurities).
- Products from chemical reactions that occur during processing (e.g., vulcanization and antioxidant activity).

The relative contribution of the above to any given sample of rubber fumes is dependent on the ingredients present in each rubber formulation.

The information published demonstrates that rubber fumes are not a single chemical entity, but a highly complex mixture of chemical species that varies depending upon the factors described above. Hence, the analytical methods that evaluate only the aerosol portion of rubber fumes (i.e., the visible fraction), like MDHS 47/2:1999 [8], and provide no information on the gas and vapour portion (i. e., invisible fraction) are unsuitable for evaluation of the full chemical composition of the fumes. It is for this reason that combinations of sampling media (e.g., different types of adsorbent tubes, impingers, filters) have had to be used in order to efficiently capture rubber fumes in factory atmospheres and enable representative compositional data to be obtained.

The studies that have been carried out have shown that when only the composition of a rubber compound is known, it may be difficult to compile an accurate predictive list of the species that are likely to be present in its process fumes because there are too many variables that influence their composition. It is necessary to consider the process conditions because these could cause the generation of specific species.

It is possible to have reasonable knowledge of the chemical composition of rubber fumes by adopting a case-by-case strategy based on three main steps.

1. Knowledge of the chemical composition of the rubber compound(s), the process technology in use and the associated process parameters – see ISO definition for rubber fumes in ISO/TS 17796:2013 [6] in Section 7.1.
2. Evaluation of the visible and invisible fractions of the fumes using qualitative methods.
3. Quantification evaluation (e.g., by GC–MS) of the components whose presence has been detected using qualitative methods.

Such an approach has been applied to improve knowledge of the chemical composition of rubber fumes and to discontinue use of dangerous substances in the production processes. The recent introduction of ISO/TS 17796:2013 [6] should assist with this work.

References

1. M.J. Forrest and B.G. Willoughby in *Proceedings of RubberChem 2004*, 9–10th November Birmingham, UK, Rapra Technology Ltd, Shawbury, UK, 2004, Paper No.1.
2. R.B. Simpson in *Rubber Basics*, Rapra Technology Limited, Shawbury, UK, 2002.
3. J.F. Funt in *Mixing of Rubber*, Smithers Rapra Technology Limited, Shawbury, UK, 2009.
4. V.A. Shershnev, *Rubber Chemistry and Technology*, 1982, **55**, 537.
5. S. Majumdar in *Rubber Vulcanisation Processes: An Overview*, Chemical Weekly, 26th August 2008, No.2, p.211.
6. ISO/TS 17796:2013 – Rubber – Trapping and identification of volatile components of rubber fumes with active sampling on a poly(2,6-diphenylphenylene oxide) type sorbent, using TD and gas chromatography method with mass spectrometric detection.
7. ISO/TS 21522:2017 – Rubber process fumes components – Quantitative test methods.
8. MDHS 47/2:1999 – Determination of rubber process dust and rubber fumes (measured as cyclohexane-soluble material) in air.

9.	M.J. Forrest, *Progress in Rubber, Plastics and Recycling Technology*, 2015, **31**, 4, 219.

10.	ISO/DTR 21275:2017 – Rubber – Comprehensive review of the composition and nature of process fumes in the rubber industry.

11.	N.M. Levin in *Proceedings of the 144th International Rubber Conference '93*, 26–29th October, Orlando, FL, USA, American Chemical Society, Rubber Division, Akron, OH, USA, 1993, Paper No.11.

12.	J. Asplund, *Kautschuk, Gummi, Kunststoffe*, 1995, **48**, 4, 276.

13.	B.G. Willoughby in *Proceedings of Hazards in the Rubber Industry*, 28–29th September, Rapra Technology Ltd, Shawbury, UK, 1999, Paper No.7.

14.	F. Pilati, S. Masoni and C. Berti, *Polymer Communications*, 1985, **26**, 280.

15.	J. Pospisil in *Aromatic Amine Antidegradants, Developments in Polymer* Stabilisation – 7, Elsevier Applied Science, London, UK, 1984.

16.	M.M. Donskaya, S.M. Kavun, A.V. Krokhin, V.G. Frolokova and Y.A. Khazanova, *International Polymer Science and Technology*, 1994, **21**, 3, T/38.

17.	C.R Blanden and S.A. Isherwood in *Proceedings of Health and Safety in the Plastics & Rubber Industries*, 15–16th September, York, UK, 1987, p.3/1–3/10.

18.	A.A. Dost, D. Redman and G. Cox in *Proceedings of International Rubber Exhibition and Conference*, 7–10th June, Manchester, UK, 1999, Paper No.5.

19.	B.G Willoughby in *Rubber Technology – 4*, Eds., A. Whelan and K.S. Lee, Elsevier Applied Science, Amsterdam, The Netherlands, 1987, p.253.

20.	L.J. Maisey in *Proceedings of SRC-81 Conference*, Helsinki, Finland, 21–22nd May, 1981, p.215.

21.	B.G. Willoughby in *Proceedings of Health and Safety in the Plastics and Rubber Industries*, Warwick, UK, 29th September– 1st October, 1980, Paper No.12.

22.	J. A. Worwood in *Proceedings of the 3rd International Conference on Health and Safety in the Plastics & Rubber Industry*, 15–16th September, York, UK, 1987, p.4/1–4/9.

23	F. De Vocht, K. Straif, N. Szeszenia-Dabrowska, L. Hagmar, T. Sorahan, I. Burstyn, R. Vermeulen and H. Kromhout, *Annals of Occupational Hygiene*, 2005, **49**, 8, 691.

24.	H. Khalfoune in *Proceedings of International Rubber Conference 2013*, Paris, France, 20–22nd March, 2013, Paper No.54.

25.	U. Giese in *Proceedings of Hazards in the European Rubber Industry*, 28–29th September, Manchester, UK, 1999, Paper No.9.

26.	P. Kuhn-Stoffers, U. Giese, R.H. Schuster and G. Wunsch, *Kautschuk, Gummi, Kunststoffe*, 1997, **50**, 5, 380.

27.	T. Will and U. Giese, *Kautschuk, Gummi, Kunststoffe*, 1996, **49**, 3, 200.

28.	Le Huy Mai and U. Giese in *Proceedings of International Rubber Conference 2013*, Paris, France, 20–22th March, 2013, Paper No.74.

29.	*Internal Report from the Nordic Rubber Industry 1991*, Summary Kautsch, Gummi, Kunststoffe, 1993, **46**, 858.

30.	T. Rogszewska and co-workers, *Polymer Journal Occupational Medicine*, 1989, **2**, 4, 366.

31.	R.H. Schuster, F. Nabholz and M. Gmünder, *Kautschuk, Gummi, Kunststoffe*, 1990, **43**, 2, 95.

32.	F. De Vocht, R. Vermeulen, I. Burstyn, W. Sobala, A. Dost, D. Taeger, U. Bergendorf, K. Straif, P. Swuste and H. Kromhout, *Occupational and Environmental Medicine*, 2008, **65**, 6, 384.

33.	H. Kromhout, P. Swuste and J.S.M. Boleij, *Annals of Occupational Hygiene*, 1994, **38**, 1, 3.

34.	V.O Frolikova, M.M. Donskaya, L.I. Yalovaya, A.M. Pichugin and I.I. Vishnyakov, *International Polymer Science and Technology*, 2009, **36**, 8, 29.

35.	V.O. Frolikova, M.M. Donskaya, L.I. Yalovaya, A.M. Pichugin and I.I. Vishnyakov, *Kauchuk I Rezina*, 2008, **5**, 20.

36.	Z.N. Nudel'man, *International Polymer Science and Technology*, 2001, **28**, 2, T/38.

37. V. Cocheo, M.L. Bellomo and G.G. Bombi, *American Industrial Hygiene Association Journal*, 1983, **44**, 7, 521.

38. M.I. Novokovskaya, V.B. Saltanova, K. Yu and Shaposhnikov, *Kauchuk I Rezina*, 1976, **6**, 48.

39. B.G. Willoughby in *Rubber Fume: Ingredient/Emission Relationships*, Rapra Technology Ltd, Shawbury, UK, 1994, p.105.

40. K.G. Ashness, G. Lawson, R.E, Wetton and B.G. Willoughby, *Plastics Rubber Processing Applications*, 1984, **4**, 69.

41. B.G. Willoughby and K.W. Scott, *Rubber Chemistry and Technology*, 1998, **71**, 310.

42. B. Willoughby in *Proceedings of the 8th Scandinavian Rubber Conference*, 10–12th June, Copenhagen, Denmark, 1985, p.593.

43. R.W.B Smith and B.G Willoughby in *Proceedings of the 3rd Health and Safety in the Plastics and Rubber Industries*, 15–16th September, York, UK, 1987, p.4/1–4/9.

44. D. Becklin, T. Herman, S. Ponto and B. Rozynov, *Rubber and Plastics News*, 1995, **25**, 2, 15/8.

45. D. Becklin, T. Herman, S. Ponto and B. Rozynov in *Proceedings of the 147th ACS Spring Meeting*, 2–5th May, American Chemical Society, Rubber Division, Philadelphia, PA, USA, 1995, Paper No.31, p.14.

46. B.V Rozynov, R.J. Liukkonen, D.O. Becklin, A.L. Noreen and S.D. Ponto, *ACS Polymer Preprints*, American Chemical Society, Division of Polymer Chemistry, 2000, **41**, 1, 692.

47. J. Zietlow and R.H. Schuster in *Proceedings of the 144th International Rubber Conference '93*, 26–29th October, American Chemical Society, Rubber Division, Orlando, Fl, USA, 1993, Paper No.10, p.8.

48. G. Lucas and U Giese, *Kautschuk, Gummi, Kunststoffe*, 2008, **61**, 4, 180.

49. Z.N. Nudel'man and V.L. Antonovskii, *International Polymer Science and Technology*, 1994, **21**, 1, p.T/52.

50. Z.N. Nudel'man and V.L. Antonovskii, *Kauchuk I Rezina*, 1993, **6**, 14.

51. Y. G. Chikishev, N.A. Klyuev, G.A. Vakhtberg and V.G. Zhil'Nikov, *Kauchuk I Rezina*, 1975, **12**, 42.

52. R. Wommelsdorff, U. Giese, C. Thomas and A. Hill, *Kautschuk, Gummi, Kunststoffe*, 1994, **47**, 8, 549.

53. J. Sakdapipanich and K. Insom, *Kautschuk, Gummi, Kunststoffe*, 2006, **59**, 7–8, 382.

54. V.P. Hoven, K. Rattanakaran and Y. Tanaka, *Rubber Chemistry and Technology*, 2003, **76**, 5, 1128.

55. A. J. Aarts and K.M Davies in *Proceedings of Rubbercon 92 – A Vision for Europe*, 15–19th June, Plastics and Rubber Institute, Brighton, UK, 1992, p.455.

56. R.H. Schuster, H. Linde H and G. Wuensch, *Kautschuk, Gummi, Kunststoffe*, 1991, **44**, 3, 222.

57. H.Linde in *Untersuchung von bei der Vulkanisation enstehenden Gasen und Daempfen*, Universitat Hannover, Germany, 1990 [Dissertation].

8 Compliance with food-contact regulations

8.1 Introduction

For rubber products that are intended for use in certain sectors of industry, there are regulations and guidelines that they must comply with before they can be placed on the market. Three of the most highly regulated of these sectors are:
- Food-contact materials and articles
- Pharmaceutical products
- Medical devices

This chapter describes the tests that rubber products undergo to ensure that they pass food-contact regulations. Chapter 9 addresses the analytical work carried out during the conduct of extractables and leachables (E&L) studies aimed at addressing the requirements of the pharmaceutical and medical sectors.

Regulations that govern the use of rubber products for food-contact applications in Europe and the United States (US) have been in existence for many years; for example, the German BfR Recommendations were first published in the 1950s. Over the last 30 years, the number of countries and regions that have published their own food- contact regulations have grown to the extent that few developed and emerging nations do not have them. Information on the food-contact regulations that apply to rubber products in the European Union (EU), US and other parts of the world is available in publications such as the ones edited by Rijk and Veraart [1] and Baughan [2].

One feature common to all regulations, and most guidance documents, is an inventory list, sometimes called a 'positive list', of substances that are allowed to be used in the manufacture of the food-contact material. A recent comparison of a number of regulatory or non-regulatory inventory lists for food-contact materials was published by Geueke and co-workers [3]. The authors estimated that >6,000 substances appear on these lists and stated that some have been linked to chronic diseases and others lack sufficient toxicological evaluation. The aim of their study was the identification of known food-contact substances that were also considered to be 'chemicals of concern' and the following three inventory lists were chosen for the investigation:
- The 2013 Pew Charitable Trusts database of direct and indirect food additives used legally in the US.
- The 2014 EU list for plastic food-contact materials and articles; the 'Union List' provided in Annex 1 of Regulation (EU) 10/2011 and updated by subsequent amendments.
- The 2011 non-plastics food-contact substances database published by the European Food Safety Authority.

https://doi.org/10.1515/9783110640281-008

The conclusion of the study was that 175 chemicals of concern were identified in the three lists and information about their applications, regulatory status and potential hazards was provided in the article.

In Section 9.2, a description is provided of the potential origins for the substances, both added intentionally (i.e., ingredients) and non-intentionally (e.g., reaction products) that have the potential to migrate from rubber products and, hence, should be addressed by E&L study programmes. The substances that have the potential to migrate into food from rubber products have the same origins and so the information in this description is also very relevant to this chapter. One important difference between an E&L study and food migration work, however, is that more direction is provided to the analyst, *via* the regulations, of the specific substances that need to be identified and quantified. For readers interested in obtaining more information on how these regulations compare with one another, an article discussing the parallels and differences between the food, medical and pharmaceutical regulations has been published by Creese [4].

A review of the regulations that apply to rubbers intended for food-contact applications has been published by Forrest [5]. In addition to providing information on the regulations, this publication also provides information on the testing that has to be carried out to meet them. It also presents the information generated by the work carried out during the course of four research projects for the UK Food Standards Agency (FSA) [6–9].

A copy of the final report for all of these projects can be obtained *via* the FSA website: http://www.foodstandards.gov.uk.

A review of the testing of rubber products (e.g., seals, gaskets, hoses) for food-contact applications such as transportation, pipe work, pumps, storage vessels, handling, preparation and manufacturing, has been provided by Isa [10]. This review includes the testing of rubber products for the migration of additives (e.g., plasticisers, process oils) according to national and international standards and regulations.

8.2 Rubbers for food-contact applications

Many types of rubber can be used to manufacture food-contact materials and articles. These include:
- Natural rubber (NR).
- Nitrile rubber (NBR).
- Ethylene propylene rubbers [i.e., ethylene propylene monomer and ethylene propylene diene monomer (EPDM)].
- Fluorocarbon rubber (FR).
- Silicone rubber.
- Thermoplastic elastomers (TPE).
- Others – including butyl rubber, polychloroprene (CR), acrylic and hydrin.

There is also a wide range of products that are manufactured from these rubber compounds. These products make contact with food in a variety of end-use conditions (i.e., contact times, areas and temperatures).

Although rubber products can be used to a limited extent for the packaging of food (e.g., seals within bottle closures), and cooking of food, they mainly come into contact with food during its manufacture, processing, storage and transportation. Information on the rubber products that make contact with food (e.g., types of rubber compounds and conditions of contact) was collected by Rapra Technology during the course of a FSA project [6]. Some of the information collected during an industry survey is shown in Tables 8.1 and 8.2. Specifically, Table 8.1 shows examples of rubber products and components that make contact with food and Table 8.2 provides representative examples of typical applications for food-contact rubbers.

Table 8.1: Examples of rubber products and components that contact food.

Location/Usage	Examples of products and components
Food transportation	Conveyor belts, hoses, rubber skirting and rubber paddle lips
Pipe work components	Seals, gaskets, flexible connectors and butterfly valves
Pumps	Progressive cavity pump stators, diaphragm pumps
Plate heat exchangers	Gaskets
Machinery/storage vessels	General seals and gaskets
Cans/bottles	Bottle seals and can seals
Food handling/preparation	Gloves and feather pluckers
Food manufacturing	Silicone sweet moulds, rubber squeeze rollers
Food wrapping	Meat and poultry nets

Reproduced with permission from M.J. Forrest in *Food Contact Materials – Rubbers, Silicones, Coatings and Inks*, Smithers Rapra Technology Ltd, Shawbury, UK, 2009, p.15. ©2009, Smithers Rapra Technology Ltd [11]

With regards to the analysis of the low-molecular-weight (MW) substances in rubbers that have the potential to migrate into food, the development of new instrumentation has provided the rubber analyst with much greater capability. A prime example is the fact that the past 15 years have seen the proliferation of liquid chromatography (LC)–mass spectrometry (MS) instruments to the extent that they have now essentially replaced conventional high-performance liquid chromatography (HPLC) instruments in the majority of laboratories. The LC–MS technique compliments gas chromatography (GC)–MS and enables the analyst, for the first time, to routinely generate data on both the thermally-labile and thermally-stable compounds present in rubber products, food and food simulants up to and beyond the absorption limit of the gastrointestinal tract (1,000 Da).

In addition, development work continues to advance analytical instrumentation so that there are improvements in important areas such as MW range, detection limits, software-assisted peak deconvolution, analysis speed, accuracy of library

Table 8.2: Representative contact conditions for typical applications of food-contact rubbers.

Component	Contact area – individual components (cm³)	Contact area – component assemblies (cm³)	Typical contact time	Maximum contact time	Contact temperature (°C) General	Contact temperature (°C) Extreme
Food[1] transportation	Up to 50,000	–	<1 h	–	<85	–
Pipework components	<1,000	<10,000	<1 h	2 weeks[2]	<140	250[3]
Pumps	<10,000	–	<1 h	2 weeks[2]	<85	–
Plate heat exchanger gaskets	<1,000	Up to 56,000	<3 min	–	<140	250[3]
General seals and gaskets	<1,000	Up to 30,000	<1 h	12 weeks[2]	<85	–
Packaging seals and closures	<1,000	–	Up to 5 years	–	Ambient	140[4]
Miscellaneous	<1,000	–	<1 h	4 weeks[5]	<85	250[6]

1 Potential contact areas could be up to 1,500,000 cm³ for some of the longer conveyor belts
2 May have extended contact times during shutdowns and beer-keg seals may have contact times of up to 12 weeks
3 Refining of vegetable oil
4 During pasteurisation/sterilisation
5 Meat and poultry nets
6 Meat and poultry nets if not removed before cooking
Reproduced with permission from M.J. Forrest in *Food Contact Materials – Rubbers, Silicones, Coatings and Inks*, Smithers Rapra Technology Ltd, Shawbury, UK, 2009, p.17. ©2009, Smithers Rapra Technology Ltd [12]

searching, and species selectivity. The advent of multi-hyphenated techniques such as GC×GC–time-of-flight (ToF)–MS and LC–MS/MS are examples of this. These instruments, with their greater resolving power and selectivity, are also improving the direct analysis of food products. The large range of low-MW compounds in these food products have caused problems in the past and caused workers to rely quite heavily on food simulants.

These developments in instrumentation, together with the research work being carried out on rubbers (e.g., the FSA projects mentioned above and in Section 8.1), has enabled more accurate conformity checks to be performed on compounds, as well as continuing to add to the understanding of the migration behaviour of rubber-related substances.

Food-contact materials are a very heavily regulated sector with, as mentioned in Section 8.1, all of the developed countries in the world having food-contact regulations. To set the scene, Section 8.3 provides a brief overview of the regulations for food-contact rubber products in Europe and the US.

8.3 Overview of the European Union and United States Food and Drug Administration regulations

8.3.1 European union legislation

At the moment, there is no harmonised EU legislation for rubber food-contact materials and articles. There is only a specific regulation published in 1993 concerning nitrosamines in baby's dummies [13]. However, all rubber materials and articles need to comply with the EU Framework Regulation (EC) 1935/2004 and the EU Good Manufacturing Practice Regulation (EC) 2023/2006 so that in normal use they will not transfer their constituents to food in quantities that could endanger health or cause unacceptable changes in the composition of food or deterioration in its organoleptic properties (i.e., taste, texture, aroma, or appearance).

8.3.2 German regulations

Within Germany, the food-contact legislation for natural and synthetic rubbers is described in BfR Recommendation XXI '*Commodity Articles Based on Natural and Synthetic Rubber*'. There are separate requirements for silicone rubbers and these are contained within BfR Recommendation XV.

In the case of Recommendation XXI, four usage categories and a special category are defined as follows:

- Category 1 (test conditions: 10 days at 40 °C): Rubber articles which come into contact with food for periods of >24 h to several months (e.g., storage containers, container linings, seals for cans and bottles).
- Category 2 (test conditions: 24 h at 40 °C): Rubber articles which come into contact with food for <24 h (e.g., food conveying belts, tubes and hoses, sealing rings for cooking pots, lock seals for milk can lids).
- Category 3 (test conditions: 10 min at 40 °C): Rubber articles which come into contact with food for <10 min (e.g., milk liners and milking machine tubes, roller coatings and conveyor belts (fatty foods only in both cases), gloves and aprons for food handling).
- Category 4 (no migration testing required): Rubber articles which are used only under conditions where no migration into food is to be expected (i.e., if the articles come into contact with the food for a very short time or only over a very small area). Examples of rubber products in this category include: conveyor belts, suction and pressure hosing for moving and loading/unloading dried food, and pump parts.
- Special Category (test conditions: 24 h at 40 °C): Rubber articles directly associated with the consumption of food and which are being, or can expected to be,

taken into the mouth (e.g., toys according to BfR Recommendation XLVII, teats, soothers, gum shields, balloons).

The following food simulants are used in connection with the German regulations: distilled water, 10% ethanol and 3% acetic acid. There is no simulant stated in the BfR Recommendation XXI for fatty food, although the document mentions the possibility of one being available in the future. The permissible overall migration limits (OML) vary according to the category and simulant (Table 8.3).

Table 8.3: OML for BfR Recommendation XXI.

Food simulant	Category and OML (mg/dm²)			
	1	2	3	Special
Distilled water	50	20	10	10 or 50#
10% Ethanol	50	20	10	–
3% Acetic acid	150 (50)	100 (20)	50 (10)	–

(value) = permissible organics within total
#dependent on product type
NB: there is no migration limit for category 4 (see above)
Reproduced with permission from M.J. Forrest in *Food Contact Materials – Rubbers, Silicones, Coatings and Inks*, Smithers Rapra Technology Ltd, Shawbury, UK, 2009, p.28.
©2009, Smithers Rapra Technology Ltd [14]

To carry out the overall migration test to the requirements of BfR Recommendation XXI, test pieces of 50 × 50 mm to give a total area of 50 cm² (both surfaces) are immersed in 100 ml of the appropriate simulant for the intended end use, the test performed and then the simulant dried down quantitatively.

The BfR regulations also include specific compositional limits and specific migration limits (SML) for a number of substances, including:
1. N-Nitrosamines and nitrosatable substances
2. Amines (all categories)
3. Milking liners and milking tubes
4. Formaldehyde
5. Acrylonitrile (ACN)
6. Zinc dibenzyldithiocarbamate

In addition to migration testing and compositional requirements, BfR Recommendation XX1 also contains inventory lists for the various end-use categories, and some of the substances within these lists have restrictions concerning their maximum level within a rubber compound. Pysklo and co-workers [15] compared the BfR inventory list of ingredients approved for food-contact rubbers with the equivalent

Polish list. The latter was originally based on the German list and this exercise showed that it was in need of updating.

The BfR Recommendation XV for Silicones covers silicone oils, silicone resins and silicone rubbers. The chapter on silicone rubbers has an inventory list stipulating acceptable starting materials and the additives that may be used in processing and manufacture; some of the latter having restrictions on their maximum levels.

Separate restrictions are stated in Recommendation XV, where silicone rubber is to be used for teats, dummies, nipple caps, teething rings or dental guards. Dummies and bottle teats must also comply with the requirements laid down in the German Commodities Regulation (Bedarfsgegenständeverordnung).

The amount of volatile organic material in the finished silicone rubber product is restricted to a maximum of 0.5%, as is the total extractable material. Test methods are referenced in Recommendation XV for these determinations as well as a test that needs to be carried out for residual, un-decomposed peroxides, which should be negative.

8.3.3 Requirements in France

French requirements for food-contact elastomers (excluding silicones) are given in the *Arrete* of 9th November 1994, which was published in the *Journal Officiel de la République Française*, 2nd December 1994 (pages 17029–17036). Four usage categories (A to D) and a special category (designated T) are described in this document together with an inventory list detailing permitted ingredients in each category. There is an OML set at 10 mg/dm^2 (60 mg/kg), the same as for plastics, and specific restrictions also apply, such as an SML for primary and secondary aromatic amines of <1 mg/kg.

8.3.4 Requirements in the United Kingdom

The UK legislation on food-contact materials is published as a number of Statutory Instruments. The use of rubbers in contact with food is covered by the legislation included in Statutory Instrument 1987 No.1523 '*Materials and Articles in Contact With Foodstuffs*'. This states that any food-contact material should not be injurious to the health of the consumer and that any contamination should not have an adverse effect on the organoleptic properties of the food. The absence of any inventory list for compounding ingredients means that UK rubber compounders usually refer to either the US Food and Drugs Administration (FDA) or the BfR regulations depending on the market to be addressed.

There are separate rules in the UK for the use of rubber in contact with potable water. These are given in the UK water-fitting bylaws scheme and include tests for the following:

- Taste
- Appearance
- Growth of aquatic micro-organisms
- Migration of substances that may be of concern to public health
- Migration of toxic metals

The test methods for the above are given in the British Standard, BS 6920, Parts 1 to 4.

8.3.5 United States Food and Drug Administration regulations

In the US, the FDA produces a guidance for industry document entitled 'Preparation of Food Contact Notifications and Food Additive Petitions for Food Contact Substances: Chemistry Recommendations'. This is in addition to the Code of Federal Regulations (CFR) Title 21, Parts 170 to 199 Food and Drugs, which contains the FDA food-contact regulations. Copies of both the guidance documents and the regulations can be assessed within the appropriate sections of the FDA website: http://www.fda.gov.

The CFR is published annually on 1st April and the main requirements for rubber products for use with food are covered in Part 170, specifically:
- Rubber articles intended for repeat use: Part 177.2600
- Closures with sealing gaskets for food containers: Part 177.1210.

Part 177.2600 is the principal section that needs to be addressed and which covers the majority of rubber compounds and products. In addition to the sections that need to be considered when compounding a food-contact rubber, there are other sections that are sometimes applicable, such as Parts 180.22 and 181.32, that place a restriction (i.e., 0.003 mg/inch^2) on the amount of ACN monomer that can migrate from NBR rubbers and blends that contain NBR rubber (e.g., polyvinyl chloride (PVC)/NBR blends).

In addition to listing acceptable compounding ingredients in these two 'rubber sections' of Part 177, the FDA regulations also allow the use in rubbers of 'prior sanctioned ingredients' (Part 181), additives that are generally recognised as safe (Parts 184 to 186) and, providing that the specified restrictions allow, substances listed in other Parts. Hence, it is the case with the FDA regulations that not all the ingredients that can be used in food-contact rubber products are to be found in the same place, and compounders and assessors may need to address a number of Parts in CFR Title 21. For example, relevant substances listed in Part 182 include zinc oxide (ZnO) (182.8991), zinc stearate (182.8994) and calcium silicate (182.2227), substances listed in Part 184 include calcium carbonate, calcium stearate and calcium oxide, and those in Part 186 include kaolin clay (186.1256) and iron oxide (186.1374). With respect to exclusions, the FDA regulations specifically prohibit the use of the following, relatively common, compounding ingredients in food-contact rubbers:

- Part 189.220 – polymerised 1,2-dihydro-2,2,4-trimethylquinoline
- Part 189.250 – mercaptoimidazoline and 2-mercaptoimidazoline

A review and comparison of the FDA and BfR regulations (Section 8.3.2), with particular emphasis on the inventory lists of approved ingredients, has been carried out by Pysklo [16]. A schedule of the commercial rubber compounding ingredients which meet the requirements of the two regulations is provided in the article.

The migration testing that needs to be carried out on rubber products is described in Part 177.2600. Test pieces are cut from the rubber test product to provide a known surface area (cut edges are included in the calculation) and immersed in an appropriate amount (e.g., 100 ml) of food simulant, either hexane for fatty food-contact applications or distilled water for aqueous food-contact applications. The samples are refluxed for 7 h in pre-cleaned glassware and then removed and placed into fresh simulant and refluxed for a further 2 h. The test pieces are then removed and both the 7- and 2-h test portions evaporated separately to dryness in conditioned crucibles and the residues weighed. Blank determinations on equivalent volumes of the food simulant used are also performed. In order to be acceptable for food use, the rubber has to pass the requirements shown in Table 8.4.

Table 8.4: FDA Part 177.2600 migration limits for repeat-use articles.

Food type	First 7 h	Succeeding 2 h
Fatty foods – hexane extractables under reflux	175 mg inch^{-2}	4 mg inch^{-2}
Aqueous foods – distilled water extractables under reflux	20 mg inch^{-2}	1 mg inch^{-2}

Reproduced with permission from M.J. Forrest in *Food Contact Materials – Rubbers, Silicones, Coatings and Inks*, Smithers Rapra Technology Ltd, Shawbury, UK, 2009, p.22. ©2009, Smithers Rapra Technology Ltd [17]

To summarise, providing that the ingredients in a rubber product are listed within the applicable parts of CFR Title 21 as being approved, the water (for aqueous food use) and hexane (for fatty food use) extractables are within the prescribed limits, and any applicable restrictions for its composition (e.g., migration of residual monomer) or end-use conditions [e.g., food type(s), contact time and temperature] are met, then the product is considered suitable for food use.

8.3.6 Council of Europe resolutions

The Council of Europe (CoE) is an international organisation with over 45 member countries that has a committee of Experts on Materials and Articles Coming into Contact with Food that meets under the auspices of the Partial Agreement in the Social and Public Health Field. Over the period 1989 to 2005, this committee published nine resolutions concerning a wide range of food-contact materials and articles, from 'Colourants' [Resolution AP(89)1] to 'Cork Stoppers and Other Cork Materials' [AP(2004)2]. The resolutions that are applicable to this section, the 'CoE Resolution on Rubber Products' [AP(2004)4] and the 'CoE Resolution on Silicone Products' [AP(2004)5], were published in 2004.

Since 2008, there has been a change within the CoE, and its activities relating to food- contact materials have been transferred to the European Directorate for the Quality of Medicines & HealthCare (EDQM). The EDQM is a Directorate of the CoE and is an organisation that protects public health in the areas of food, medicines, and organ transplants. It has its own laboratory and publishes well-respect standards, such as the European Pharmacopoeia, which is legally binding within the member states of the CoE.

Once adopted, the food-contact material resolutions and supporting documents drawn up by the CoE are not legally binding, but member states, of which the UK is one, are expected to take note of them.

8.3.6.1 CoE resolution on rubber products

The 'CoE Rubber Resolution' [AP(2004)4] on food-contact elastomers has an inventory list of additives within it and a small section that deals with breakdown products – nitrosamines and amines. The inventory list is described as Technical Document No.1 – List of substances to be used in the manufacture of rubber products intended to come into contact with foodstuffs. An FSA project [8] was commissioned to study the breakdown and reaction products from the curatives and antidegradants present in this inventory list.

In addition to Technical Document No.1, there are four other documents associated with this CoE Resolution:
- Technical Document No.2: Guidelines concerning the manufacture of rubber products intended to come into contact with foodstuffs.
- Technical Document No.3: Good manufacturing practices of rubber products intended to come into contact with foodstuffs.
- Technical Document No.4: Test conditions and methods of analysis for rubber products intended to come into contact with foodstuffs.
- Technical Document No.5: Practical guide for users of Resolution AP(2004)4 on rubber products intended to come into contact with foodstuffs.

The resolution places rubber products into one of three categories:
- Category I comprises the following rubber products for which migration testing is required:
  - Feeding teats.

- Rubber products to come into contact with baby food, for which the R-total is ≥0.001 (a definition of R-total is given below).
- Category II comprises rubber products for which R-total is ≥0.001 and for which migration testing is required.
- Category III comprises rubber products for which R-total is <0.001 and for which migration testing is not required, except for rubber products containing nitrosamines, nitrosatable substances or aromatic amines and Category III substances having an SML in Technical Document No.1.

These three categories take into account the wide variety of applications for which rubber products are used and the fact that migration may vary with the application. The level of migration for rubber products may be estimated by taking into account four factors: R_1, R_2, R_3 and R_4. Categories are based on the intended use or on the result of the multiplication of the four factors ($R_1 \times R_2 \times R_3 \times R_4 = R$ total).

The factors R_1, R_2, R_3 and R_4 can be defined as follows:

'R_1 refers to the relative contact area (A_R) between rubber products and food or beverage, expressed in cm^2 of rubber surface per kg of food or beverage. For a relative area smaller or equal to 100 cm^2/kg foodstuffs, R_1 has a value calculated according to the formula: $R_1 = A_R/100$. For a relative surface larger than 100 cm_2/kg, R_1 always has the value of 1.00.

R_2 refers to the temperature during the contact period of the rubber product with the food or beverage. At a temperature lower than or equal to 130 °C, R_2 has a value calculated according to the formula: $R2 = 0.05e^{0.023T}$. Where 'e' is the base of the natural or Napierian logarithms and T is the contact temperature, expressed in °C. For temperatures higher than 130 °C, R_2 always has the value 1.00.

R_3 refers to the time, t, expressed in hours, during which a rubber product is in contact with the food or beverage. For a contact time shorter than or equal to 10 hours, R_3 has a value calculated according to the formula: $R_3 = t/10$. For a contact time of more than 10 hours, R_3 has the value 1.00.

R_4 refers to the number of times, N, that one and the same rubber product, or part of that rubber product comes into recurrent contact with a quantity of food or beverage. If the number of contact times is greater than 1000, then R_4 is calculated according to the formula: $^{10}log R4 = 6 - 2 \, ^{10}log N$. If the number of contact times is smaller than or equal to 1000, then R_4 always has the value 1.00.'

The resolution also states that, among other things:
1. Rubber products in Categories I and II should not transfer their constituents to foodstuffs or food simulants in total quantities >60 mg/kg food or food simulant (i.e., an OML).

2. Rubber products of Categories I and II should comply with the restrictions laid down in Technical Document No 1. In addition these rubber products should comply with the requirements set out in Table 1 of the resolution, except rubber teats, which should comply with EU Directive 93/11/EEC.

3. Rubber products intended for repeated use should be subjected to tests according to EU Directive 2002/72/EC Annex 1 [Note: this EU Directive has now been replaced by EU Regulation (EU) 10/2001].

4. Rubber products belonging to Category III do not require migration testing, unless otherwise specified.

5. Verification of compliance with the quantitative restrictions should be carried out according to the requirements laid down in Technical Document No.2. This document provides more detail regarding the following:

a) Further definitions and data surrounding the 'R' values.

b) Examples of rubber products that fall into Categories II and III and the calculations applied to them.

c) Migration tests.

8.3.6.2 CoE resolution on silicone materials

There is a separate CoE Resolution, designated AP(2004)5, which addresses silicone materials for food contact. The resolution defines the silicone product group comprising silicone rubbers, silicone liquids, silicone pastes and silicone resins. Blends of silicone rubber with organic polymers are covered by the resolution where the silicone monomer units are the pre-dominate species by weight.

Silicones that are used as food additives (e.g., as defoamers) in the manufacture of substances such as wine are not covered by this resolution, but polysiloxanes used as emulsifiers are. The resolution gives an OML of 10 mg/dm^2 of the surface area of the product or material, or 60 mg/kg of food, depending on the intended end-use of the silicone rubber product (see below). There are restrictions on the types of monomers that can be used to produce the silicone polymers and there is an inventory list – Technical Document No.1 – of substances used in the manufacture of silicone products (including rubbers) for food-contact applications.

An OML of 60 mg/kg of food or food simulant is stipulated for rubber products that are regarded as being in Categories I and II, as defined in the '*CoE Resolution on Rubber Products*' AP(2004)4 (Section 8.3.6.1). The choice of food stimulant and the conditions that are used for the overall migration experiment (i.e., time and temperature) should be appropriate bearing in mind the conditions that the rubber product will see in service. Guidance for the designing of these tests is given in Technical Document No.4 of the Resolution.

8.4 Analysis work to characterise food-contact rubbers

8.4.1 Introduction

Although there are strict rules and guidelines about how a food-contact rubber can be formulated and manufactured, the complexity of rubber technology, particularly the chemistry associated with vulcanisation, has meant that considerable effort has been made over the last 15 years to fully characterise the substances that have the potential to migrate into food. As mentioned in Section 8.1, the UK FSA funded a series of research projects from 1994 [6−9] to achieve this goal and this activity, and the efforts of other funding bodies and research institutions, also encouraged other work in this area. This section provides an overview of the analytical techniques that have been developed to fingerprint the chemical substances that are present in food- contact rubbers (Section 8.4.2) and the work that has been carried out to identify and quantify substances that have been targeted specifically (Section 8.4.3).

8.4.2 Fingerprinting potential migrants from rubber compounds

8.4.2.1 Use of gas chromatography−mass spectrometry to fingerprint food-contact rubber samples

It is often useful to produce a qualitative or semi-quantitative fingerprint of the low-MW species in a rubber compound that have the potential to migrate into food. GC−MS is often used for this purpose due to its high resolving power (important with rubbers due to their complexity) and the identification power of the mass spectrometer. In order to obtain data on as large a range of species as possible, it is often advisable to use both headspace GC−MS (the solid rubber samples being heated to ≈150 °C) and direct injection GC−MS, where an extract solution of the rubber, produced using a non-selective solvent (e.g., dichloromethane or acetone), is injected straight into the instrument. In the latter case, semi-quantitative data can be obtained on the species that are detected by use of a single calibrant compound such as eicosane.

Work by Forrest and co-workers [18] using typical food-contact rubber compounds has shown that, on average, between 20 and 30 compounds can be detected by conventional GC−MS using these types of approaches. The commercialisation of multi-dimensional GC−MS instruments has provided the analyst with greater resolving power, coupled with improved detection limits and enhanced deconvolution software. When a two-dimensional GC−MS system (GC×GC−ToF−MS) was applied to the same food-contact rubber compounds by the research team, its use enabled the detection of over 100 compounds.

To illustrate the GC−MS results obtained by Forrest and co-workers [18] using the two types of GC−MS system, the data presented on a food-grade FR has been selected. The formulation of the peroxide-cured food-grade FR (designated 49V) that was compounded and vulcanised for 11 min at 170 °C for this exercise is shown in Table 8.5.

Table 8.5: Formulation of food-grade FR 49V.

Ingredient	phr
Viton™ GBL-200 (peroxide cross-linkable fluorocarbon tetrapolymer)	100
ZnO	3
Medium thermal N990 (CB filler)	30
TAC	3
Luperco 101-XL [2,5-dimethyl-2,5(di-*tert*-butyl-peroxy)hexane]	4

CB: Carbon black
TAC: Triallyl cyanurate
Reproduced with permission from M.J. Forrest, S. Holding and D. Howells, *Polymer Testing*, 2006, **25**, 63. ©2006, Elsevier Ltd [19]

An acetone extract, obtained by ultrasonicating 0.3 g in 10 ml of solvent for 30 min, from this FR compound was analysed using the GC×GC–ToF–MS experimental conditions listed below, and the three-dimensional chromatogram that resulted from the analysis is shown in Figure 8.1.

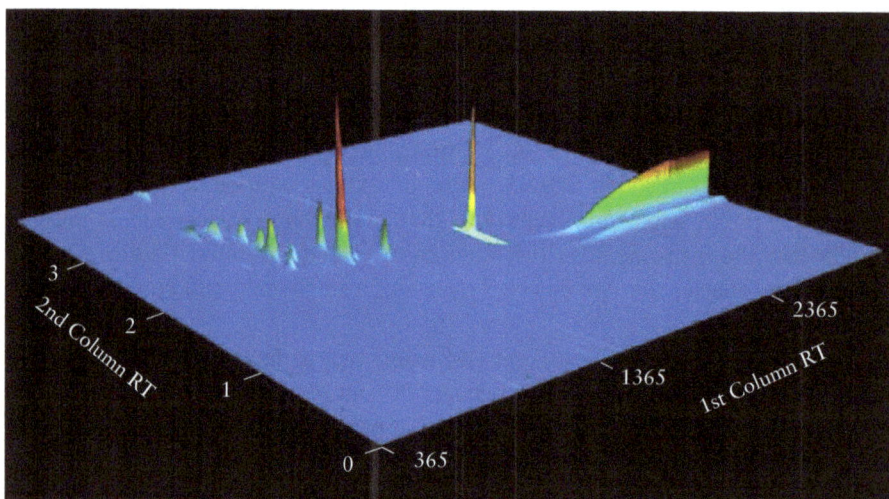

Figure 8.1: GC×GC–ToF–MS total ion current (TIC) chromatogram of the acetone extract of peroxide-cured FR 49V (RT: retention time). M.J. Forrest, S. Holding and D. Howells *in Proceedings of the High Performance and Specialty Elastomers*, 20–21st April, Geneva, Switzerland, Rapra Technology Ltd, Shawbury, UK, 2005, Paper No.2. ©2005, Rapra Technology Ltd [20].

- Instrument: Agilent 6890 gas chromatograph with LECO Pegasus III GC×GC–ToF–MS.
- Injection: PTV injection set at 10 °C above the primary oven. Injection volume of 1 µl.
- Primary column: J and W scientific HP-5MS (30 m × 0.250 mm with film thickness of 0.25 µm).
- Secondary column: SGE BPX-50 (1.8 m × 0.100 mm with film thickness of 0.10 µm).
- Carrier Gas: helium, 1.0 ml/min at constant flow.
- Primary oven programme: 40 °C for 10 min and the 10 °C/min to 320 °C. Hold at 320 °C for 15 min.
- Secondary oven programme: 50 °C for 10 min and then 10 °C/min to 330 °C. Hold at 330 °C for 15 min.
- Modulator offset: 30 °C.
- Modulator frequency: 4 s.
- Hot time of modulator: 0.30 s.
- MS setting: 30–650 Da scanned at 76 spectra/s.

As demonstrated by Figure 8.1, the GC×GC–ToF–MS instrument provides a high degree of resolution and, hence, separation of the species, resulting in a large number being detected and identified in the acetone extract. Over 20 diagnostic compounds related directly to specific formulation ingredients were identified and over 150 compounds in general. These numbers compared with only 10 and 36, respectively, for the conventional GC–MS analysis of the same acetone extract, illustrating the much greater profiling power of the GC×GC–ToF–MS system.

8.4.2.2 Use of liquid chromatography–mass spectrometry to fingerprint food-contact rubber samples

In-house databases have been developed as a result of laboratories using LC–MS systems on rubber samples. Hence, the inclusion of this technique into the fingerprinting process has complemented the data generated by GC–MS by contributing information on thermally-labile and/or relatively-large (e.g., oligomeric) potential migrants.

Acetone, acetonitrile and diethyl ether solutions of rubber extracts can be analysed directly by LC–MS. This was demonstrated by Sidwell [21], who described how LC–MS was used to provide additional information on the species present in an ether extract of a food-grade EPDM rubber. The formulation of this EPDM compound, which was designated 'EPDM 2', is provided in Table 8.6.

Some of the results presented by Sidwell [21], from both LC–MS and GC–MS analyses, are reproduced here. For example, Figure 8.2 shows the TIC GC–MS trace for the examination of the diethyl ether extractable species from a sample of the vulcanised sulfur-cured EPDM 2 compound that had not been post-cured, a

Table 8.6: Formulation of food-grade EPDM.

Ingredient	phr
Keltan® 720	100
ZnO	5
Stearic acid	1
Self-reinforcing CB N762	50
Strukpar 2280 paraffinic oil	8
Struktol® WB16	1
Sulfur	2
MBT	1.5
Tetramethylthiuram disulfide	0.8
TDEC	0.8
Dipentamethylene thiuram hexasulfide	0.8

MBT: 2-Mercaptobenzothiazole
TDEC: Tellurium diethyldithiocarbamate
Reproduced with permission from M.J. Forrest, S. Holding and
D. Howells, *Polymer Testing*, 2006, **25**, 63.
©2006, Elsevier Ltd [19]

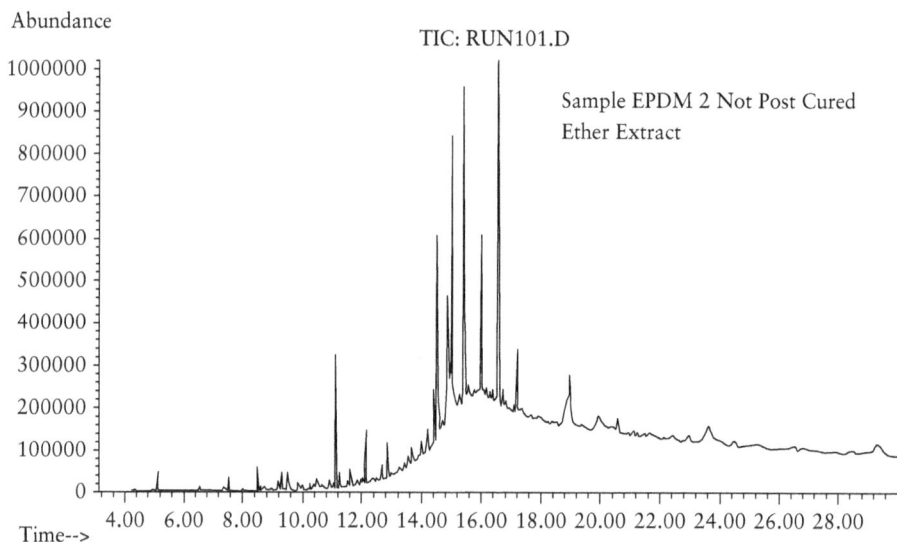

Figure 8.2: GC–MS TIC chromatogram of the diethyl ether extract of EPDM 2. Reproduced with permission from M.J. Forrest in *Food Contact Materials – Rubbers, Silicones, Coatings and Inks*, Smithers Rapra Technology Ltd, Shawbury, UK, 2009, p.37. ©2009, Smithers Rapra Technology Ltd [22].

procedure that is sometimes used to 'clean-up' food-contact rubber products. The species detected in this GC–MS chromatogram are shown in Table 8.7 and, importantly, all of these species are of a relatively low MW.

Table 8.7: Species extracted from EPDM 2 by diethyl ether and identified by GC–MS (Figure 8.2).

Peak time (mins)	Assignment	Origin of the species
5.11	Isothiocyanato-ethane	From cure system
7.52	Dicyclopentadiene	Monomer
8.50	N-formylpiperidine	From cure system
9.19	Tetramethylthiourea	From cure system
9.27	Benzothiazole	From cure system
11.92	2-(Methylthio)-benzothiazole	From cure system
12.16	N-ethyl-2-benzothiazolamine	From cure system
14.48	N-(2-hydroxyethyl)-dodecanamide	From process aid
14.88	Pyrene	From CB
15.09	N,N-dimethylpalmitamide	From process aid
15.42	N-(2-hydroxyethyl)-dodecanamide	From process aid
16.57	Di(2-ethylhexyl)phthalate	Contaminant
17.18	4-Ethyl-2-propyl-thiazole	From cure system

Reproduced with permission from M.J. Forrest in *Food Contact Materials – Rubbers, Silicones, Coatings and Inks*, Smithers Rapra Technology Ltd, Shawbury, UK, 2009, p.38. ©2009, Smithers Rapra Technology Ltd [23].

In an attempt to obtain more information on this diethyl ether extract, it was also analysed by LC–MS using a C_{18} reverse-phase gradient elution separation with atmospheric pressure chemical ionisation (APCI). Using the APCI head in the positive mode gave the LC–MS TIC chromatogram shown in Figure 8.3. The last three peaks showed ions of masses 538, 566 and 594, respectively, and are believed to relate to the presence of tellurium dithiocarbamates in the extract (i.e., ions + $2H^+$ from the ionised protonic solvent), which originated from the TDEC accelerator. Information on the proposed chemical structure of these three species is presented below:
– Mass 538 – tellurium^{4+} with two dimethyldithiocarbamate groups and two diethyldithiocarbamate groups (tellurium^{4+} being detected separately).
– Mass 566 – tellurium^{4+} with three dimethyldithiocarbamate groups and one diethyldithiocarbamate groups.
– Mass 594 – tellurium^{4+} with four diethyldithiocarbamate groups.

Clearly, the presence of such high-MW species in the diethyl ether extract was not detected by the GC–MS analyses (Table 8.7).

MSD2 TIC, MS File (J0329\LCMS0003.D)   APCI, Pos, Scan, Frag: 190

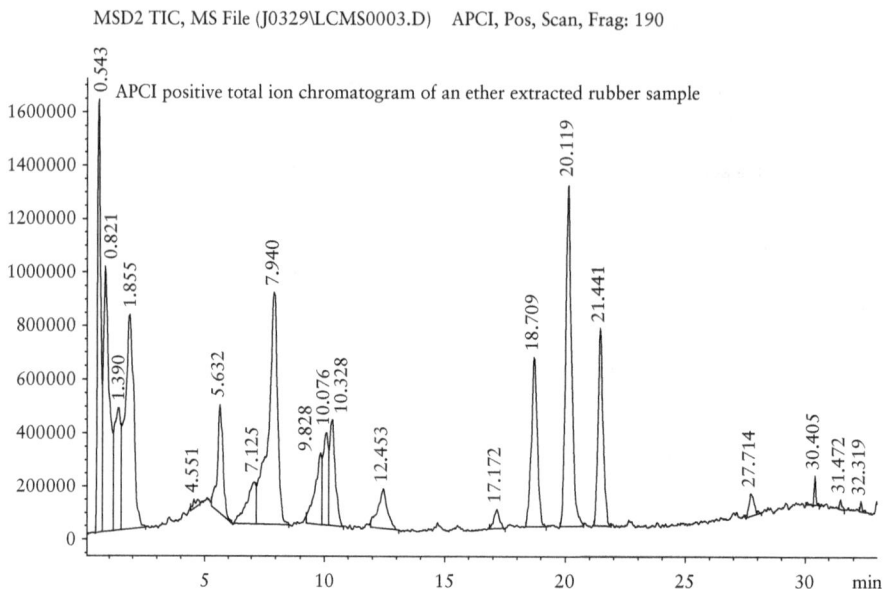

Figure 8.3: LC–MS APCI (positive mode) TIC chromatogram of the diethyl ether extract of EPDM 2. Reproduced with permission from M.J. Forrest in *Food Contact Materials – Rubbers, Silicones, Coatings and Inks*, Smithers Rapra Technology Ltd, Shawbury, UK, 2009, p.38. ©2009, Smithers Rapra Technology Ltd [23].

Other studies have demonstrated the power of LC–MS-based techniques to fingerprint low-MW species in food-contact rubbers. For example, Fichtner and Giese [24] showed how both LC–MS and LC–MS/MS can be used to profile the low-MW species present in the rubbers and plastic components used in food and pharmaceutical industries. The advantages that these techniques offer over HPLC were emphasised in their article, as is the work that will be required in the future to ensure that these techniques are exploited to their full potential.

8.4.3 Determination of specific substances in food-contact rubbers

The main classes of potential specific migrants as well as the analytical techniques used to detect and quantify them are described in Sections 8.4.3.1–8.4.3.5.

8.4.3.1 Monomers
Monomers are either in the gaseous state or relatively volatile liquids. Hence, GC and GC–MS-based techniques are used to determine them in both rubber compounds and food simulants/food products. To simplify the analysis, a static

headspace sampler is often used to isolate the monomer from the sample matrix because an extract-based procedure often presents chromatographic problems, with the extraction solvent and co-extractants often obscuring the analyte.

8.4.3.2 Plasticisers and process oils

These additives are essentially viscous, high-boiling-point liquids and so the most appropriate technique to use is LC–MS. A range of synthetic plasticisers such as phthalates, adipates, mellitates and sebacates can be detected using APCI ionisation. Process oils are hydrocarbon mineral oils and require either an atmospheric pressure photoionisation head (which can ionise non-polar species) or, if the oil has sufficient aromatic character, the use of in-line ultraviolet or fluorescence detectors. A fluorescence detector is particularly sensitive in the detection of polyaromatic hydrocarbon compounds in such oils.

8.4.3.3 Cure system species, accelerators and their reaction products

This class of additive can present problems as it is often thermally labile, reactive and, in some cases, has a degree of ionic character (e.g., dithiocarbamate salts). In some cases, the reaction products [e.g., methyl aniline from N,N′-di-*ortho*-tolyl guanidine and cyclohexylamine from N-cyclohexyl-2-benzothiazole sulfenamide (CBS)] are stable and so GC and GC–MS can be used. Peroxides are popular curatives for food-use rubbers and the stable, breakdown products of these can be easily detected by GC–MS.

In a number of cases, LC–MS is a more appropriate technique than GC–MS. It is also easier to use LC–MS with most of the approved food simulants because, being compatible with the mobile phase, they can be injected directly into the instrument,

Nitrosamines are derived from the secondary amines that are the breakdown products of a number of commonly used accelerators (Section 7.5.1 and 8.5.3). These potentially carcinogenic species can be determined at low-ppm levels by the use of a combined GC–thermal energy analyser (TEA) instrument. Samples can be prepared from rubber compounds by either solvent extraction or food migration studies and then, after a concentration step, injected into the GC. The separated nitrosamines enter a catalytic pyrolyser where nitrosyl radicals are generated. These react with ozone introduced into the system to form a new radical that is chemiluminescent as it returns to the ground state. The emitted light generated by this loss of energy is detected and quantified. A number of factors have been found to affect the levels of nitrosamines found in a particular rubber sample [25].

8.4.3.4 Antidegradants and their reaction products

This class of additive is less thermally labile and reactive than the preceding one and GC-based methods can be used for a number of its members. However, due to the relatively large number of high-temperature processing steps used with rubbers (i.e., mixing, extruding/calendaring, moulding and, in some cases, post-curing), a number of low-volatility, oligomeric antidegradants are commercially available and the higher oligomers of these far exceed the MW limits of GC. LC–MS methods, therefore, have to be used and the technique has proved to be of great value in determining a range of amine and phenolic types.

8.4.3.5 Oligomers

Low-MW oligomers have the potential to migrate from rubbers and those that have a MW of ≤1,000 Da are of particular interest because, like other chemical substances in this weight range, they have the potential to be absorbed by the human gastrointestinal tract. A number of chromatographic techniques have the capability to detect and quantify oligomers, including GC–MS, LC–MS and supercritical fluid chromatography (SFC). Their effectiveness depends upon the chemical nature of the oligomers (e.g., structure and thermal stability) and the weight range being targeted.

8.5 Analysis work to detect and quantify specific migrants in food simulants and food

8.5.1 Introduction

National regulations and CoE Resolutions stipulate concentration limits for certain species within food-contact rubber compounds (Section 8.3) and they also have SMLs for certain migrant compounds.

Analytical work is, therefore, required on a quality-assurance basis to ensure that a food-approved rubber compound remains fit for purpose (e.g. by checking the monomer level) and to ensure that compounds having SMLs do not exceed them in food simulants or food samples prepared using appropriate contact conditions.

These tests are, therefore, used to target specific chemical compounds for which there is a toxicological concern. The test methodology varies according to the regulations being applied, but some species (e.g., nitrosamines and nitrosatable substances) appear regularly due to the degree of concern associated with them. Other specific migrants that are often targeted include:
- Aromatic amines
- Other amines (e.g., cycloaliphatic amines)
- Peroxides and their breakdown products
- Formaldehyde
- Monomers (e.g., ACN)
- Accelerators (e.g., zinc dibutyldithiocarbamate, CBS, MBT)

This list is not complete because it is recognised that rubber products contain two important classes of ingredients (antidegradants and curatives) that are reactive and so produce a large range of reaction products and breakdown products. Work carried out at Rapra for the FSA [8] showed that there are >1,000 of these products originating from the 200 compounds in the 2004 CoE Rubber Resolution AP(2004)4 inventory list.

For convenience, in this section, the work that has been carried out in this area has been divided into two sections:
- The chemical type of the potential migrant – Section 8.5.2.
- The specific food-contact product – Section 8.5.3.

8.5.2 Chemical type of the potential migrant

8.5.2.1 Alkylphenol and bisphenol A
Concerns over their potential to function as endocrine disruptors led to a Japanese study on the levels of alkylphenols in 60 rubber products [26]. Such compounds are used as starting materials in the manufacture of a number of rubber additives, particularly oligomeric phenolic antioxidants. The work concentrated on four compounds: p-*tert*-butylphenol (PTBP), p-*tert*-octylphenol (PTOP), p-nonylphenol (NP) and bisphenol A. The results showed the presence of PTOP in three samples in the range 2.2 to 37 µg/g, NP in 15 samples in the range 2.6 to 513 µg/g, and no PTBP or NP in any samples. Some specific migration experiments for NP were also carried out using water, 20% ethanol and n-heptane. The levels were found to vary from 0.004 to 1.519 µg/ml, with the highest results being obtained with the n-heptane.

Hakkarainen covered the use of solid-phase microextraction for the analysis of migratable substances from food-contact materials in a chapter within a book on chromatography [27]. In particular, he described how the technique can be used to target a number of substances, including:
- Butylated hydroxyltoluene (BHT) in bottled drinking water.
- Bisphenol-type contaminants.
- Acetaldehydes and terephthalic acid in polyethylene terephthalate bottles.
- Butyl tin compounds in food and beverage packaging.

8.5.2.2 Peroxide breakdown products
Peroxides can be used to cure silicones and a number of other rubbers. Work carried out by Novitiskaya and Ivanova [28] on a peroxide-cured NBR detected between 0.82 and 6.41 mg/l of the breakdown product diisopropyl benzene in an aqueous food simulant (distilled water). Work has also been carried out at

Smithers Rapra on the migration of substances from silicone rubbers, and among the substances detected were peroxide breakdown products (Section 8.5.3).

8.5.2.3 Dimethylsiloxanes and other components from silicone rubbers

A test report produced by the Fraunhofer Institute of Food Technology and Packaging [29] commented on the migration of siloxanes from three silicone rubbers – a high- temperature curing material, a room temperature curing material, and a cured liquid silicone rubber. Five food simulants (isooctane, ethanol, ethanol/ water, ethyl acetate and olive oil) were used and one of the experimental parameters investigated was the degree to which the thickness of the sample effected over- all migration. In the case of the hydrophobic solvents, it was found to be more important than the polarity of the simulant. As expected, the results obtained with ethanol/water mixtures showed that the amount of migrating oligomeric material reduced markedly with increasing water content, a virtually zero result being ob- tained above the level of 30%. The migrants were characterised by SFC using both a flame ionisation detector and MS detection. A homologous series of methyl-termi- nated linear siloxane oligomers of ≤ 20 $SiMe_2O$ units was identified.

A review of the use of a number of analytical techniques [e.g., infrared, GC–MS, nuclear magnetic resonance (NMR), atomic absorption spectroscopy] to identify and quantify polydimethylsiloxanes (PDMS) in a wide range of matrixes (e.g., food, pharmaceuticals and cosmetics) was published by Mojsiewicz-Pienkowska and Lu- kasiak [30]. Their study also considered the toxicological issues surrounding PDMS.

Smithers Rapra have carried out studies of the migration of silicone oligomers from food-contact silicone elastomers as part of the FSA project A03046 [9, 31]. They found that a range of cyclic and linear siloxane oligomers could be detected. Further information of the findings of this research is provided in Section 8.5.3.

8.5.2.4 Accelerators and antidegradants

An investigation has been carried out and reported by Kazarinova and Ledovskikh [32] on the migration of the accelerator diphenyl guanidine and its reaction prod- ucts from rubber compounds into food simulants. The concentration of migrants was found to be influenced by the fillers present in the rubber compounds, with non-black vulcanisates giving the highest values.

The results of a study presented in 1999 by Kretzchmar [33] suggested that when guanidine accelerators are used in sulfur vulcanisation with phenylenedi- amine-based antioxidants, carcinogenic aromatic amines and toxic isothiocyanates may be formed. Similarly, when N-(1,3-dimethyl butyl)-N′-phenyl-p-phenylenedi- amine is used as an antiozonant, it may decompose and react in aqueous food types to form aromatic amines. This study also included the effects that compound formulation and curing conditions can have on these reactions.

Barnes and co-workers [34] developed a LC–MS method to identify vulcanisation agents and their breakdown products in food and drink samples. A large sample of 236 retail foodstuffs were analysed for the presence of MBT and its breakdown product benzothiazole (BT). The accelerators 2,2'-dithiobis(benzothiazole) (MBTS) and CBS, which are commonly used in food-contact rubbers, were also searched for, and MBT and BT are also known to be breakdown products of these two compounds. The detection limit for these species was found to dependent on the food product type and ranged from 0.005 to 0.043 mg/kg. MBT, BT, MBTS or CBS was not detected by the analytical work in any of the samples above these values.

Kruger and co-workers [35] studied the migration of primary aromatic amines (e.g., aniline) and secondary amines into food simulants that came into contact with styrene-butadiene rubber (SBR) rubber compounds containing a series of phenylenediamine antiozonants having thioether and sulfoxide groups. The results that the team obtained were discussed in comparison with those obtained using N-phenyl-N'-(1,3-dimethylbutyl)-p-phenylenediamine as an antioxonant in the SBR compound.

8.5.2.5 Oligomers
The wide availability of LC–MS instruments means that they are now rivalling SFC for the analysis of oligomers. The MW range of LC–MS instruments can be extended to ≥4,000 Da, and this capability makes them ideal for the characterisation of oligomers. For example, it has been shown [9] that silicone oligomers can be detected by this technique in food simulants and some of the results obtained are shown in Section 8.5.3.

8.5.3 Food-contact products

8.5.3.1 Teats and soothers
Historically, a variety of rubbers has been used in the manufacture of teats for baby feeding bottles, including NR, silicone and styrene-butadiene block copolymers. The rubbers that were vulcanised using sulfur-based cure systems presented particular concerns in this product area due to the potential for nitrosamines to be present. The EU published legislation on the level of nitrosamines permitted in babies' teats and soothers in 1993 in the form of a European directive (Section 8.3.1).

The publication of this directive came at a time when there were concerns over the presence of nitrosamines in the processing environment within the rubber industry (Section 7.5.1) and in rubber products, as well as from studies conducted on teats and soothers. An example of the latter was carried out by Mizuishi and co-workers [36]. They focused on the migration of nitrosamines from teats and soothers made from silicone, NR, synthetic polyisoprene and SBR, and showed that several nitrosamine compounds were detected and that, predictably, the products made from NR yielded the highest levels.

In addition to the targeting of nitrosamines, analytical work of a more general nature has been carried out on teats and soothers. For example, a survey of the extractables present in rubber teats was published in 1991 [37]. The samples were extracted with diethyl ether or acetone and the extracts analysed by GC and GC–MS. Data was obtained on 49 rubber teats commercially available in the Netherlands and a number of compounds not permitted in the Dutch regulations were identified: dibenzylamine, acetophenone, zinc dibenzyldithiocarbamate, 4,4'-thio-*bis*(2-*tert*-butyl-5-methyl) phenol and *bis*(2-hydroxy-3-*tert*-butyl 5-ethylphenyl)methane.

A survey from the Netherlands in 2003 [38] looked at the migration of N-nitrosamines, N-nitrosatable substances and MBT from 19 samples of teats and soothers. In addition to these species, screening work was carried out for any other potential migrants. In finding that the majority of teats and soothers were manufactured from silicone rubber, this study highlighted how the industry was moving away from conventional, sulfur-cured synthetic rubbers and towards cleaner rubbers that could not contain substances like nitrosamines. However, NR products were present in the sample and only one was found to be above the permissible limit, and that was for nitrosatable substances at 0.23 mg/kg. MBT was found in only one of the NR products, and this was below the migration limit of 0.3 mg/teat.

Dopico-Garcia and co-workers [39] evaluated the stability of some phenolic antioxidants and the oxidation product of one phosphate antioxidant in four food simulants (distilled water, 3% acetic acid, 10% ethanol and olive oil) at different temperatures and time intervals. The results showed that, in general, the antioxidants were more stable in olive oil than in the aqueous food simulants. Of the aqueous simulants, distilled water and 10% ethanol allowed the highest stability at low temperatures. Some differences were observed according to MW with the 3% acetic acid simulant. The low-MW antioxidants had good stability in this simulant, even at high temperatures, but the high-MW examples decomposed rapidly at low temperature.

Di Feng and co-workers [40] analysed non-target compounds in silicone rubber teats with a view to providing the means to carry out a safety evaluation and early-warning mechanism for these and similar polymeric materials. They used two analytical approaches to extract non-target compounds in 30 silicone rubber teats: purge and trap and solid-phase microextraction. This strategy enabled 140 compounds in 12 categories to be separated and detected by GC–MS. Three qualitative methods were used to identify the compounds: spectral matching using a library, retention index, and confirmation by use of standards. Consideration was given to how many of these 140 compounds represented a potential danger to infant health, and 53 were selected based on particular criteria, such as their concentration and potential toxicity. These 53 compounds included aromatic compounds, siloxanes, BHT, trimethyl silanol, N,N-dibutylformamide and BT.

Sung and co-workers [41] used a LC–MS/MS method as an alternative to the GC– TEA method recommended by the European Committee on Standardisation for the simultaneous determination of eight N-nitrosamines released into artificial saliva from rubber teats and soothers. The team found that the method could be validated, using N-nitrosodipropylamine-d14 as the internal standard, with relatively-good analytical results, including a sufficiently low detection limit of 0.1 to 2 µg/kg of sample. The results also showed that the LC–MS/MS method was sufficiently rugged and successful to be used for the routine analysis to demonstrate compliance with European Directive 93/11/EEC (Section 8.3.1).

8.5.3.2 Rubber adhesives
Rubber adhesives, both natural and synthetic, are used widely in the manufacture of multi-laminates commonly employed as food packaging. For this reason, it is important to identify the compounds that have the potential to migrate from the laminate into food. Nerin and co-workers [42] identified 29 compounds in two rubber adhesives, some of them with high toxicity levels according to the theoretical model of Cramer (e.g., benzene 4-cyanocyclohexene and benzene isothiocyanate). The team carried out work to determine the partition of these compounds between adhesive and different substrates. The diffusion in both media was found to be very variable depending upon the properties of the compounds and substrates used. The results of the study also showed that only three compounds (benzene 4-cyanocyclohexene, BHT and 2-cyclopentyl-1,3,5-trimethylbenzene) migrated into food and these were below the SML set by the EU, and the values recommended by Cramer.

8.5.3.3 Meat netting
NR has been the traditional material for elastomeric meat netting to hold joints of meat together during packaging and cooking for many years. This has led to a number of studies on the levels of N-nitrosamines, nitrosatable and other compounds. Work has been carried out in the US by Marsden and Pesselman [43] using a typical product from NR latex in contact with a 50% ethanol simulant for 150 min at 152 °C. The results showed that the most abundant zinc carbamate compound was zinc dibenzyldithiocarbamate (860 µg/g netting) and the most abundant secondary amine was dimethylamine (8.8 ng/g netting).

A collaborative survey of ten meat-netting samples obtained from four manufacturers was carried out in the Netherlands [44] by workers from the Inspectorate for Health Protection and Veterinary Public Health as well as the National Institute of Public Health and the Environment. All ten samples consisted of both NR and vegetable fibres and, in addition to nitrosamines and N-nitrosatable substances, the samples were screened for other potential migrants. Nitrosamines were detected in concentrations up to 2 mg/kg of netting and the two N-nitrosatable compounds dimethylamine and dibenzylamine were found in up to 0.4 mg/kg of netting. These values were not considered to be of concern to public health because of the ratio of meat netting to food product. The other potential migrants identified included alkanes, alkenes, acids, antioxidants, plasticisers and sterols, several of which were not authorised for food contact in the Netherlands, but were allowed in other countries.

8.5.3.4 Rubber gloves for handling food

Datta and Gonlag [45, 46] looked at a number of issues involving latex gloves, including nitrosamines, zinc, allergy problems and the food-contact legislation and regulations that exist in Europe. Additional information concerning the allergy problems that can be associated with latex gloves is provided in Section 11.2.15.

A study was published in 2001 [47] that reported GC–MS results from solvents extracts (e.g., n-heptane and n-hexane extracts) obtained on a range of gloves, including those produced from NR and NBR. A range of accelerator and plasticiser-type species were identified, but it was apparent in the case of the nitrile samples that a relatively large number of extracted compounds could not be identified by GC–MS, with no match being found in commercially available libraries.

Another study was then carried out on NBR gloves by members of the same team with a view to improving the overall quality of the data [48]. Six compounds, which were common to a number of the nitrile gloves used in the original work, were isolated from an n-hexane extract by silica gel chromatography, and then these compounds were identified by NMR and high-resolution MS.

8.5.3.5 Thermoplastic rubbers

Although the market share for TPE in food-contact situations is relatively low, they find niche applications where their ease of processing and physical properties enables the replacement of vulcanised rubber. Such applications can include conveyor belting, diaphragms and kitchen utensils. Because of their intermediate status, it is often the case that legislation for both rubbers and plastics has to be consulted to enable a full regulatory assessment to be made. A study by Sidwell [49] addressed the approach to the regulation of TPE in Europe and the US. He also reviewed the use of mathematical modelling to assess the migration of substances from TPE into food.

A review of the use of thermoplastic rubbers for the food packaging, medical and pharmaceutical industry was presented by Siepmann at the 2012 'Thermoplastic Elastomers' conference [50]. In this review it was highlighted that these materials were demonstrating an annual growth of ≈7.5% in these regulated industries, and they were being chosen as substitutes for vulcanised rubbers, PVC or silicones.

8.5.3.6 Natural and synthetic rubber products

As part of the work carried out on the FSA research project that targeted breakdown products and reaction products in food-contact rubbers [8], GC–MS and LC–MS were used to identify substances that had migrated from the rubbers into food simulants and food products. For the preparation of the migration samples, food simulants, food products and test conditions were selected that were representative of the types of end-use conditions that the different rubber compounds could experience in service. Table 8.8 shows some of the results that were obtained using GC–MS to analyse the selected food products.

Table 8.8: Identification of migrants into different food products from a variety of rubber compounds by GC–MS.

Rubber compound and condition*	Food product	Examples of migrants detected by GC–MS**
Hydrin 20D aged	Olive oil	4,4'-Thiobis[2-(1,1-dimethylethyl-5-methyl-phenol)]
EPDM 239C unaged	Beer	4-Oxopentanoic acid, n-butyl ester
	Olive oil	TAC
EPDM 240C unaged	Olive	9,10-Dihydro-9,9-dimethyl-acridine
		N-phenyl-1-naphthylamine
NR 504A aged	Olive oil	N-isopropyl-N'-phenyl-p-phenylenediamine
		4,4'-Thiobis[2-(1,1-dimethylethyl-5-methyl-phenol)]
EPDM 242C unaged	White wine	N-butyl-1-butanamine
	Olive oil	N-phenyl-1-naphthylamine
SBR 195S aged	Apple juice	N-isopropyl-N'-phenyl-p-phenylenediamine
	Olive oil	N-isopropyl-N'-phenyl-p-phenylenediamine
Butyl 51B unaged	Olive oil	Aniline
		N-phenyl-octadecanamide
CR 201E aged	Olive oil	N-phenyl-1-naphthylamine

*Some cured rubber samples were aged by placing in an oven prior to the migration test to
 reflect ageing in service
**The foodstuffs were extracted (e.g., with a pentane/diethyl ether mixture) to provide
   samples for analyses
Adapted from M.J. Forrest in *Food Contact Materials – Rubbers, Silicones, Coatings and Inks*,
Smithers Rapra Technology Ltd, Shawbury, UK, 2009, p.85 [51]

An example of the data obtained on the food simulants using LC–MS, in both positive and negative APCI mode, is shown in Table 8.9. This table lists the migrants that were detected in 3% acetic acid that had contacted a sample of a cured SBR rubber compound, designated 195S, for 4 h at 100 °C.

It can be seen from the assignments in these two tables that the migrants detected by GC–MS and LC–MS were either breakdown/reaction products of antidegradants or accelerators, or were examples of these types of compounding ingredients in their unchanged form. These two sets of data are only a selection of the results that were obtained during this 3-year study. A full set of results is presented in the final report for the project [8], which is available on the FSA website (Section 8.1). In addition, Sidwell reviewed some of the findings of this research in

Table 8.9: Identification of migrants from the SBR 195S into 3% acetic acid by LC–MS.

APCI mode of detection	Retention time (min)	Mass (m/e)	Migrant
Positive	18.6	185.0	N-phenyl-1,4-phenylenediamine
Positive	20.2	397.2	MBT zinc salt
Positive	23.8	227.1	N-isopropyl-N'-phenyl-p-phenylenediamine
Negative	16.8	165.9	2-Mercaptobenzothiazole

Adapted from M.J. Forrest in *Food Contact Materials – Rubbers, Silicones, Coatings and Inks*, Smithers Rapra Technology Ltd, Shawbury, UK, 2009, p.311 [52]

a paper presented to Rapra's '*Food Contact Polymers*' conference in 2009 [53]. Some of the examples described in the paper included the use of GC–MS and LC–MS to detect curative breakdown products and volatile substances from NR and NBR compounds, and extractable siloxane oligomers from food-grade silicone rubbers.

8.5.3.7 Silicone rubber products

Silicone rubbers are in wide use as food-contact materials for various applications. This fact is reflected in the amount of food-related research and investigative work that has been published. A review on the use of silicone products for food-contact applications was published by Forrest in 2006 [54]. This publication includes an extensive literature search. Some of the relevant literature that has been published since 2006 is presented below.

Smithers Rapra carried out studies on the migration from food-contact silicone elastomers as part of the FSA project A03046 [9, 31]. A combination of GC×GC–ToF–MS and LC–MS was used to examine, detect, and identify the migrants in a number of food simulants, including olive oil and 95% ethanol. The results revealed that the migrating species could include:
- Silicone oligomers:
  - Cyclic siloxanes
  - Linear siloxanes (methyl and hydroxyl terminated)
- Cure system breakdown products:
  - Peroxide breakdown products (e.g., ketones and aldehydes)
  - Metal catalysts (e.g., platinum compounds)
- Oxidation products:
  - Cyclic oligomers – from main chain scission
  - Aldehydes – oxidation of alkyl groups (e.g., formaldehyde from methyl groups)

Examples of the data obtained are shown in Figures 8.4 (the GC×GC–ToF–MS TIC chromatogram of the acetonitrile extract of the olive oil food simulant sample) and

Figure 8.5 [the LC–MS APCI (positive mode) TIC chromatogram of the 95% ethanol food simulant sample]. A wide range of species peaks, including those of siloxane oligomers, can be observed in Figure 8.4 due to the high resolving power of the GC×GC–ToF–MS instrument. In Figure 8.5, the sequence of species peaks are due to a series of oligomers from n = 7 (i.e., heptamer) to around n = 20 with a difference of 74 mass units between each oligomer, equal to the dimethylsiloxane repeat unit.

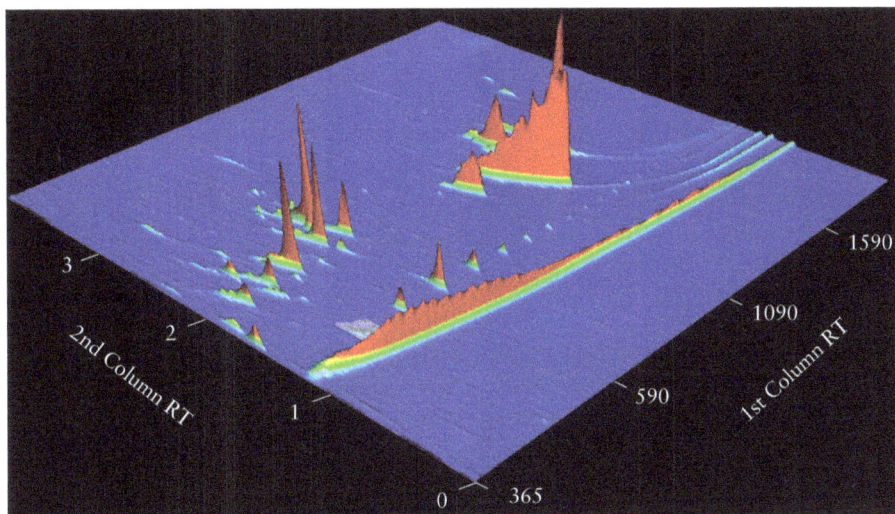

Figure 8.4: GC×GC–ToF–MS TIC chromatogram – olive oil food simulant sample (acetonitrile extract). Reproduced with permission from J. Sidwell in *Proceedings of the Food Contact Polymers 2007*, 21–22nd February, Brussels, Belgium, Smithers Rapra, Shawbury, UK, 2007, Paper No.5. ©2007, Smithers Rapra [31].

Helling and Simat [55] of the Saxon Institute for Public and Veterinary Health presented an overview of the migration behaviour of silicone rubbers used as baking moulds. The migration results were obtained using proton NMR analysis of a range of foodstuffs that had contacted the silicone rubbers. The study also contained a critical assessment of the suitability of food simulants for evaluation of the migration from silicone rubbers into cakes and muffins. Other aspects of the subject addressed by the study included:
- Evaluation of migration from silicone rubbers into cakes, meatloaf and crème brûlée.
- Migration from silicone rubbers into pizza during long-term usage.
- Migration of fat from food into silicone rubbers.
- Influence of temperature treatment of silicone moulds of migration of siloxanes into food.

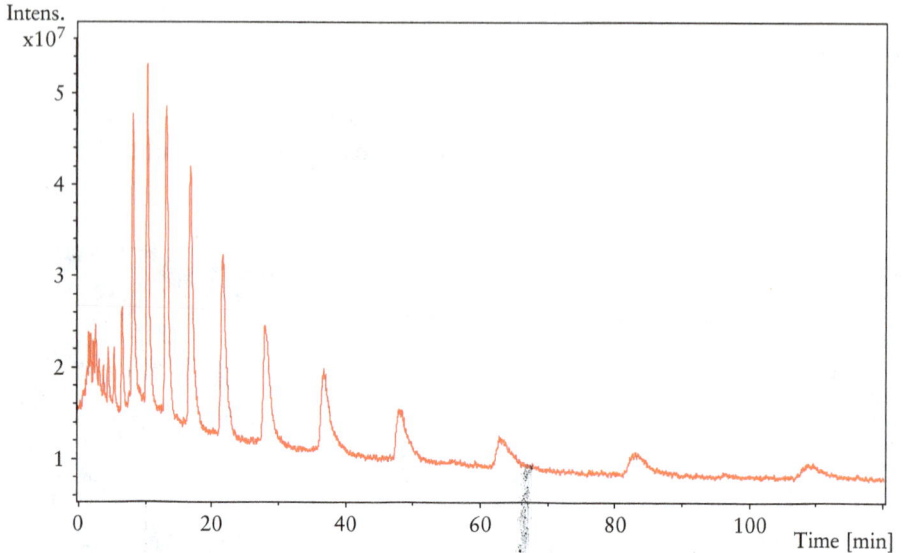

Figure 8.5: LC–MS APCI (positive mode) TIC chromatogram – 95% ethanol food simulant sample. Reproduced with permission from J. Sidwell in *Proceedings of the Food Contact Polymers 2007*, 21–22nd February, Brussels, Belgium, Smithers Rapra, Shawbury, UK, 2007, Paper No.5. ©2007, Smithers Rapra [31].

In another study, Helling and Simat [56] reported the results obtained on ≈100 silicone rubber food-contact products (e.g., silicone teats and moulds) with respect to migration behaviour, loss of volatile substances, amount of platinum catalyst, take-up of fat from the foods, and the hygienic considerations surrounding the absorption of fat. In the case of silicone teats, a high level of chemical and physical stability was observed, with only minor levels (0.26%) of fat absorption. The sterilisation of silicone rubber teats using a microwave up to 100-times was not found to have a significant effect on its migration behaviour in contact with milk or its physical properties. In the case of silicone moulds, a high level of oligomers was found to migrate and the levels exceeded the current limits set by a number of EU national regulations and the CoE Resolution (i.e., 60 mg/kg). The level of oligomer migration from the moulds was found to decrease with repeated use, but the amount of absorbed fat from the food (e.g., flaky pastry and soufflés) continued to increase up to 1.2%.

The results of other studies on the migration performance of silicone rubber baking moulds have been published by Helling and other workers at the Saxon Institute for Public and Veterinary Health. For example, in an article published in 2010 in *Food Additives & Contaminants* [57], the researchers reported on the migration results obtained using two food simulants, 95 and 50% ethanol, and also foodstuffs that had different fat contents (e.g., creamed cake and meatloaf). Data were generated on the substances that had migrated using the proton NMR method that the team had previously validated and employed for this purpose (see below). Other areas addressed by the article included the effect of tempering on siloxane migration, the determination of volatile organic compounds (restricted in regulations such as the BfR Recommendation XV), and the take-up of fat potentially leading to organoleptic and hygiene issues. Another study [58] on silicone baking moulds published by the team in the same journal described how the substances that migrated were characterised using proton NMR, LC–MS and GC–MS, and were identified as cyclic organosiloxane oligomers and linear partly hydroxyl-terminated organosiloxanes. These migration studies were carried out using cakes that came into contact with a range of food simulants (olive oil, isooctane, ethanol and Tenax™) and a validated proton NMR method enabled quantification of the migrating siloxanes.

Helling and co-workers [59] reported on the work they carried out on the migration properties of silicone materials during typical long-term commercial and household use applications. This work was a continuation of earlier studies, examples of which are provided in the paragraphs and references above, in that pre-characterised silicone products were assessed during these long-term trials. One case in particular involved silicone moulds that were used in a commercial pizza bakery on a daily basis up to 1,700 times. The following properties were monitored by the team over this period:

- Migration behaviour
- Uptake of fat
- Amount of volatiles and extractables
- Physical properties (e.g., elongation and tensile strength)

The migration of substances from the silicone mould into the food was determined using proton NMR, and was found to decrease during the experimental exercise from values between 11 to 18 mg/kg to below the limit of detection (<1 mg/kg). No evidence was found for the formation of new migratable siloxanes by degradation during the study period, and the loss of the siloxanes was compensated for by the uptake of fat and other lipophilic food constituents. Also, the release of volatile organic compounds decreased from 0.44% for the new moulds to 0.14% for the oldest ones (i.e., those with ≈1,700 uses – equivalent to 400 h at 180 °C). The physical properties of the mould remained reasonably constant during the study. Similar results for the volatile compounds and physical properties were obtained for a long-term household study that utilised baby teats sterilised in a microwave oven. However, there was a difference with regard to the extractable siloxane content of new and old samples due to the weak extracting power of milk, with no significant differences being detected.

References

1.   *Global Legislation for Food Packaging Materials*, Eds., R. Rijk and R. Veraat, Wiley-VCH Verlag GmbH & Co., Weinheim, Germany, 2010.
2.   *Global Legislation for Food Contact Materials*, Volume 278, Ed., J.S. Baughan, Woodhead Publishing Series in Food Science, Elsevier Science and Technology, Amsterdam, The Netherlands, 2015.
3.   B. Geueke, C.C. Wagner and J. Muncke, *Food Additives & Contaminants: Part A*, 2014, **31**, 8, 1438.
4.   M. Creese, *Pharmaceutical Manufacturing and Packing Sourcer*, 2012, **February**, 30.
5.   M.J. Forrest in *Food Contact Materials – Rubbers, Silicones, Coatings and Inks*, Smithers Rapra Technology Ltd, Shawbury, UK, 2009.
6.   Project FS2219 – Migration Studies – Food Contact Materials, FSA, UK, 1994–1997.
7.   Project FS2248 – Further research on chemical migration from food contact rubbers and other elastomers FSA, UK, 1999–2001.
8.   Project A03038 – An investigation of the breakdown products of curatives and antidegradants used to produce food contact elastomers, FSA, UK, 2002–2004.
9.   Project A03046 – Chemical migration from silicones used in connection with food contact materials and articles, FSA, UK, 2004–2005.
10. R.C. Isa, *Malaysian Rubber Technology Developments*, 2011, Special Edition, **11**, 11.
11. M.J. Forrest in *Food Contact Materials – Rubbers, Silicones, Coatings and Inks*, Smithers Rapra Technology Ltd, Shawbury, UK, 2009, p.15.
12. M.J. Forrest in *Food Contact Materials – Rubbers, Silicones, Coatings and Inks*, Smithers Rapra Technology Ltd, Shawbury, UK, 2009, p.17.
13. European Union Directive 93/11/EEC: Release of N-nitrosamines from rubber teats.
14. M.J. Forrest in *Food Contact Materials – Rubbers, Silicones, Coatings and Inks*, Smithers Rapra Technology Ltd, Shawbury, UK, 2009, p.28.
15. L. Pysklo, T. Kleps and K. Cwiek-Ludwicka, *Elastomery*, 2002, **6**, 4–5, 39.
16. L. Pysklo, *Elastomery*, 2003, **7**, 1, 26.
17. M.J. Forrest in *Food Contact Materials – Rubbers, Silicones, Coatings and Inks*, Smithers Rapra Technology Ltd, Shawbury, UK, 2009, p.22.
18. M.J. Forrest, S. Holding and D. Howells in *Proceedings of the High Performance and Specialty Elastomers*, 20–21st April, Geneva, Switzerland, Rapra Technology Ltd, Shawbury, UK, 2005.
19. M.J. Forrest, S. Holding and D. Howells, *Polymer Testing*, 2006, **25**, 63.
20. M.J. Forrest, S. Holding and D. Howells in *Proceedings of the High Performance and Specialty Elastomers*, 20–21st April, Geneva, Switzerland, Rapra Technology Ltd, Shawbury, UK, 2005, Paper No.2.
21. J.A. Sidwell in *Proceedings of RubberChem*, 11–12th June, Munich, Germany, Rapra Technology Ltd, Shawbury, UK, 2002.
22. M.J. Forrest in *Food Contact Materials – Rubbers, Silicones, Coatings and Inks*, Smithers Rapra Technology Ltd, Shawbury, UK, 2009, p.37.
23. M.J. Forrest in *Food Contact Materials – Rubbers, Silicones, Coatings and Inks*, Smithers Rapra Technology Ltd, Shawbury, UK, 2009, p.38.
24. S. Fichtner and U. Giese, *Kautschuk, Gummi, Kunststoffe*, 2004, **57**, 3, 116.
25. B.G. Willoughby and K.W. Scott in *Nitrosamines in Rubber*, Rapra Technology Ltd, Shawbury, UK, 1997.
26. A. Ozaki and T. Baba, *Food Additives & Contaminants*, 2003, **20**, 1, 92.

27. M. Hakkarainen in *Chromatography for Sustainable Polymeric Materials, Renewable, Degradable and Recyclable*, Eds., A. Albertsson and M. Hakkarainen, Springer-Verlag, Berlin, Germany, 2008.

28. L.P. Novitiskaya and T.P. Ivanova, *Gigiena i Sanitariya*, 1989, **5**, 88.

29. O. Piringer and T. Bucherl in *Extraction and Migration Measurements of Silicone Articles and Materials Coming into Contact with Foodstuffs*, FhG Test Report, 1994.

30. K. Mojsiewicz-Pienkowska and J. Lukasiak, *Polimery*, 2003, **48**, 6, 401.

31. J.A Sidwell in *Proceedings of Food Contact Polymers*, 21–22nd February, Brussels, Belgium, Smithers Rapra, Shawbury, UK, 2007, Paper No.5.

32. N.F. Kazarinova and N.G. Ledovskikh, *Kauchuk I Rezina*, 1978, **1**, 26.

33. H-J. Kretzchmar in *Proceedings of Hazards in the European Rubber Industry*, 28–29th September, Manchester, UK, Rapra Technology Ltd, Shawbury, UK, 1999, Paper No.6.

34. K.A. Barnes, L. Castle, A.P. Damant, W.A. Read and D.R. Speck, *Food Additives & Contaminants*, 2003, **20**, 2, 196.

35. R.H. Kruger, C. Boissiere, K. Klein-Hartwig and H-J. Kretzschmar, *Food Additives & Contaminants*, 2005, **22**, 10, 968.

36. K. Mizuishi, M. Takeuchi, H. Yamanobe and Y. Watanabe, *Annual Report of Tokyo Metropolitan Research Laboratory of Public Health*, 1986, **37**, 145.

37. J.B.H. van Lierop in *Food Policy Trends in Europe: Nutrition*, Technology, *Analysis and Safety*, Eds., H. Deelstra, M. Fondu, W. Ooghe and R. Van Havere, Woodhead Publishing Ltd, Cambridge, UK, 1991.

38. K. Bouma, F.M. Nab and R.C. Schothorst, *Food Additives & Contaminants*, 2003, **20**, 9, 853.

39. M.S. Dopico-García, J.M. López-Vilariño and M.V. González-Rodriguez, *Journal of Applied Polymer Science*, 2006, **100**, 1, 656.

40. D. Feng, H. Yang, D. Qi and Z. Li, *Polymer Testing*, 2016, **56**, 91.

41. J.H. Sung, I.S. Kwak, S.K. Park, H.I. Kim, H.S. Lim, H.J. Park and S.H. Kim, *Food Additives & Contaminants: Part A*, 2010, **27**, 12, 1745.

42. C. Nerin, J. Gaspar, P. Vera, E. Canellas, M. Azner and P. Mercea, *International Journal of Adhesion and Adhesives*, 2013, **40**, 1, 56.

43. J. Marsden and R. Pesselman, *Food Technology*, 1993, **47**, 3, 131.

44. K. Bouma and R.C. Schothorst, *Food Additives & Contaminants*, 2003, **20**, 3, 300.

45. R. Datta and A.T. Gonlag, *Gummi Fasern Kunststoffe*, 2003, **56**, 12, 768.

46. R. Datta and A.T. Gonlag, *Gummi Fasern Kunststoffe*, 2004, **57**, 6, 310.

47. C. Wakui, Y. Kawamura and T. Maitani, *Journal of the Food Hygienics Society of Japan*, 2001, **42**, 322.

48. M. Mutsuga, C. Wakui, Y. Kawamura and T. Maitani, *Food Additives & Contaminants*, 2002, **19**, 11.

49. J.A. Sidwell in *Proceedings of Thermoplastic Elastomers*, 8–9th November, Brussels, Belgium, Smithers Rapra Technology Ltd, Shawbury, UK, 2011.

50. D. Siepmann in *Proceedings of Thermoplastic Elastomers*, 13–14th November 2012, Berlin, Germany, Smithers Rapra Limited, Shawbury, UK, 2012, Paper No.19.

51. M.J. Forrest in *Food Contact Materials – Rubbers, Silicones, Coatings and Inks*, Smithers Rapra Technology Ltd, Shawbury, UK, 2009, p.85.

52. M.J. Forrest in *Food Contact Materials – Rubbers, Silicones, Coatings and Inks*, Smithers Rapra Technology Ltd, Shawbury, UK, 2009, p.311.

53. J.A. Sidwell in *Proceedings of Food Contact Polymers*, 21–22nd April, Brussels, Belgium, Smithers Rapra Technology Ltd, Shawbury, UK, 2009, Paper No.7.

54. M.J. Forrest in *Silicone Products for Food Contact*, Rapra Review Report No.188, Smithers Rapra Technology Ltd, Shawbury, UK, 2006.

55. R. Helling and T.J. Simat in *Proceedings of Silicone Elastomers*, 7–8th October, Hamburg, Germany, Smithers Rapra, Shawbury, UK, 2009, Paper No.9.
56. R. Helling and T.J. Simat in *Proceedings of Silicone Elastomers*, 30th–31st March, Cologne, Germany, Smithers Rapra, Shawbury, UK, 2011, Paper No.11.
57. R. Helling, K. Kutschbach and T.J Simat, *Food Additives & Contaminants*, 2010, **27**, 3, 396.
58. R. Helling, A. Mieth, S. Altmann and T.J. Simat, *Food Additives & Contaminants*, 2009, **26**, 3, 395.
59. R. Helling, P. Seilfried, D. Fritzche and T. Simat, *Food Additives & Contaminants: Part A*, 2012, **29**, 9, 1489.

9 Extractable and leachable studies on rubber products

9.1 Introduction

The tests that have to be carried out on rubber products intended for food-contact applications, many of which target migrating substances, are covered in Chapter 8. This chapter describes the extractables and leachables (E&L) studies carried out on rubber products to obtain information to support regulatory submissions for the medical and pharmaceutical industries.

Although these two types of exercises often target and characterise the same set of primarily, low-molecular weight (MW) substances, there are significant and important differences in the methodologies and protocols. Creese [1] took this as his theme and compared and contrasted the difference between the two sets of requirements. There are also significant regulatory differences in that a specific set of regulations cover food-contact materials and articles and others, for example, the European Union Medicinal Products Directive 2001/83/EC, apply to products that contain pharmaceutical products.

For a number of years, manufacturers of rubber products that were intended for use in pharmaceutical applications (e.g., seals in single-use inhalers) and for certain classes of medical devices (e.g., those that were not invasive) tended to use food- approved rubber products as one of the starting points in their selection process for an appropriate product. In recent years, a number of guidance documents and standards have been published, such as those by the US Food and Drug Administration (FDA) [2], the Product Quality Research Institute (PQRI) [3], the United States Pharmacopeia (USP) [4–6] and the International Council for Harmonisation of Technical Requirements for Pharmaceuticals for Human Use (ICH) [7]. The objective of these guidance documents is to enable manufacturers and end-users to carry out specific E&L test programmes to establish if the rubber products are acceptable for their intended end-use [8]. This is a dynamic area and, in 2016, representatives of the PQRI reported to the E&L USA Conference on their toxicology teams' approach to thresholds for leachables in parenteral drug products [9]. Included in the presentation was mention of a database of 606 chemicals that had been shown to potentially leach from container closure systems. These chemicals had been compiled, analysed and sorted into classes using a modified Cramer approach and a final classification scheme had been developed following an initial risk assessment of the most toxic leachables. A database referred to as the 'E&L materials and safety database' has been compiled by the Extractables and Leachables Safety Information Exchange [10].

One of the fundamental differences between the extractables and leachables parts of an E&L work programme is that solvents (e.g., isopropanol) that

https://doi.org/10.1515/9783110640281-009

simulate drug products are usually employed for the extractable study, whereas the actual drug products are often used for the leachables study. It is also usual practice to use 'worst case' conditions in terms of temperature and extraction process (e.g., refluxing) in extractables work to ensure that everything that has the potential to be extracted and, hence, ultimately leach into a pharmaceutical product in service, is removed from the rubber. Leachable work is usually carried out using the complete drug packaging and under conditions that, although still representing the worst case, are much closer to the actual storage temperature and time.

The leachables that can be determined from a rubber, or other component of the pharmaceutical packaging (e.g., plastic, label, adhesive and so on), can be regarded as an imperfect sub-set of the extractables profile. It is an imperfect sub-set because there is the potential for substances that are present in the extractables profile to be chemically changed into new compounds during the leachables study, for example, by a reaction with other components of the drug packaging or with the drug itself (e.g., the 'active component' or other constituents within its formulation).

With regard to the definitions that apply to 'extractable' and 'leachable', general ones that have proved useful in describing the two groups of substances are detailed below.

9.1.1 Extractables

Extractables are chemical compounds that migrate from any product (e.g., a drug) contacting containing/closure material(s) when exposed to a solvent under exaggerated conditions such as the use of a challenging solvent and/or high temperature and time. These compounds can contaminant the drug product. Extractables are an intrinsic property of the contact material.

9.1.2 Leachables

Leachables are chemical compounds, typically a sub-set of extractables, which migrate into a drug formulation from any product (e.g., a drug) contact material(s) as a result of direct contact under normal process conditions or accelerated storage conditions:

- A property of the contact material, drug product and conditions of use.
- Leachables are *generally* a sub-set of extractables, but additional substances can be generated by reactions that occur between leachable substances from the contact material(s) and species within the drug media.

The two groups can be represented pictorially as shown in Figure 9.1.

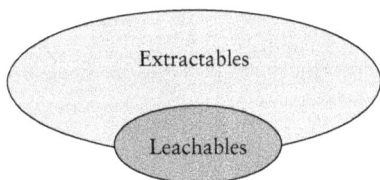

Figure 9.1: Schematic representation of E&L.

As with the majority of analytical investigations, prior knowledge of the specimen to be tested is invaluable in assisting the analyst in choosing the correct and most appropriate approach. With regulatory work that uses comprehensive documents that provide in-depth detail on the methodology that must be used to demonstrate compliance (e.g., food-contact regulations) this prior knowledge is not so important. However, with E&L work, the guidance documents that are available assist the analyst in putting together a programme of work, but this is often complemented by input concerning the composition of the rubber so that an effective 'study plan' can be devised. Section 9.2 provides an overview of the origins of the substances and species that have the potential to contribute to the E&L profiles of rubber products.

An introduction to the subject of E&L has been provided by Feilden [11] in the form of an presentation in 2012 to the E&L USA Conference and also in a book published in 2011 [12]. In addition, further useful background information is provided by Bailey [13] that outlines a strategy for evaluating E&L to minimise risk when changing materials in primary packaging components.

9.2 Origins of substances that can contribute to extractable and leachable profiles of rubbers

The complex compositional nature of vulcanised rubber products means that there are numerous potential sources for the substances that can contribute to E&L profiles and these have been the subject of a review by Forrest [14] and Hulme and Forrest [15]. Also, Norwood and co-workers [16] provided a general overview of the E&L issues that can be associated with materials, particularly rubber products, used in contact with therapeutic products.

The species that can leach from rubber products can be reduced by undertaking measures associated with their compounding and their manufacture, and these aspects have been addressed by Toynbee [17] and Janssen [18].

An overview of the potential origin of the substances that can contribute to an E&L profile is provided in Sections 9.2.1–9.2.5 and this overview underlines how complex the situation is because, in addition to species associated with the polymerisation stage, the polymer itself and the additives incorporated into the final compound, there are chemical changes within the rubber that can take place for a number a reasons. Such chemical changes can include:

1. Reaction products from intermediates and byproducts in the steps leading up to the formation of the polymer.
2. Reaction products arising from the curing chemistry.
3. Products formed during antidegradant and stabiliser activity.
4. Thermal decomposition products of the polymer and additives.

Because it is often these same, relatively low-MW substances, that have the potential to migrate into food (Chapter 8) and that contribute to vulcanisation fume (Chapter 7), this review also has relevance to these subject areas and provides some important background chemistry.

9.2.1 Polymer

9.2.1.1 Polymerisation residues, monomers, oligomers and impurities

Polymerisation residues are largely unavoidable during polymer production, but can be controlled. For pharmaceutical, medical-grade polymers, stringent quality assurance controls are put in place and are often highlighted by the manufacturers and are a big selling point – although at a cost premium.

Polymerisation residues can originate from many sources, but may include:

- Monomers and oligomers
- Catalyst remnants
- Polymerisation solvents and surfactants
- Non-polymerisable impurities
- Impurities from the manufacturing plant and polymerisation vessel
- MW modifiers (e.g., chain transfer agents)

Among the polymerisation modifiers used in emulsion polymerisation processes [e.g., for styrene-butadiene rubber (SBR) or nitrile rubber (NBR)] are long-chain mercaptans (thiols) or disulfides. One such example is a mixture of C_{12} mercaptans with high tertiary (i.e., $R^1R^2R^3CSH$) content. This substance has been found to break down in the rubber during curing to the respective alkenes [19]:

$$\text{mixed dodecyl mercaptans} \rightarrow \text{mixed dodecenes}$$

Another example of a disulfide used for molecular control in polymerisation reactions is the thiuram disulfide compound used in the production of chloroprene–sulfur copolymers [20]. Copolymerising sulfur in this way improves the crystallisation resistance of the resulting polychloroprene (CR), and these types of CR (e.g., Neoprene® G types) contain residual thiuram disulfide that can react further in the subsequent curing reaction.

In natural rubber (NR), a number of species may remain from the biosynthesis or subsequent treatment steps (Chapter 11). Examples include a number of fatty acids (e.g., formic, acetic, propionic, lactic succinic and malic) [21] and the presence of these create a number of possibilities for the formation of amides.

All polymers have a MW distribution, the width of which is dependent upon the type of polymerisation technique used in their manufacture (e.g., free radical, ionic or condensation). This means that at the end of a polymerisation there will be both residual monomer species remaining in the polymer and oligomers composed of a variable number of monomer units. The relatively small MW of these monomers and oligomers means that they have the potential to migrate into food and pharmaceutical products in their unchanged form, or they can undergo reactions in the rubber and the reaction products can have the potential to migrate.

In some rubbers, for example, SBR, it is possible to detect residual monomers (e.g., styrene) at 100 ppm. In other cases (e.g., NBR), where the monomer is regarded as being more toxic, the permissible level of residual monomer is much lower, for example, acrylonitrile (ACN) has a maximum limit of 50 ppb for materials intended for food-contact use. Residual monomers, such as ACN, may also play a part in chemical reactions if additives such as accelerators are present. An example is the involvement of ACN in thiuram reactions during NBR vulcanisation [22] and the new products that can be formed in this way can include:

– 2-Cyanoethyl diethyldithiocarbamate from tetraethylthiuram disulfide (TETD).
– 2-Cyanoethyl dimethyldithiocarbamate from tetramethylthiuram disulfide (TMTD) or tetramethylthiuram monosulfide.

With regard to oligomers, the situation with pharmaceutical/medical products is different from food-contact studies due to the greater number of ways that species can enter into the blood stream and contact bodily fluids (e.g., absorption through a nasal membrane or direct, long-term implantation) and so a different range of MW may be of interest depending upon their chemical nature.

9.2.1.2 Degradation products of the polymer

Some polymers will begin to 'unzip' to their monomers if heated (e.g., those containing styrene such as SBR and styrene-butadiene-styrene). However, if such monomers are found, not all of them may have arisen from degradation because, as shown in Section 9.2.1.1, the residual monomer from the polymerisation may also be present.

In some cases, the presence of dissimilar units in the backbone can compromise stability and lead to the formation of new substances. For example, in the case of rubber copolymers:

– Acetic acid from ethylene-vinyl acetate copolymers.
– Hydrogen fluoride from vinylidene fluoride copolymers (e.g., FKM-type fluorocarbon rubbers).

In considering polymer breakdown due to heating, thermo-oxidative attack during processing or in service cannot be ruled out. In fact, thermo-oxidation may also occur to the products of thermal degradation, or an initial oxidative step may trigger the thermal degradation, so the two processes may not be easily separated. The products of thermo-oxidation may include aldehydes and ketones.

9.2.2 Additives in the unaltered state

9.2.2.1 Plasticisers and fillers

Plasticisers are used to soften a rubber product (e.g., by reducing the proportion of filler and polymer) and/or improve its low-temperature flexibility by reducing their glass transition temperatures. Such additives also assist in the processing and manufacture of rubber products by improving their rheological properties.

There are many types of plasticisers that can be used in rubber products, including hydrocarbon oils, esters and phosphates, and these substances are all relatively thermally stable and non-reactive, so will be present in E&L profiles in their unchanged form. Because they can be present at relatively high levels (e.g., 15% w/w) they may also contribute to the chemical species in these profiles due to the presence of the starting compounds that they are manufactured from. A typical example is 2-ethylhexanol, a starting material for the manufacture of di(2-ethylhexyl)phthalate.

Fillers, when added to rubbers, have the opposite effect to plasticisers in that they harden the compound, but they are also used for other important reasons (e.g., to increase tensile strength and tear strength). They are usually inorganic compounds [e.g., calcium carbonate ($CaCO_3$), silica and aluminium silicates] or different forms of carbon black (CB). Hence, they are essentially insoluble in the solvents and drug preparations used E&L studies, but may still have an impact (e.g., ppm levels of silica) on the profiles, particularly the 'extractables' profile. Impurities may also be present in fillers (e.g., hydrocarbons from CB, Section 3.6.4) and whether these contribute to an E&L profile depends upon their chemical nature and how well they are bound into the filler matrix. Overall, fillers can usually be expected to have much less of an impact on E&L than plasticisers.

9.2.2.2 Antidegradants, process aids and pigments

Although antidegradants (e.g., hindered phenolic compounds and amine compounds) are designed to be reactive and stabilise the rubber during processing by undergoing reactions, such as those involved in the chain breaking of free-radical reactions and the dissociation of hydroperoxide groups, they are added at a level that will mean that a high proportion of the initial stabiliser is present in a rubber in its unchanged form so that it can continue to protect the rubber in service. For this reason, this class of additive, and the ones stated below, can contribute to E&L profiles.

Flow promoters and lubricants improve the processibility of rubbers by reducing the friction between the rubber compound and the processing equipment. Examples of these types of additives are stearate compounds (e.g., calcium stearate and zinc stearate).

Although rubber compounds are usually black due to the very common use of CB filler, they can be other colours (e.g., white or green) and these compounds will contain pigments. Both organic pigments (e.g., azo dyes) and inorganic pigments (e.g., titanium dioxide) can be used in rubber compounds.

9.2.3 Breakdown and reaction products of cure system components

Curatives are designed to break down in rubbers during vulcanisation to initiate chemical reactions that will result in covalent bonds being formed as crosslinks between polymer molecules. This breakdown ensures that only low levels of the original curative compound(s) are present in the rubber. Indeed, in order to obtain the best service-life properties, and to 'clean-up' the rubber product, a post-cure operation is often employed after an initial curing cycle to ensure that the majority (i.e., >95%) of the curative has reacted within the rubber. Some examples of the principal curative systems used in rubbers, and the breakdown products/reaction products that may be formed as a result of their use, are given below.

9.2.3.1 Peroxides
The decomposition of peroxides proceeds at measurable rates usually at temperatures above 100 °C. The generic products of this decomposition, depending upon the structure of the original peroxide compound, include: alcohols, alkenes and ketones. The actual compounds formed, and their relative amounts, will depend on the peroxide used and the environment within which it has reacted (e.g., cure temperature, type of rubber).

For simplicity, the various dialkyl peroxides can be described in terms of two generic formulae:

$$X-O-O-X \qquad X-O-O-Y-O-O-X \tag{9.1}$$

where X is a monovalent organic group and Y is a divalent organic group.

In the case of most commercial peroxides, X is either a *tert*-butyl group or the cumyl group (e.g., di-*tert*-butyl peroxide, dicumyl peroxide). In the former case, the principal decomposition products include *tert*-butanol, iso-butene and acetone. In the latter case the principal decomposition products include cumyl alcohol, α-methyl styrene and acetophenone.

Divalent groupings of the type Y decompose in a similar fashion, but not necessarily to give the relatively simple molecules for which toxicological information is documented. Probably the simplest commercial example of this group is 1,1-*bis*(*tert*-butylperoxy)cyclohexane, in which the divalent grouping breaks down principally to cyclohexanone.

Diacyl peroxides decompose more readily than dialkyl peroxides and this breakdown is also amenable to catalysis (e.g., by tertiary amines such as N,N-dimethylaniline). The mode of breakdown is also different from that of dialkyl peroxides in that the acyloxy radical that is initially formed decomposes rapidly to an aryl radical by the loss of carbon dioxide. The principal byproducts are, therefore, those of this second radical and for aromatic types of these peroxides they include single-ring compounds (by hydrogen abstraction) and double-ring compounds (by radical coupling).

The thermal breakdown characteristics and products of different types of diacyl peroxide have been described by Davies [23]. Examples of the abstraction and coupling products of two common peroxides are as follows:
- Dibenzoyl peroxide – benzene (abstraction); biphenyl (coupling).
- *Bis*(2,4-dichlorobenzoyl) – *m*-dichlorobenzene (abstraction); 2,2′,4,4′-tetrachlorobiphenyl (coupling).

9.2.3.2 Sulfur donor-type cure systems

Asymmetric cleavage in molecules of the type X-S-S-X (to give X radicals) is believed to play a part in the sulfur donor curing reactions of agents such as 4,4′-dithiodimorpholine (DTDM) and TMTD [24]. Hydrogen abstraction or coupling may then follow to give a range of products and, using DTDM and TMTD to illustrate this, examples of these products can include:

$$DTDM \rightarrow morpholine \quad TMTD \rightarrow tetramethylthiourea$$

9.2.3.3 Accelerators used in sulfur-type cure systems

Many studies have addressed the mechanisms and byproducts of accelerator action in sulfur cures. Both sequential and competitive processes are involved, and the nature and yield of byproducts may be highly dependent on the proportions of the ingredients used in the rubber compound and on its thermal history. Amines and

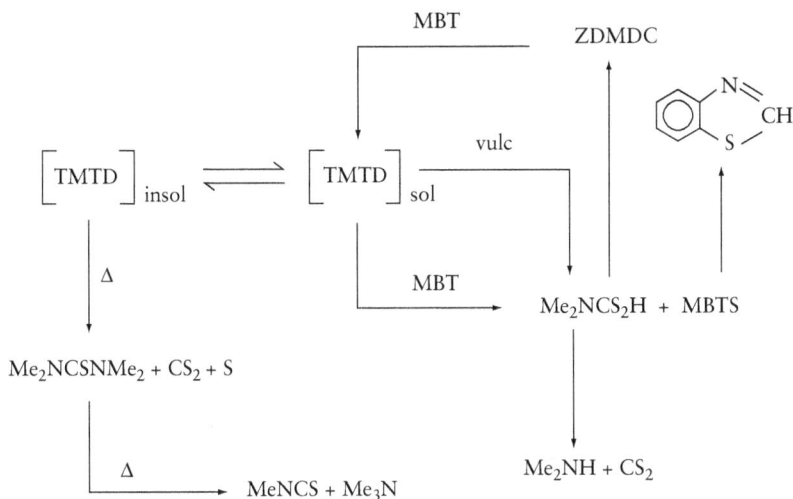

Figure 9.2: Some potential reactions associated with the accelerator TMTD [MBTS: 2,2′-dithiobis (benzothiazole) and ZDMC: zinc dimethyldithiocarbamate]. Reproduced with permission from M.J. Forrest and B. Willoughby in *Proceedings of the RubberChem Conference*, 9–10th November, Birmingham, UK, Rapra Technology Ltd, Shawbury, UK, 2004, Paper No.1. ©2004, Rapra Technology Ltd [26].

organo-sulfur compounds may be expected to be formed and the scheme shown in Figure 9.2 illustrates just some of the possibilities when TMTD is used as an accelerator [25]. The reaction scheme in Figure 9.2 also includes reactions with 2-mercaptobenzothiazole (MBT), an accelerator in its own right, but also a breakdown product of thiazole and sulfenamide accelerators, which are often used in cure systems in conjunction with TMTD.

The scheme shown above includes possible interactions with MBT. It should be recognised that MBT is an active agent in sulfenamide [e.g., N-cyclohexyl-2-benzothiazole sulfenamide (CBS), 2-(4-Morpholinothio)benzothiazole (MBS)] acceleration [27] and in these cases both MBT and MBTS are formed. The compound MBT is important because it is a 'special case' compound for E&L studies as defined by organisations like the PQRI (Section 9.3).

Amines are also formed in these types of reactions [e.g., cyclohexylamine from CBS and tertiary butylamine from N-*tert*-butyl-2-benzothiazole sulfenamide (TBBS)]. Indeed, amine formation extends beyond the thiuram and sulfenamide examples because many accelerator compounds are manufactured from amines and some reversion to these starting compounds appears possible. Examples of some of the amines formed in accelerator action are listed in Table 9.1 [26]. These are distinguished in terms of primary (RNH_2) or secondary (R_2NH) amine character.

Amines are also highly reactive and so they would be expected to react further. Possible reactions of this sort include:

Table 9.1: Primary and secondary amines from accelerators and curing agents.

Rubber compounding ingredient	Amine formed	Boiling point of amine (°C)
TMTD	Dimethylamine (2°)	7.4
TBBS	*tert*-Butylamine (1°)	44
TETD	Diethylamine (2°)	56
N,N-diisopropyl-2-benzothiazole sulfenamide	Di-isopropylamine (2°)	84
DTDM	Morpholine (2°)	128
N-oxydiethylene-2-benzothiazyl sulfenamide	Morpholine (2°)	128
CBS	Cyclohexylamine (1°)	135
TBTD	Dibutylamine (2°)	159
DPG	Aniline (1°)	184

(1°) = Primary amine (2°) = Secondary amine
DPG: Diphenyl guanidine
TBTD: Tetrabutylthiuram disulfide
Reproduced with permission from M.J. Forrest and B. Willoughby in *Proceedings of the RubberChem Conference*, 9–10th November, Birmingham, UK, Rapra Technology Ltd, Shawbury, UK, 2004, Paper No.1. ©2004, Rapra Technology Ltd [26]

- Amide formation with fatty acids:

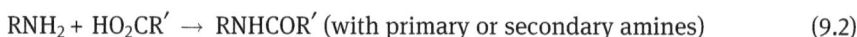

$$RNH_2 + HO_2CR' \rightarrow RNHCOR' \text{ (with primary or secondary amines)} \qquad (9.2)$$

- Nitrosamine formation with nitrosating agents (NOX):

$$R_2NH + NOX \rightarrow R_2N - N = O + HOX \text{ (with secondary amines only)} \qquad (9.3)$$

Note: nitrosamines are another PQRI 'special case' compound, Section 9.3.

Fatty acids for reaction (i) can include the components of commercial stearic acid (mainly octadecanoic, hexadecanoic and tetradecanoic) and any introduced into the rubber compound as impurities. Sources of NOX for reaction (ii) may include the nitrite salt bath medium used in continuous vulcanisation processes or impurities introduced by other ingredients of the rubber compound (e.g., CB).

Other examples of potential amine reactions include symmetrical and unsymmetrical thiourea formation from exchange reactions in mixed accelerator systems.

9.2.4 Breakdown and reaction products of antidegradants used in rubbers

9.2.4.1 Antioxidants
Antioxidants can be subdivided by their mode of action into 'primary' or 'secondary' types [28]. Primary antioxidants function by 'chain breaking' mechanisms

whereas secondary antioxidants function by a form of 'sacrificial oxidation'. The former are phenols and aromatic amines, whereas the latter are species such as phosphites and thioesters. In practice, both may produce oxidised products, and examples of some of the possibilities are listed in Table 9.2 [14].

Table 9.2: Some possible oxidation products of antioxidants.

Antioxidants	Possible reaction product
Butylated hydroxytoluene	2,6-Di-*tert*-butyl-1,4-benzoquinone
Trisnonylphenyl phosphite	Trisnonylphenyl phosphate
Phenyl-beta-naphthylamine	1,2-Napthoquinone-2-anil
6PPD	N-(1,3-dimethylbutyl)-N'-phenyl-1,4-*p*-benzoquinone diimine

6PPD: N-(1,3-dimethylbutyl)-N'-phenyl-phenylenediamine
Reproduced with permission from M.J Forrest in *Proceedings of Extractables & Leachables for Pharmaceutical Products*, 14–15th September, London, UK, Smithers Rapra Technology Ltd, Shawbury, UK, 2010, Paper No.1. ©2010, Smithers Rapra [14]

Another reaction to consider in this context is thermal decomposition of the antioxidant. For example, chemical compounds which are the products of Freidel–Crafts reactions may undergo a catalysed reversion which selectively de-alkylates the aromatic ring. Potential catalysts include acids, metal ions or amides [29]. Antioxidants made by Freidel–Crafts chemistry include hindered phenols and alkylated (or arylated) diphenylamines.

9.2.4.2 Antiozonants used in rubber products

Some rubbers are particularly prone to attack by ozone (e.g., diene-type rubbers such as NBR) and so antiozonants have to be incorporated into the compounds to ensure good ageing resistance in service. Antiozonants act by a sacrificial reaction with ozone [30]. A byproduct of their chemistry in rubber is side-chain oxidation and this can lead to the release of ketones as illustrated below for two antiozonants [isopropyl *p*-phenylenediamine (IPPD) and 6PPD] in common use in the rubber industry:

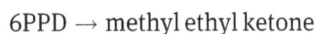

IPPD → acetone

6PPD → methyl ethyl ketone

9.2.5 Carbon black fillers and the formation of nitrosamines

The involvement of CB in nitrosamine formation has been controversial and the co-agents are believed to be the oxides of nitrogen adsorbed on the surface of the CB.

Their action in the formation of nitroamines is notably temperature-dependent and has been investigated by Willoughby and Scott [31]. This effect arises because not all the oxides are active in nitrosation. Dinitrogen trioxide (N_2O_3) and dinitrogen tetroxide (N_2O_4) are NOX, but nitrogen dioxide (NO_2) is not. The respective reactions with secondary amines can be depicted as:

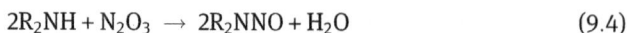

$$2R_2NH + N_2O_3 \rightarrow 2R_2NNO + H_2O \tag{9.4}$$

$$R_2NH + N_2O_4 \rightarrow R_2NNO + HNO_3 \tag{9.5}$$

The origin of the secondary amine from the use of certain accelerators is discussed in Section 9.2.3.3.

Given that N_2O_4 and NO_2 are interconvertible (N_2O_4 dissociates to NO_2 on heating and reforms on cooling), it is perhaps not surprising that nitrosamine formation in CB-filled, accelerated sulfur vulcanisates is dependent on thermal history [31].

9.3 Overview of extractable and leachable testing of rubber products

Mention has been made in Section 9.1 of the guidance documents and protocols concerning E&L that have been published in recent years. One of these, in addition to a number of other publications of relevance in this field, has been produced by the PQRI and they have produced a summary of their recommendations that address the following areas:

- Safety thresholds
- Analytical evaluation threshold (AET)
- Integration of safety evaluation
- Componentry
- Controlled extraction studies
- Leachables studies
- Routine extractables testing

The last of these areas is where three important special case compounds (i.e., poly-aromatic hydrocarbons, MBT and nitrosamines) are listed, along with the recommendation that they should be evaluated 'by specific analytical techniques and technology defined thresholds for leachable studies and routine extraction testing'. These three compounds are particularly relevant to rubber compounds (Section 9.2) and are often cited in the literature (Section 9.5).

Information on the potential origins of E&L from rubber products is provided in Section 9.2. It can be seen that these compounds tend to be low- or relatively low-MW species. Section 9.4 provides a case study detailing the approach used to carry out an extractables study on a rubber product (e.g., a seal) of the type that could be used in pharmaceutical packaging, a drug dispensing device or a medical device, and illustrates the possible substances that could be found. Once such an extractables profile has been obtained, the next step involves its examination by a toxicologist, who would pay particular attention to the substances detected at the highest concentrations. This toxicological assessment would highlight substances that should be specifically targeted as 'leachables' during the next stage in the process. In this leachables study, the analytical programme would be devised to include analytical techniques and protocols optimised and validated to provide accurate quantifications of these substances, often in the drug media. These data would then be used, along with the end-use information associated with the medicinal product, to calculate the maximum human daily exposure from normal clinical use to establish if safe daily intake levels of the leachables were being adhered to [33, 33]. A keynote paper was presented by Robison [34] of the FDA at the 2016 E&L USA Conference in which the design of E&L studies and the toxicological risk assessment of E&L in primary and secondary packaging were addressed. Case studies were covered in the presentation that related to the following three products and a guide to the best practices provided:

- Pre-filled auto-injector
- Flexible bags for use with intravenous drug products
- Pre-filled syringe

A number of articles have addressed the design of analytical strategies to assess the safety of E&L of pharmaceutical packaging, such as those produced by Van Dongen [35], Schulz [36] and Feilden [37, 38].

The complexity of the rubber products that were traditionally used in sealing applications in the medical/pharmaceutical industry [e.g., ethylene propylene diene monomer (EPDM) and nitrile] meant that potential migrants could be present in complex mixtures, which represented a challenge to the analyst (Section 9.4). Because of the complexity of conventional rubber compounds, there has been a trend over recent years for manufacturers to produce 'clean' rubbers for pharmaceutical and medical applications. These are designed to address the concerns of the industry by containing novel, 'low extractables' cure systems and fewer additives that can contribute to E&L profiles, as well as avoiding classes of materials (e.g., those that are animal-derived) and additives that could produce species of high concern, (e.g., MBT or nitrosamines). An example of these specialist clean rubbers is the bromobutyl rubbers, marketed by ExxonMobil Chemical Company, which are cured using proprietary cure systems.

With regard to changes in materials in the search of lower levels of E&L, other possibilities are available to manufacturers of packaging and sealing materials. One of these is to use thermoplastic elastomers (TPE) and high-performance TPE. Examples of these materials are commercially available and can be used for disposable devices, such as pre-filled syringes and vial stoppers [39–41]. Another option is to use elastomeric cyclic olefin copolymers (COC) as alternatives to TPE. These COC rubbers are regarded as having exceptional toughness and impact properties, as well as the transparency, dielectric properties, low E&L and chemical resistance that have come to be associated with this class of material. They can be used for a broad range of applications in the medical, pharmaceutical packaging, food packaging, optics and electronics sectors [42]. Despite the introduction of this new generation of clean rubbers, there is still a need at times, in order to satisfy particular criteria, for more traditional rubbers.

Irrespective of the type of rubber product being studied, the fact that a mixture of compounds will result from the extractables phase of the work means that gas chromatography (GC)–mass spectrometry (MS), both headspace GC–MS and solution injection (also called 'direct injection') GC–MS, is an ideal analytical technique for these mixtures with its low detection limit, high resolving power, and good identification and quantification capabilities. The ability of headspace GC– MS, particularly dynamic headspace GC–MS, to make an important contribution in the E&L screening of pharma and medical grade polymers has been reviewed by Jeanguyot [43].

Some problems, however, can be encountered with GC–MS and these occur if an analyte is thermally labile (degradation problems), highly polar (adsorption problems) or of a reasonably high MW (poor elution from the column). Because of these considerations, liquid chromatography (LC)–MS is employed alongside GC–MS, although care has to be taken because the mobile phase/separating column combination is crucial to ensuring good chromatographic performance with this technique. As yet, there are no comprehensive commercial libraries available to assist with identification and so there is not the same ease of identification that exists with GC–MS, for which comprehensive 70-eV fragmentation voltage libraries have been available for many years. The lack of a standardised approach means that this is likely to remain the case for a long time to come. Identification is, therefore, based on the analysis of standards and an understanding of the chemical processes (e.g., formation of adduct ions) that occur within the mass spectrometer.

Recent developments in instrumentation have provided the analyst with even more powerful tools for this type of work. In the case of GC–MS, developments in the GC oven area have enabled the GC×GC–time-of-flight (ToF)–MS instrument to be developed. This technique is essentially a two-dimensional chromatographic operation where both a non-polar and a polar column can be used in series for all the peaks in the chromatogram *via* a modulation device. Hence, mixed peaks (i.e., those that contain two or more species due to co-elution occurring with the first column), or an area of the chromatogram where a series of peaks are unresolved, can be separated into their individual components by use of the second column. ToF–MS is complimentary to this technology due to its very fast data collection rate and this, coupled with enhancements in the software package, provides an additional peak deconvolution capability. This can sometimes be referred to as the 'third dimension' in these instruments.

In the case of LC–MS instruments, there have been a number of developments that have made sophisticated versions of the technique accessible to test house laboratories. For example, the LC–trap, which is essentially an LC–MS×MS instrument in which fragment ions (or molecular ions) resulting from the first analysis, can be fragmented a second time to provide a further series of fragments. This provides the benefits of enabling structural information to be obtained on the original ions (good for identifying unknowns) and definitive assignments in complex LC chromatograms where peak resolution can be a problem and mixed spectra result. These benefits are important in LC–MS because molecular ions (or molecular adduct) ions are often formed and these are not always specific enough [e.g., where two compounds (or adducts) have the same MW]. The second fragmentation pattern, together with the initial ion(s) and the peak retention time, will assist the analyst in obtaining an unequivocal identification.

Another important development is the greater availability of LC–MS instruments equipped with a high-resolution accurate mass (HRAM) mass spectrometer, which again enables a greater number of substances in extractables studies to be identified and increases confidence in these assignments (Section 2.5.5). The performance of these instruments can also be improved by the upgrading of the LC component to one that has ultra-performance liquid chromatography (UPLC) status, the acronym for the technique then becoming UPLC–HRAM/MS. UPLC–HRAM/MS instruments are highly regarded for the rapid and sensitive screening of non-volatile organic compounds, such as antioxidants, lubricants and larger polymer degradation products. As mentioned in Section 2.5.5, there are options in the mode of operation of the MS component of LC–MS systems. Using these instruments in both the atmospheric pressure chemical ionisation (APCI) and electron ionisation (EI) mode offers the chance of generating the most complete profile because the former is more sensitive to compounds with low-to-medium polarity, whereas the latter is more sensitive to highly polar compounds.

Ion mobility-MS is another novel approach to screening for E&L components from packaging materials. Cabovska and Riches [44] and Cabovska [45] provided an introduction to this technique along with collisional cross-section (CCS). Ion mobility- MS measures the drift time of an ion and the application of a calibration provides the CCS, a key physiochemical property of a species. According to the authors, the advantages of CCS include a reduction in the number of false-positive data, an additional dimension of separation to obtain cleaner spectra in complex samples, and an improvement in reliability for the assignment of isomers, which are not separated chromatographically.

As is apparent from the above information, this is a very active area for developments and a recent presentation by D'Silva of ThermoFisher Scientific [46] provided an overview of the analytical capabilities that are available for E&L studies, together with an introduction to a new technology, called Q Exactive GC–MS from

AstraZeneca, for identifying and quantifying unknown impurities in support of pharmaceutical applications.

To complete the picture when carrying out screening work for extractables on a rubber product, a number of other analytical tests are usually employed. For example, a semi-quantitative 36-element scan is carried out on solvent extracts (e.g., hexane or isopropanol) of the product by inductively coupled plasma (ICP). The non-volatile residue (NVR) of these solvent extracts is also obtained by quantitatively drying down in an oven, and the bulk chemical characteristics of this NVR are obtained by recording an infrared (IR) spectrum using a Fourier–Transform infrared (FTIR) spectrometer.

To summarise, the analytical techniques that are usually brought together in a complementary way to characterise the extractables that are present in rubber products are listed, together with their contributions to the data obtained, in Table 9.3.

Table 9.3: Analytical techniques used in an extractables study and their contribution.

Analytical technique	Substances targeted
Headspace–GC–MS	Volatile organic compounds
GC–MS*	Semi-volatile organic compounds
LC–MS and LC– ultraviolet	Non-volatile organic compounds
ICP–OES or ICP–MS	Heavy metals and other elements
Ion chromatography	Anions and cations
IR	General chemical profile of dried, and weighed extracted material (i.e., NVR)

* Solution injection GC–MS
OES: Optical emission spectrometry

Additional examples of the general and specific analytical approaches used to undertake E&L studies to ensure that a comprehensive assessment results are provided in Section 9.5. This section also contains a number of published examples of the E&L work that has been carried out on rubber products, and provides specific information on the substances that the authors have identified and quantified.

9.4 Case study – potential and actual extractables from an nitrile rubber compound

9.4.1 Composition of the nitrile rubber compound

Section 9.2 discussed the origins of E&L species in rubber products. To provide a reasonably in-depth illustration of some of these substances, and how analytical techniques such as GC–MS and LC–MS can be employed to identify them in a screening study, an example of a relatively complex rubber formulation is provided in this section and the formulation of this rubber compound is shown in Table 9.4.

Table 9.4: Formulation for the NBR compound.

Ingredients	phr
Breon N36C60 [acrylonitrile-butadiene copolymer (nitrile) rubber]	100
ZnO	5
Stearic acid	2
HAF N330 (CB filler)	15
Translink 77 (calcined and surface-modified clay filler with vinyl functional surface modification)	15
Sulfur (MC grade)	1.5
DPG (curative)	0.15
MBTS (curative)	1.5
Rhenogran MPTD70 (curative)	0.29
Rhenofit PAN (antidegradant)	1
Wingstay® 29 [mixture of styrenated diphenylamines (antidegradant)]	1.43

HAF: High abrasion furnace
MPTD: Dimethyl diphenyl thiuram disulfide
PAN: N-phenyl-1-naphthylamine
Reproduced with permission from M.J Forrest in *Proceedings of Extractables & Leachables for Pharmaceutical Products*, 14–15th September, London, UK, Smithers Rapra Technology Ltd, Shawbury, UK, 2010, Paper No.1. ©2010, Smithers Rapra [14]

The NBR compound in Table 9.4 is a traditional one that meets European food-contact legislation with respect to its composition and could be used to produce gaskets and seals for pharmaceutical packaging and medical devices although, as mentioned in Section 9.3, these industries have started to move away from compounds like these in recent years due to their desire to use cleaner rubbers. However, due to its relative complexity, it provides a useful illustration of the extractables data that can be obtained using a combination of GC–MS and LC–MS.

9.4.2 Potential extractables

Given the ingredients in the NBR formulation (Table 9.4, Section 9.4.1), and considering the theoretical background to the origins of extractable species in rubber compounds presented in Section 9.2, the potentially extractables species, including reaction products and breakdown products, that might be detected by a combination of GC–MS and LC–MS are shown in Table 9.5.

Table 9.5: Possible extractable species from the cured NBR compound that could be detected by a combination of GC–MS and LC–MS.

Ingredient	Description	Potential extractable species
Breon N36C60	NBR polymer	ACN monomer, oligomers and dodecenes
Stearic acid	Activator	Mixture of carboxylic acids (C_{14}, C_{16} and C_{18}) Amides by reaction with amines listed within this table
DPG	Accelerator	Unchanged ingredient N,N'-diphenylthiourea Trisphenyl amino-1,3,5-triazine Diphenylcarbodiimide Aniline Ammonia N-phenyl-N-methyl-N'-phenylthiourea N,N'-diphenylurea Diphenylamine Phenyl isothiocyanate Phenyl isocyanate
MBTS	Accelerator	Unchanged ingredient MBT Mercaptobenzothiazole zinc salt Dibenzothiazyl monosulfide Benzothiazole (BT) Hydrogen sulfide Aniline Carbon disulfide Phenyl mercaptan
MPTD	Accelerator	Unchanged ingredient Methylphenyldithiocarbamic acid methylphenylammonium salt Methylphenyldithiocarbamic acid zinc salt N-methylaniline Carbon disulfide Carbonyl sulfide Sym-dimethyldiphenylthiourea N-methyldiphenylamine N,N-dimethylaniline Methyl isothiocyanate Phenyl isothiocyanate N-nitroso-N-methylamine 2-Cyanoethyl methylphenyl-dithiocarbamate
PAN	Antidegradant	Unchanged ingredient 1,2-Naphthoquinone-1-anil

(continued)

Table 9.5 (continued)

Ingredient	Description	Potential extractable species
Styrenated diphenylamines	Antidegradant	Unchanged ingredient
		Monostyrenated diphenylamine
		Distyrenated diphenylamine-N-oxide
		Diphenylamine
		Aniline
		Styrene

Note: the full chemical names for the acronyms of the ingredients are shown in Table 9.4 (Section 9.4.1)
Reproduced with permission from M.J Forrest in *Proceedings of Extractables & Leachables for Pharmaceutical Products*, 14–15th September, London, UK, Smithers Rapra Technology Ltd, Shawbury, UK, 2010, Paper No.1. ©2010, Smithers Rapra [14]

9.4.3 Gas chromatography–mass spectrometry analysis of the nitrile rubber sample

As mentioned in Section 9.3, it is common practice to use both headspace GC–MS and solution injection GC–MS to ensure as wide a range of extractable substances is

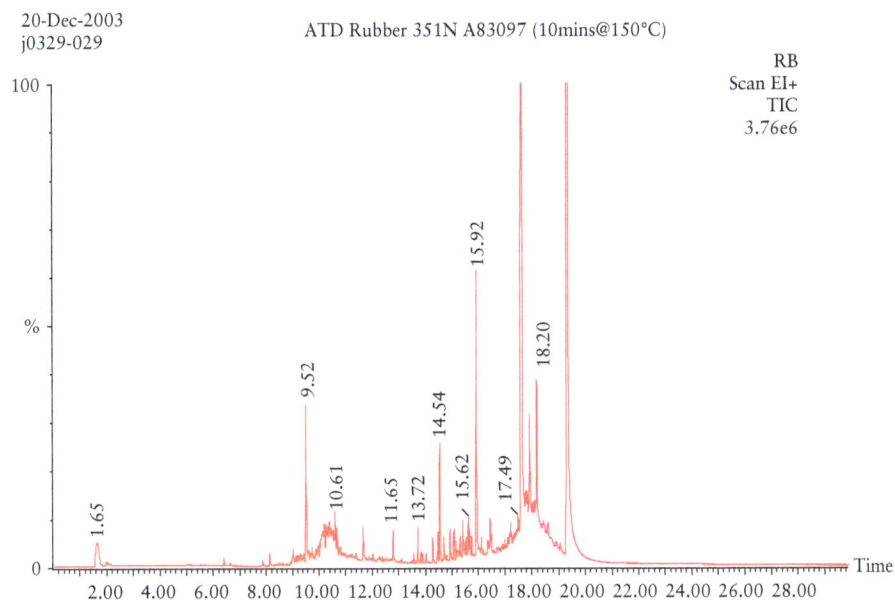

Figure 9.3: Headspace GC–MS chromatogram for the NBR sample (ATD: automated thermal desorption; TIC: total ion current). Reproduced with permission from M.J Forrest in *Proceedings of Extractables & Leachables for Pharmaceutical Products*, 14–15th September, London, UK, Smithers Rapra Technology Ltd, Shawbury, UK, 2010, Paper No.1. ©2010, Smithers Rapra [14].

detected as possible. Once a cured test sheet of the NBR compound in Table 9.4 was produced, samples for GC–MS analysis were prepared as shown below.

9.4.3.1 Headspace gas chromatography–mass spectrometry
A small amount (e.g., 0.2 g) of the sample was cut finely, placed into a headspace vial and then sealed. The vial was heated at 150 °C for ≈30 min and then the headspace injected into the GC–MS.

9.4.3.2 Solution injection gas chromatography–mass spectrometry
A small amount of the sample (e.g., 0.3 g) was cut finely, placed into a vial and 2 ml of acetone added. The vial was then placed into an ultrasonic bath and agitated for 30 min.

   The headspace GC–MS chromatogram and solution injection (acetone extract) GC–MS chromatogram for the vulcanised NBR sample are shown in Figures 9.3 and 9.4, respectively.

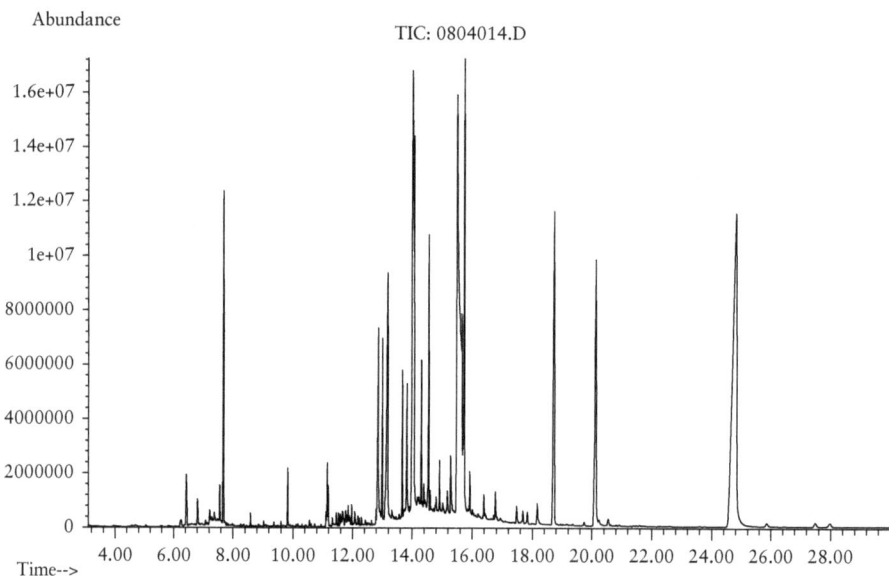

Figure 9.4: Solution injection (acetone extract) GC–MS chromatogram for the NBR sample. Reproduced with permission from M.J Forrest in *Proceedings of Extractables & Leachables for Pharmaceutical Products*, 14–15th September, London, UK, Smithers Rapra Technology Ltd, Shawbury, UK, 2010, Paper No.1. ©2010, Smithers Rapra [14].

9.4.4 Analysis of the nitrile rubber sample by liquid chromatography–mass spectrometry

An extract of the NBR sample was obtained using a solvent (methanol) that was compatible with the mobile phase (methanol/water gradient) of the LC–MS and injected into the instrument. As mentioned in Section 9.3, the use of LC–MS compliments GC–MS in that species which are thermally labile (e.g., MBTS) and/or of a high MW (e.g., species in oligomeric antidegradants like Wingstay® 29) can be identified. Because of the lack of commercial LC–MS libraries, identification was based on the analysis of standards and an understanding of the chemical processes (e.g., formation of adduct ions with the mobile phase) that occur within the mass spectrometer source.

The LC–MS chromatogram for the methanol extract of the NBR sample is shown in Figure 9.5.

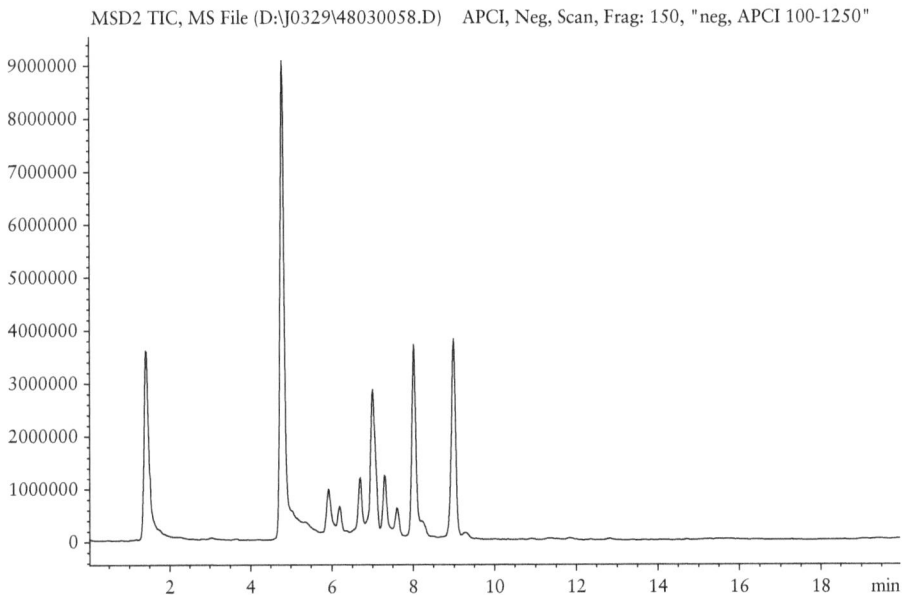

Figure 9.5: LC–MS chromatogram for the methanol extract of the NBR sample. Reproduced with permission from M.J. Forrest in *Proceedings of Extractables & Leachables for Pharmaceutical Products*, 14–15th September, London, UK, Smithers Rapra Technology Ltd, Shawbury, UK, 2010, Paper No.1. ©2010, Smithers Rapra [14].

9.4.5 Summery of the extractables detected by gas chromatography–mass spectrometry and liquid chromatography–mass spectrometry analysis

The data obtained on the 'extractables' from the NBR sample by both GC–MS and LC–MS revealed the presence of three main types of diagnostic (i.e., of traceable origin and hence useful) species:
– Original compounding ingredients in the unchanged form.
– Impurities and substances from compounding ingredients.
– Breakdown products and reaction products.

To illustrate these results, some selected examples of these three categories are shown below:
1. Examples of the original ingredients included:
    – MBTS.
    – PAN.
    – Styrenated diphenylamines (adducts of 1:1, 2:1 and 3:1 stoichiometry).
2. Examples of impurities in the original ingredients included:
    – ACN (residual monomer in the NBR polymer).
    – Vinyl cylohexene and cyanocyclohexene (polymerisation intermediates).
    – Alkylbenzenes (possible impurities in the CB).
    – 1-Octanethiol/dioctyl disulfide (polymerisation residues – substances used to control MW).
3. Examples of breakdown products/reaction products included:
    – Aniline.
    – N-phenylpalmitamide (aniline/$C_{15}CO_2H$ reaction product).
    – N-phenylstearamide (aniline/$C_{17}CO_2H$ reaction product).
    – BT.
    – Carbon disulfide.
    – Diphenylamine.
    – Dodecenes.
    – 2-Mercaptobenzothiazole.
    – N-methylaniline.

Not all of the theoretically predicted breakdown products (Section 9.4.2) were found by GC–MS and LC–MS. It may be that some others were present in amounts below the detection limits of the GC–MS and LC–MS analyses, or that the conditions for their formation were not favourable for this particular NBR compound with its particular thermal history.

In some cases, species were seen that are recognised derivatives of some of the predicted breakdown products. One example is the 1:1 condensation product of an aromatic primary amine ($ArNH_2$) with the solvent that was used to prepare the extract, acetone, i.e.:

$$ArNH_2 + O = CMe_2 \;\rightarrow\; ArN = CMe_2 + H_2O \qquad\qquad (9.6)$$

The actual species that was detected in this case was isopropylidene aniline, which is also called acetone 'anil'.

As is normal with this type of analytical work, a number of non-diagnostic, commonly detected species, were present in the extractables data (e.g., aliphatic hydrocarbons) that were of uncertain origins, hence the term 'non-diagnostic'.

One point worth mentioning is that this is essentially a trace component analysis, and the presence of components from external sources (e.g., vulcanisation fume species from other rubber samples being cured and/or stored nearby) in the assignment list should not be surprising. Thus, while the low-MW species within the rubber may have an impact on the environment around the rubber, that environment may also have an impact on the low-MW species within the rubber.

As mentioned in Section 9.3, it would be common practice during an extractables study to use other analytical techniques in order to provide a fuller picture of the substances that could potentially leach into a medicinal product during storage or in service. Such additional work would include:

–   Semi-quantitative multi-element (e.g., 36 element) analysis of the extraction media by ICP.
–   Determination of the NVR of the extraction media by quantitative drying down in pre-weighed glassware (e.g., a petri-dish).
–   Analysis of the NVR residue obtained above by FTIR. This could be carried out using transmission or attenuated total reflectance techniques or, if insufficient material was available for these approaches, an FTIR microscope would be used.

9.5 Determination of extractables and leachables

This section contains examples of E&L investigations and studies on rubber products and samples that have been published. It also contains articles and other publications that illustrate how the analytical techniques presented in Table 9.3, and discussed in Section 9.3, have been shown to be of great value in the determination of low-MW substances from rubber samples and so confirms their relevance to E&L work.

Ruberto [47] described how carrying out a comprehensive materials assessment prior to conducting an E&L study can be a useful tool to predict migrants, design a suitable controlled extraction study, facilitate data analysis, verify vendor supply, and evaluate supply chain integrity. Also included in this presentation made to the 2012 E&L Conference were typical extractables profiles that can be obtained from plastics and elastomers, and sources of unexpected migrants.

The PQRI publications and guidance documents for E&L work were introduced in Sections 9.1 and 9.3. Researchers within the pharmaceutical industry and associated networks have commented on how these relate to specific commercial situations. For example, Ball and co-workers [48] discussed the derivation and application of the PQRI safety thresholds for leachables in orally-inhaled nasal drug products. In their study, experimental data was included, the conversion of the thresholds to analytical thresholds in an E&L characterisation process described, and a summary of best practice recommendations provided.

Both GC×GC–ToF–MS and LC–MSxMS have been shown to be very useful in a number of publications, including those produced by Forrest and co-workers [49, 50] for the characterisation of low-MW constituents in rubber products. Their strengths in this area, particularly when they are combined, mean that they can make a significant contribution to E&L studies. Weil [51] of the LC–MS manufacturer Agilent Technologies discussed the impact of mobile phase, chromatography, ionisation mode (i.e., EI, APCI and atmospheric pressure photoionisation), and the MS interface on data obtained during E&L investigations. Also included in the study was specific information concerning the following aspects of the LC–MS analytical system:

- Effect of mobile phase buffers on ionisation and in-source fragmentation.
- Influence of ion sampling on detection limits.
- Suitability of automated data mining with statistical analysis as a simplified E&L differential analysis and multiple class comparison method.

Christiaens [52] considered the use of water for injection as an extraction solvent as well as the relationship between E&E studies and physico-chemical compendial tests. Also included within the presentation to the E&L for Pharmaceutical Products Conference in 2009 were data generated by GC–MS studies and, in relation to rubber samples, the results of a comparison between the sample preparation conditions in compendial tests.

Christiaens [53] also determined the amount of halobutyl and non-halogenated oligomers in pharmaceutical rubbers. The origin and formation of these species was examined and as well as the differences between their postulated chemical structures and actual structures. With respect to the determination of these substances, the analytical techniques and methods that could be used for their detection in E&L studies were described, and consideration was given to the toxicity of these substances and the need for analytical standards to assist with detection and validation work.

Kirchmeyer [54] carried out an in-depth analysis of the inorganic extractables from rubbers intended for pharmaceutical applications in response to the publication of the ICH Q3D 'Guideline for Metal Impurities' and the announcement that USP 232 'Elemental Impurities – Limits' and USP 233 'Elemental Impurities – Procedures' will replace USP 231 'Heavy Metals'. For the analytical work, atomic absorption spectrometry (zinc only), ICP–OES and ICP–MS were used and testing was carried out on water extracts from a range of starting materials (e.g., $CaCO_3$, talc, calcined clay, zinc oxide and silica) and rubber compounds that contained these ingredients. The results obtained using standard elemental test methods (e.g., International Organizations for Standardizations' ISO 8871-1:2003) demonstrated the ability of ICP–OES and ICP–MS to determine the inorganic E&L from raw materials and rubber compounds, and so could be used for the testing of rubbers intended for pharmaceutical applications. The results also demonstrated how the washing process used for the rubber samples could be responsible for increasing the concentration of some water-soluble substances and that brominated butyl rubber samples had very low levels of the elements listed in USP 232.

A number of authors have reported on E&L work carried out on rubber products used in the packaging of different types of pharmaceutical products. For example, Nielsen [55] presented at paper at the 2015 E&L Conference which addressed the identification of leachables in a topical product during its development using, as a case study, an ointment in a prototype packaging composed of polymers and rubbers. In the case of ophthalmic drug products, Houston [56] in a paper to the 2016 E&L USA Conference addressed the challenges, important study parameters, key variables and potential pitfalls in ophthalmic drug product E&L testing and looked at how assessments of product packaging changes could be conducted in an effective and time-efficient manner. Akapo [57] described a systematic approach to identify the E&L from three polymeric components used in the dispensing device for aqueous- based parenteral medications:

- Chlorobutyl rubber (vial stopper)
- COC plastic (syringe barrel)
- Bromobutyl rubber (stopper/tip cap)

The study also provided a discussion on the estimation of AET for the three materials, stated the results that were calculated in these cases, namely 0.15 µg/vial for the vial stopper and 0.03 µg/syringe for the syringe components (i.e., barrel, tip cap and plunger stopper), and provided details of the analytical methods that had been developed to monitor the levels of the leachables. Also, in regard to parenteral products, Janssen [58] reported on the impact that the extraction conditions and time had on the extractable results obtained on a chlorobutyl rubber compound. His study assessed the influence of a number of the variables, including:

- Surface area of the rubber sample.
- Extraction technique (e.g., reflux, Soxhlet, ultrasonication).
- Extraction time (0.5 to 8.0 h).
- Extraction solvent (isopropanol, water at pH 11 and water at pH 3).

A range of analytical techniques, similar to that shown in Table 9.3 (Section 9.3), was used to identify and quantify the extractable species. The results showed that extending the extraction time (e.g., from 1 to 8 h) had an influence on the quantitative data, but with evidence that the 8-h period was still not exhaustive. With respect to qualitative data, extending the extraction time was found to have less impact, with a similar set of extractables being detected.

The materials involved in the manufacture of e-cigarettes have been the subject of E&L testing. Kesselring [59] provided an overview of the regulatory requirements for medical devices together with a brief overview of their critical parts and e-cigarettes. In his study, test results for e-cigarettes extracted with solvent and gas were provided, as were data for medical devices made from silicone, SBR, polyurethane, polyethylene and polyvinyl chloride. Finally, safety concerns over e-cigarettes were considered.

Section 9.1 provided a definition of leachables and this presented the possibility that a sub-set of substances could be generated due to a reaction between leachables from the contact material and substances within the drug product. One of the scenarios that could allow this type of event to occur is when thiuram disulfide-type accelerators are used in the sulfur cure system for rubber stoppers. By undergoing thiol–disulfide exchange reactions, such compounds can undergo rapid reactions between themselves and thiols to yield a mixture of disulfide products. Corredor and co-workers [60] investigated the possibility that thiuram disulfides from rubber stoppers undergo exchange reactions with drug products containing pendant thiol groups. The results of their investigation enabled them to report the formation and identification of mixed thiuram disulfides of TMTD, TBTD and captopril (a thiol- containing drug). To detect these substances they developed a reversed-phase LC–MS method that utilised an ODS-AQ column (from YMC Co Ltd.) at 35 °C in a mobile phase consisting of a) (acetonitrile 20: water 80: trifluoroacetic acid: 0.1) and b) (acetonitrile 100: trifluoroacetic acid: 0.1). MS using molecular ions at 420.9 and 337.1 m/z, respectively, identified captopril–TMTD and captopril–TBTD disulfides. Possible structures for these compounds were provided and pH was found to have an effect on their formation, with the captopril–TMTD disulfide being enhanced at pH 6.0 and reaching a maximum concentration after 4.1 h. At lower pH values (e.g., 4.0 and 2.2), the mixed captopril adduct product was detected in solution after 20 h.

Rushing [61] also reported on the E&L profile that can be obtained from a rubber stopper, in this case in combination with a glass vial, and related to the early phase of the development of drug products. The study also covered the regulatory questions and actions that need to be addressed and undertaken, and the considerations that need to be included in the design of an E&L study.

The trend away from the use of traditional rubbers that contain accelerated sulfur cure systems (e.g., NR, NBR, EPDM) for the manufacture of pharmaceutical packaging components (e.g., stoppers) was introduced in Section 9.3. Wong of

ExxonMobil Chemical Company [62] emphasised this trend to improve cleanliness in the industry due to the increasing awareness of the potential impact that impurities have on drugs (e.g., deactivation) and, hence, the health of patients. As a consequence, the level of E&L components has become an important parameter with regard to pharmaceutical stoppers being qualified. Wong pointed out that the cleanliness of the stoppers themselves was determined by the manufacturing process, the cleanliness of each ingredient within the rubber compound (i.e., its purity), and various residual ingredients (e.g., curatives), and the reaction products and breakdown products that can be generated during processing, curing and ageing. Bromobutyl can be effectively cured *via* low levels of sulfur-free and/or zinc-free curing systems, so it is becoming a very popular rubber for pharmaceutical stoppers and seals. Specific examples of clean curing systems for bromobutyl rubbers were provided in the study and the extractables profiles of cured samples produced from these compounds described.

In addition to being considered for individual container/drug combinations, information on leachables are also of considerable importance when it comes to the items and components within a manufacturing site. Wilhelm [63] reported on the scope of the leachables project at Roche's Penzberg site, the leachables assessment strategy, the opportunities for assessing a large-scale production site, and laboratory testing. In addition, the study, which was presented at 2015 E&L Europe Conference, also included the leachable results obtained for pumps, flow meters, valve membranes, tubing systems and chromatography systems.

Cecchini [64] provided an introduction to the regulatory expectations with regard to extractables from the polymeric materials used in the storage of drug products or drug delivery systems. A number of other subjects were also covered in the study, including:
– An overview of biotech product manufacturing.
– The 'decision tree' used in the initiation of an E&L evaluation.
– An E&L study for a developmental product.
– Case studies involving an extraction study of the contact materials for a downstream processing product; an E&L study on a drug substance/active pharmaceutical ingredient storage container; and an extraction study on F&F product-contact materials.

There have been recent developments regarding E&L in the biopharmaceutical industry and these have been reviewed by Menzel [65] in a presentation that dealt with the BioPhorum Operations Group's (BPOG) best practice guide [66] BPOG *'Standardised Extractables Testing Protocol, Pharmaceutical Engineering* (2014)' for leachables that might originate from single-use manufacturing components. This best practice guide included risk assessment using a standard model approach, the design of a leachable study, the test methods that should be employed and applicable analytical considerations.

Single-use systems (SUS) are used increasingly in bio-manufacturing processes due to their multitude of advantages. However, one of the main concerns for their use are their E&L profile and this was addressed by Ding [67] of Pall Life Sciences. As a SUS provider, Pall Life Sciences have systemically studied the extractables from single-use components, including filters, tubing, connectors and bio-containers. A study presented by Ding summarised the test design and strategy employed by the company. This work enabled a library of extractable substances to be generated and this facilitated product-specific E&L evaluation during process validation work. Martin [68] reported on the regulatory requirements for disposable SUS for the manufacture of bio-pharmaceuticals and pharmaceuticals in the US and Europe and outlined the recommended approach of Bio-Process Systems Alliance for suppliers and users to address E&L from single-use process contact equipment.

References

1. M. Creese, *Pharmaceutical Manufacturing and Packing Sourcer*, 2012, **February**, 30.
2. *Guidance for Industry: Elemental Impurities in Drug Products: Guidance for Industry*, Food and Drug Administration, Rockville, MD, USA, 2016. https://www.fda.gov/downloads/Drugs/Guidances/UCM509432.pdf.
3. Product Quality Research Institute in *Thresholds and Best Practices for Extractables and Leachables* in *Proceedings of the 3rd PQRI/FDA Conference on Advancing Product Quality*, 22nd March, Washington, DC, USA, 2017.
4. USP 38–NF33 – General chapters on Extractables and leachables.
5. USP 1663 – Assessment of extractables associated with pharmaceutical packaging/delivery systems.
6. USP 1664 – Assessment of drug product leachables associated with pharmaceutical/ packaging delivery systems.
7. *ICH Guideline for Elemental Impurities Q3D*, ICH Harmonised Guideline, 2014.
8. M.N. Eakins in *Proceedings of Extractables & Leachables Europe Conference*, 10–12th November, London, UK, Smithers Rapra Technology Ltd, Shawbury, UK, 2015, Paper No.3.
9. J. Bones in *Proceedings of Extractables & Leachables USA Conference*, 9–11th May, Bethesda, MD, USA, Smithers Rapra, Akron, OH, USA, 2016, Paper No.14.
10. D.J. Ball in *Proceedings of Extractables & Leachables for Pharmaceutical Products Conference*, 13–14th May, Barcelona, Spain, Smithers Rapra Technology Ltd, Shawbury, UK, 2009, Paper No.7.
11. A. Feilden in *Proceedings of Extractables & Leachables USA Conference*, 15–17th May, Baltimore, MD, USA, Smithers Rapra, Akron, OH, USA, 2012, Paper No.1.
12. A. Feilden in *Update on Undertaking Extractable and Leachable Testing*, Smithers Rapra Technology Ltd, Shawbury, UK, 2011.
13. S. Bailey in *Proceedings of Extractables & Leachables Conference*, 5–6th March, Dublin, Ireland, Smithers Rapra Technology Ltd, Shawbury, UK, 2008, Paper No.14.
14. M.J. Forrest in *Proceedings of Extractables & Leachables for Pharmaceutical Products*, 14–15th September, London, UK, Smithers Rapra Technology Ltd, Shawbury, UK, 2010, Paper No.1.

15. T. Hulme and M.J. Forrest in *Proceedings of Extractables & Leachables for Pharmaceutical Products Conference*, 27–28th September, Dublin, Ireland, Smithers Rapra Technology Ltd, Shawbury, UK, 2011, Paper No.1.

16. D.L Norwood, M. Ruberto, F.L. DeGrazio, J. Castner and W. Boehringer, *Rubber and Plastics News*, 2009, **38**, 24, 15.

17. J. Toynbee in *Proceedings of Extractables & Leachables for Pharmaceutical Products Conference*, 14–15th September 2010, London, UK, Smithers Rapra Technology Ltd, Shawbury, UK, 2010, Paper No.19.

18. R. Janssen in *Proceedings of Extractables & Leachables for Pharmaceutical Products Conference*, 27–28th September 2011, Dublin, Ireland, Smithers Rapra Technology Ltd, Shawbury, UK, 2011, Paper No.10.

19. K.G. Ashness, G. Lawson, R.E. Wetton and B.G. Willoughby, *Plastics and Rubber Processing and Applications*, 1984, **4**, 69.

20. K. J. Saunder in *Organic Polymer Chemistry*, Chapman and Hall, London, UK, 1973.

21. R.C. Crafts, J.E. Davey, G.P. McSweeney and I.S. Stevens, *Journal of National Rubber Research*, 1990, **5**, 275.

22. B.G. Willoughby in *Health and Safety, in Developments in Rubber* Technology 4, Elsevier Applied Science, London, UK, 1987.

23. A.G. Davies in *Organic Peroxides*, Butterworth's London, London, UK, 1961.

24. V.A. Shershnev, *Rubber Chemistry and Technology*, 1982, **55**, 537.

25. B.G. Willoughby in *Proceedings of the Hazards in the European Rubber Industry Conference*, Shawbury, 28–29th September, Rapra Technology Ltd, Shawbury, UK, 1999.

26. M.J. Forrest and B. Willoughby in *Proceedings of the RubberChem Conference*, 9–10th November, Birmingham, UK, Rapra Technology Ltd, Shawbury, UK, 2004, Paper No.1.

27. A.Y. Coran in *Vulcanisation, in Science and Technology of Rubber*, Ed., F.R. Eirich, Academic Press, New York, NY, USA, 1978.

28. N. Grassie and G. Scott in *Polymer Degradation and Stabilisation*, Cambridge University Press, Cambridge, UK, 1985.

29. F. Pilati, S. Masoni and C. Berti, *Polymer Communications*, 1985, **26**, 280.

30. J. Pospisil in *Developments in Polymer* Stabilisation – 7, Elsevier Applied Science, London, UK, 1984.

31. B.G. Willoughby and K.W. Scott, *Rubber Chemistry & Technology*, 1998, **71**, 310.

32. R. Hawkins in *Proceedings of Extractables & Leachables for Pharmaceutical Products Conference*, 14–15th September, London, UK, Smithers Rapra Technology Ltd, Shawbury, UK, 2010, Paper No.2.

33. R. Hawkins in *Proceedings of Extractables & Leachables for Pharmaceutical Products Conference*, 13–14th May, Barcelona, Spain, Smithers Rapra Technology Ltd, Shawbury, UK, 2009, Paper No.1.

34. T.W. Robison in *Proceedings of Extractables & Leachables USA Conference*, 9–11th May, Bethesda, MD, USA, Smithers Rapra, Akron, OH, USA, 2016, Paper No.1.

35. W. Van Dongen, *Flexible and Plastic Packaging Innovation News*, 2008, **2**, 5, 174.

36. H. Schulz in *Proceedings of Extractables and Leachables Conference*, 5–6th March, Dublin, Ireland, Smithers Rapra Technology, Shawbury, UK, 2008, Paper No.9.

37. A. Feilden in *Proceedings of Extractables and Leachables Conference*, 5–6th March, Dublin, Ireland, Smithers Rapra Technology Ltd, Shawbury, UK, 2008, Paper No.8.

38. A. Feilden in *Proceedings of Extractables & Leachables for Pharmaceutical Products Conference*, 13–14th May, Barcelona, Spain, Smithers Rapra Technology Ltd, Shawbury, UK, 2009, Paper No.5.

39. R. Varma, Rubber World, 2008, **238**, 3, 17.

40. Anon, *High-Barrier TPE for Medical Uses* in Plastics Technology, 2007, **53**, 8, 23.
41. *British Plastics and Rubber*, 2009, **February**, 22.
42. P.D. Tatarka in *Proceedings of the 188th ACS Fall Technical Meeting*, 13–15th October, Cleveland, OH, USA, ACS Rubber Division, Akron, OH, USA, 2015, Paper No.47.
43. V. Jeanguyot in *Proceedings of Extractables & Leachables for Pharmaceutical Products Conference*, 27–28th September, Smithers Rapra Technology Ltd, Shawbury, UK, 2011, Paper No.14.
44. B. Cabovska and E. Riches in *Proceedings of Extractables & Leachables Europe Conference*, 10–12th November, London, UK, Smithers Rapra Technology Ltd, Shawbury, UK, 2015, Paper No.6.
45. B. Cabovska in *Proceedings of Extractables & Leachables USA Conference*, 9–11th May, Bethesda, MD, USA, Smithers Rapra Technology Ltd, Akron, OH, USA, 2016, Paper No.17.
46. K. D'Silva in *Proceedings of Extractables & Leachables USA Conference*, 9–11th May, Bethesda, MD, USA, Smithers Rapra Technology, Akron, OH, USA, 2016, Paper No.16.
47. M.A. Ruberto in *Proceedings of Extractables & Leachables USA Conference*, 15–17th May, Baltimore, MD, USA, Smithers Rapra, Akron, OH, USA, 2012, Paper No.8.
48. D.J. Ball, D.L. Norwood and L. Nagao in *Proceedings of Extractables& Leachables Conference*, 5–6th March, Dublin, Ireland, Smithers RapraTechnology Ltd, Shawbury, UK, 2008, Paper No.3.
49. M. Forrest, S. Holding and D. Howells, *Polymer Testing*, 2006, **25**, 63.
50. M.J. Forrest, S.R. Holding, D. Howells and M. Eardley in *Proceedings of Silicone Elastomers*, 19–20th September, Frankfurt, Germany, Smithers Rapra Technology, Shawbury, UK, 2006.
51. D.A. Weil in *Proceedings of Extractables & Leachables USA Conference*, 9–11th May, Bethesda, MD, USA, Smithers Rapra Technology, Akron, OH, USA, 2016, Paper No.15.
52. P. Christiaens in *Proceedings of Extractables and Leachables for Pharmaceutical Products Conference*, 13–14th May, Barcelona, Spain, Smithers Rapra Technology, Shawbury, UK, 2009, Paper No.10.
53. P. Christiaens in *Proceedings of Extractables & Leachables Conference*, 12–13th December, Vienna, Austria, Smithers Rapra Technology Ltd, Shawbury, UK, 2012, Paper No.15.
54. H. Kirchmeyer in *Proceedings of Extractables & Leachables Conference*, 12–13th December, Vienna, Austria, Smithers Rapra Technology Ltd, Shawbury, UK, 2012, Paper No.14.
55. L.D. Nielsen in *Proceedings of Extractables & Leachables USA Conference*, 12–14th May, Bethesda, MD, USA, Smithers Rapra Technology Ltd, Akron, OH, USA, 2015, Paper No.8.
56. C.T. Houston in *Proceedings of Extractables & Leachables USA Conference*, 9–11th May, Bethesda, MD, USA, Smithers Rapra, Akron, OH, USA, 2016, Paper No.11.
57. S.O. Akapo in *Proceedings of Extractables & Leachables USA Conference*, 7–9th May, Providence, Rhode Island, USA, Smithers Rapra Technology, Akron, USA, 2013, Paper No.10.
58. R. Janssen in *Proceedings in Extractables & Leachables Europe*, 10–11th December, Smithers Rapra Technology Ltd, Shawbury, UK, 2013.
59. A. Kesselring in *Proceedings of Extractables & Leachables USA Conference*, 9–11th May, Bethesda, MD, USA, Smithers Rapra, Akron, OH, USA, 2016, Paper No.4.
60. C. Corredor, F.P. Tomasella and J. Young, *Journal of Chromatography A*, 2009, **1216**, 1, 43.
61. W. Rushing in *Proceedings of Extractables & Leachables USA Conference*, 12–14th May, Besthesda, MD, USA, Smithers Rapra Technology Ltd, Akron, OH, USA, 2015, Paper No.11.
62. W.K. Wong in *Proceedings of Extractables and Leachables for Pharmaceutical Products Conference*, 14–15th September, London, UK, Smithers Rapra Technology Ltd, Shawbury, UK, 2010, Paper No.22.
63. J. Wilhelm in *Proceedings of Extractables & Leachables Europe Conference*, 10–12th November, London, UK, Smithers Rapra Technology Limited, Shawbury, UK, 2015, Paper No.9.

64. I. Cecchini in *Proceedings of Extractables & Leachables Europe Conference*, 10–12th November, London, UK, Smithers Rapra Technology, Shawbury, UK, 2015, Paper No.7.
65. C. Menzel in *Proceedings of Extractables & Leachables Europe Conference*, 10–12th November, London, UK, Smithers Rapra Technology Ltd, Shawbury, UK, 2015, Paper No.10.
66. *Standardised Extractables Testing Protocol, Pharmaceutical Engineering*, BioPhorum Operation Group, Sheffield, UK, 2014.
67. W. Ding in *Proceedings of Extractables and Leachables for Pharmaceutical Products Conference*, 27–28th September, Dublin, Ireland, Smithers Rapra Technology, Shawbury, UK, 2011, Paper No.12.
68. J. Martin in *Proceedings of Extractables and Leachables for Pharmaceutical Products Conference*, 13–14th May, Barcelona, Spain, Smithers Rapra Technology Ltd, Shawbury, UK, 2009, Paper No.3.

10 Analysis of surface blooms and contaminants

10.1 Introduction

Rubber products and compounds can fail in service, or prior to service, for a number of reasons and some of these have been covered by other sections in this book, such as incorrect composition (Chapters 3–5).

This section deals specifically with two other important reasons for failure:
1. Surface blooms
2. Contamination

The first of these has caused problems for rubber technologists since the beginning of the industry and they can range from a displeasing aesthetic appearance to serious product failure. An example of the latter is when a surface bloom interferes with the rubber bonding to another substrate, such as metal. The second can by its nature take many forms and the form that the contamination is in (e.g., liquid, solid particle, fibre) will have a significant influence on the analytical approach and instrumentation that will be employed to identify it and, hopefully, by doing so, trace its source.

Section 10.1 provides an overview of blooming on the surface of rubbers, covering aspects of its science, such as the reasons for its formation, to the analytical techniques and methodologies that have been found to be the most effective in characterising it. Some ways in which blooms can be avoided are also presented in this section. Section 10.2 provides examples of the most commonly encountered contaminants and the ways in which they can be identified, both *in situ* and after removal from the rubber compounding ingredient, compound or product.

10.2 Surface blooms

10.2.1 Introduction

The appearance of a bloom on the surface of a rubber product or compound is a relatively common phenomenon that can lead to a number of problems. It is often, therefore, the subject of an analytical investigation in order to obtain information on its identity and, hence, the possible reason(s) for its appearance. To facilitate this investigation, as with most areas of rubber analysis, it is useful to have an understanding of what causes blooming, the substances that may be present in a bloom, and the reasons for their formation. Therefore, in addition to describing the analytical techniques and methods that have been used to characterise blooms, this

https://doi.org/10.1515/9783110640281-010

section also provides a brief overview of the background and chemistry associated with their formation.

It is useful to start with a definition and one, relatively narrow, definition of a surface bloom is '*Formation on the surface of an uncured or cured rubber of a thin coating of a compounding ingredient which has limited solubility in the rubber matrix*'.

As will be seen during the course of this section, this is a narrow definition for a number of reasons, one of which is that a bloom can be a mixture of compounding ingredients, or a reaction product of a compounding ingredient.

A bloom can appear during storage or service, and it can be an aesthetic problem which effectively categorises the product as a 'failure' as much as other deficiencies, such as a poor cure state or air blisters. Blooming can also cause bonding problems in composite systems (e.g., rubber to metal) by interfering with the wetting of the adhesive, and it can cause other problems, such as contaminating a nearby surface or altering the profile of an extractable and leachables investigation (Chapter 9).

In general, for a substance to bloom it must have a relatively low degree of solubility in a rubber matrix, and be present in excess of that solubility. This excess will be present as discrete particles dispersed throughout the matrix and the molecules within these particles will migrate (i.e., diffuse) to the surface of the rubber. This migration is dependent upon the:
- Overall polarity of the rubber matrix
- Solubility of the substance in the rubber
- Concentration of the substance in the rubber
- Temperature
- Time

At a given temperature, the solubility of a substance is influenced by its molecular weight (MW) and chemical structure. Also, in some cases, as the temperature drops the solubility of a substance will be reduced and the occurrence of a bloom made more likely. Typical examples of some of the rubber ingredients that have been known to bloom are hydrocarbon waxes and antidegradants, and other examples are provided in Section 10.2.2.1.

Blooming is not always undesirable. For example, the blooming of paraffinic waxes onto the surface of a rubber product provides a physical barrier to ozone attack and assists in the migration of antidegradants to the surface, where they can protect the rubber from ozone and oxygen. It is also possible for reaction products and/or breakdown products of rubber ingredients such as peroxide curatives, accelerators or antidegradants to bloom to the surface of vulcanised products.

Some examples of a white surface bloom on rubber components are shown in Figures 10.1–10.3. Figure 10.3 illustrates how mechanical pressure, in this case due

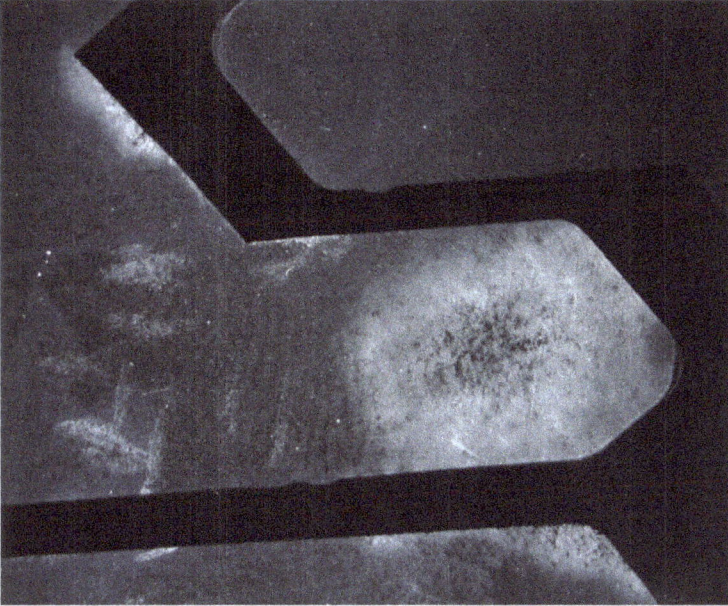

Figure 10.1: Example of a white bloom on the surface of a cured rubber product.

Figure 10.2: Example of a white bloom on the surface of uncured rubber extrudate.

Figure 10.3: Illustration of how mechanical pressure (finger impression) can encourage a bloom, in this case a white bloom, to appear on the surface of a cured rubber product.

to a finger, can cause or exacerbate a bloom due to a 'seeding' effect. Mechanical pressure may also result in localised reactions that can cause a bloom.

In addition to 'true' blooms which can be said to conform to the definition and description above, there are also a number of other examples of bloom-like phenomena and these are included in Section 10.2.2.

10.2.2 Different types of bloom or bloom-like phenomena

The different types of bloom or bloom-like phenomena can be classified and described as shown below.

10.2.2.1 True blooms

True blooms can occur if a substance within the rubber compound has limited solubility in the rubber matrix and is present in amounts that exceed this solubility. This substance can then migrate to the surface and crystallise out onto it when the product cools down after the vulcanisation process. Examples of substances that can bloom in this way include:

- Paraffinic waxes.
- Antioxidants (e.g., p-phenylenediamines).
- Accelerators [e.g., zinc diethyldithiocarbamate (ZDEC), 2-mercaptobenzothiazole].
- Elemental sulfur.
- Zinc salts of carboxylic acids (e.g., zinc stearate).
- Reaction products of sulfur cure systems (e.g., accelerators).
- Reaction products of other cure systems (e.g., peroxide cure systems).
- Retarders, lubricants and colourants.

Examples of this type of bloom may dissolve back into the rubber matrix if the product is heated.

10.2.2.2 Modified blooms
Modified blooms can result from the presence of a chemical substance in the rubber matrix which reacts deliberately, or accidentally, with the environment. Examples of substances which can create modified blooms include:
- p-Phenylenediamine antidegradants if used in excess.
- Zinc stearate when in excess – creates an initial 'oily' bloom which can then solidify if moisture is present in the atmosphere.

10.2.2.3 Pseudo-blooms
Pseudo-blooms can occur if degradation of the rubber surface has occurred to expose a filler (e.g., an inorganic filler) that has been incorporated into the rubber. This results in the appearance of a pseudo-bloom which can look like 'frosting'.

Inorganic fillers, such as calcium carbonate, are not soluble in the rubber matrix and so cannot migrate to the surface and form true blooms, but their particles can appear on the surface if the original rubber surface has been removed by degradation.

10.2.2.4 Surface contamination
An undesirable deposit on the surface of a rubber component can appear due to a number of possible events:
- Inorganic or organic substances in washing or rinsing solutions.
- Silicone-release agents giving a greasy bloom which attract airborne materials.
- Airborne dusts or mists in a factory atmosphere.
- Handling the product with contaminated gloves.

10.2.2.5 Staining or discolouration

These effects can result if substances on the surface of rubber products react with substances in other products or materials (e.g., textiles, plastics) that they come into contact with. Examples include:

- Elemental sulfur or dithiocarbamates in the presence of copper or iron.
- Zinc oxide (ZnO) in the presence of lead.
- When copper or iron are present in a contact medium.
- Reactions of breakdown products of phenolic-type antioxidants.
- Reactions of breakdown products of p-phenylenediamines.

10.2.2.6 Hazing

Hazing is rare and is due to the cloudy appearance within a transparent rubber product, or at the surface of it, causing opacity. It can appear as a result of insoluble solid particles or, rarely, droplets of a liquid. Examples include the incorrect grade or level of ZnO, calcium oxide or other poorly dispersed ingredients.

10.2.3 Analysis and identification of blooms

10.2.3.1 Simple investigative tests

There are a number of simple tests that can be applied to a rubber to obtain some information on the substance(s) that may be causing a bloom. Two examples of such tests are those that involve a change in colour taking place on the surface of the rubber, and the affect that heating a rubber product exhibiting a bloom has on the appearance of that bloom. Examples of these two classes of test are listed below.

Surface colour change:

- Reactions occurring due to exposure to light:
  - Formation of a grey/brown colour indicates an amine antioxidant.
  - Formation of a pink colour indicates a phenolic-type antioxidant.
- Patchy and inconsistent colouration over the surface:
  - Indicates that rubber ingredients on the surface have reacted with chemicals that they have contacted in the environment.
- Reactions occurring on contact with a metal:
  - Copper and iron will react with dithiocarbamates on the surface resulting in coloured salt products.

Heating the sample:

- Does the bloom disappear upon heating?
  - Yes – this result indicates that it is a true bloom and the substances have re-dissolved in the rubber.
  - No – this result indicates that it is a 'modified bloom' or a pseudo-bloom due to surface oxidation or the reaction of an ingredient on the surface.

10.2.3.2 In situ *tests*

One of the first approaches taken if a bloom needs to be investigated is to examine it using optical microscopy. This initial assessment will enable some information on the colour, structure, complexity, depth and overall nature of the bloom to be obtained. The close-up photographic images of a white bloom shown in Figures 10.1–10.3 illustrates the results that can be achieved using this approach.

It is also possible to analyse rubber blooms *in situ* on the surface of a rubber compound using various surface specific analytical techniques, but both Fourier-Transform infrared (FTIR) spectroscopy and energy-dispersive X-ray spectroscopy (EDX) are particularly useful, especially if used in combination.

With FTIR spectroscopy, an attenuated total reflectance (ATR) spectroscopy sampling accessory can be used and, although an amount of the infrared (IR) radiation will penetrate through the bloom into the surface of the rubber, it is possible to obtain a spectrum. Unfortunately, the fact that the majority of rubber samples are carbon black-filled (a substance that absorbs IR light and so reduces the strength of the reflected response) is another limitation of this technique. Use of an ATR crystal that reduces the penetration depth of the IR radiation (e.g., germanium crystal) can improve the surface specificity of the technique and improve the quality of the data obtained.

An example of where ATR–FTIR has been used to identify a bloom is illustrated by Figure 10.4. This work was conducted on a rubber glove exhibiting a bloom on its surface. There are two ATR–FTIR spectra in Figure 10.4, the top one shows an area of the glove which showed no bloom ('control' spectrum) and the one below is the spectrum recorded of an area with the bloom ('sample' spectrum). The difference in the IR data between the control and sample spectra are consistent for the accelerator ZDEC and it is possible that this ingredient has bloomed to the surface because it has exceeded its solubility within the rubber matrix.

Using the EDX technique will enable both electron micrographs of the bloom to be obtained, which will provide important pictorial images of the bloom that are useful for assessing its structure (e.g., evidence for crystals), and semi-quantitative elemental data. When carrying out this work, a baseline determination from a clean, freshly cut rubber surface is obtained in order to gauge the degree of enhancement of each element within the area of the bloom.

Table 10.1 shows the EDX data obtained on the yellow areas of bloom present on the surface of a moulded rubber component. Duplicate sets of data are displayed for both a freshly cut surface and the surface exhibiting the yellow bloom. These show a significant enhancement in the response for sulfur in the surface with the bloom. This relatively complex set of data also shows the importance of obtaining background data to aid interpretation. The likely source(s) of the elements in Table 10.1 are:

- Carbon – base rubber and any organic ingredients.
- Oxygen – any ingredients that is an oxide or contains oxygen.
- Aluminium – possibly from a silicate-type filler.

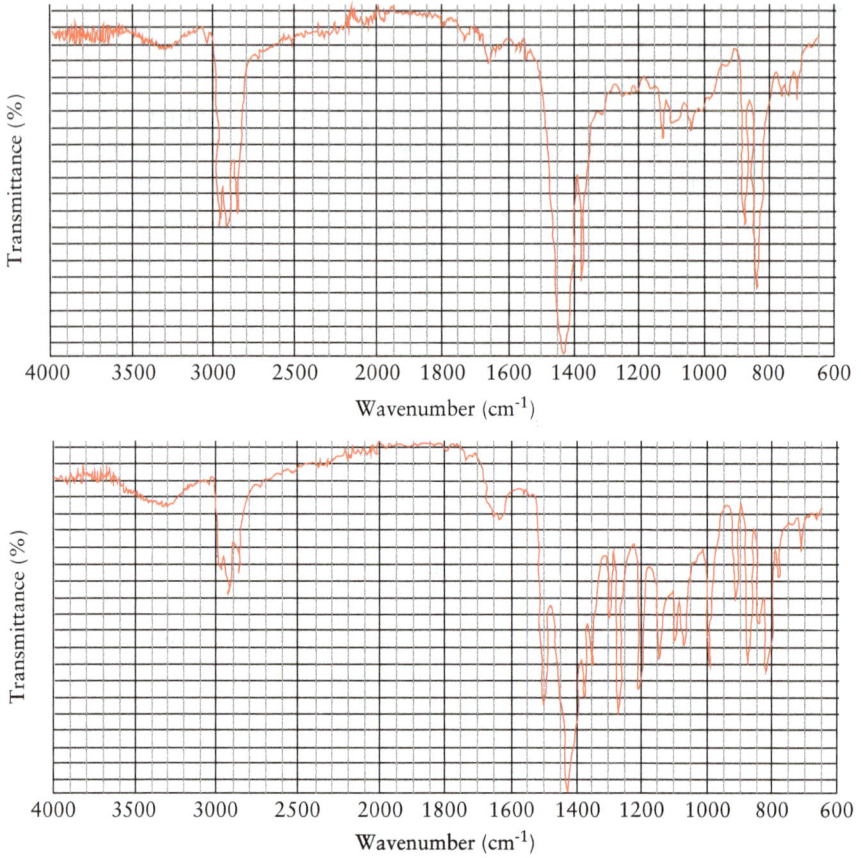

Figure 10.4: ATR–FTIR spectra of glove surfaces. (a) control surface (above) and (b) surface with ZDEC bloom (below).

Table 10.1: EDX data obtained on the yellow areas of bloom present on the surface of a moulded rubber component.

Sample	Semi-quantitative elemental data (%)						
	Carbon	Oxygen	Aluminium	Silicon	Sulfur	Titanium	Zinc
Freshly cut surface (1)	65.0	21.2	5.4	5.1	1.7	1.1	1.1
Freshly cut surface (2)	64.0	22.0	5.5	5.2	1.7	1.1	1.0
Yellow bloom (1)	46.0	2.7	0.5	1.0	49.2	0.1	0.5
Yellow bloom (2)	53.1	3.4	0.9	1.3	41.0	0.2	0.5

- Silicon – possibly from a silicate-type filler.
- Titanium – probably from the pigment titanium dioxide.
- Zinc – probably from curing co-agent ZnO.
- Sulfur – elemental sulfur and any other compounds (e.g., accelerators) that contain sulfur.

In the case of this particular bloom, the images provided by EDX were extremely useful in identifying its cause. Two of the images obtained on the yellow bloom *in situ* are shown in Figures 10.5 and 10.6. The close-up image in particular (Figure 10.6) shows crystalline structures which, together with the discovery of sulfur (Table 10.1), and its yellow colour, confirmed that the bloom on the moulded rubber component was due to elemental sulfur.

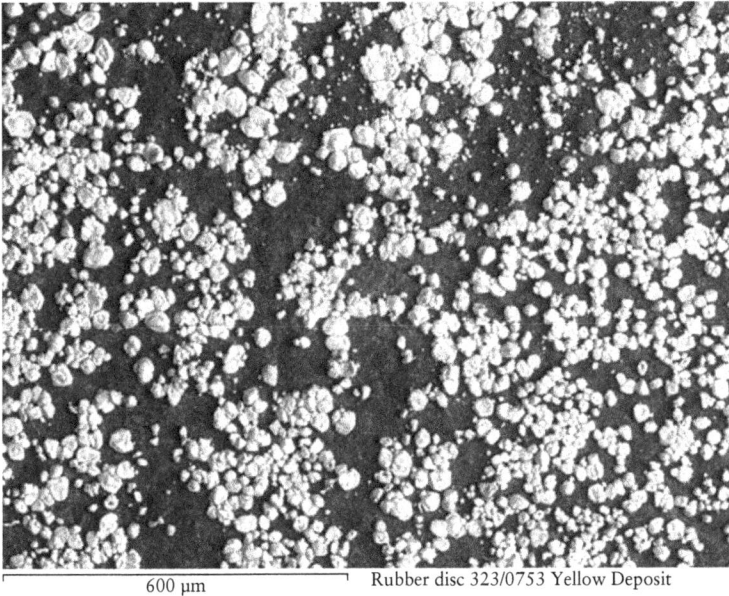

600 μm   Rubber disc 323/0753 Yellow Deposit

Figure 10.5: Scanning electron micrograph of a yellow bloom on the moulded rubber component.

Because it does not provide information on the actual compounds that the elements in a bloom are present in, it is unlikely that EDX data alone will positively identify the culprit species. As mentioned above, ATR–FTIR can encounter problems due to absorption of IR radiation. Fortunately, there is a viable alternative to the ATR–FTIR route: FTIR microspectrometry. Investigative work using this technique can be conducted using two main approaches, with IR spectral information either being obtained *in situ* using the reflectance mode (statically or in

60 μm   Rubber Disc 323/0753 Yellow Deposit

Figure 10.6: Close-up scanning electron micrograph of a yellow bloom on the moulded rubber component.

the 'mapping' configuration), or on a small amount of the bloom that has been removed from the surface of the rubber. The latter, which can be regarded as an 'indirect' approach, is dependent upon there being sufficient bloom material to function effectively but, due to the sensitivity of the technique, this is not usually a problem.

With the indirect approach, a small amount of the bloom is carefully taken from the surface of the rubber using a sharp instrument (e.g., scalpel tip), placed on the window of the sampling accessory in the FTIR microscope, and a transmission FTIR spectrum recorded. This FTIR spectrum, in combination with the EDX data that have been obtained on the bloom, is usually very effective at characterising it, even if more than one compound is present.

If problems of identification are still encountered, it is possible to remove the bloom using a dry-swab technique, which avoids contamination of the bloom constituents by solvent-soluble extractables in the rubber product, and the bloom material taken off the swab using a general-purpose solvent (e.g., chloroform). Having prepared a solution of the bloom, it is then possible to analyse it using a sensitive analytical technique, such as gas chromatography (GC)–mass spectrometry (MS). Such an approach has the benefit of utilising a chromatographic separation step and so this aids the characterisation of complex blooms.

The three-dimensional (3D) GC×GC–time-of-flight (ToF)–MS chromatogram of the species present in an 'oily' bloom removed from the surface of a diene rubber sample is shown in Figure 10.7. The assignments obtained for the species present in this chromatogram, together with its overall configuration, suggested that the bloom was a hydrocarbon oil. This conclusion was supported by the data in the FTIR spectrum obtained on the bloom (Figure 10.8) and its oily appearance on the

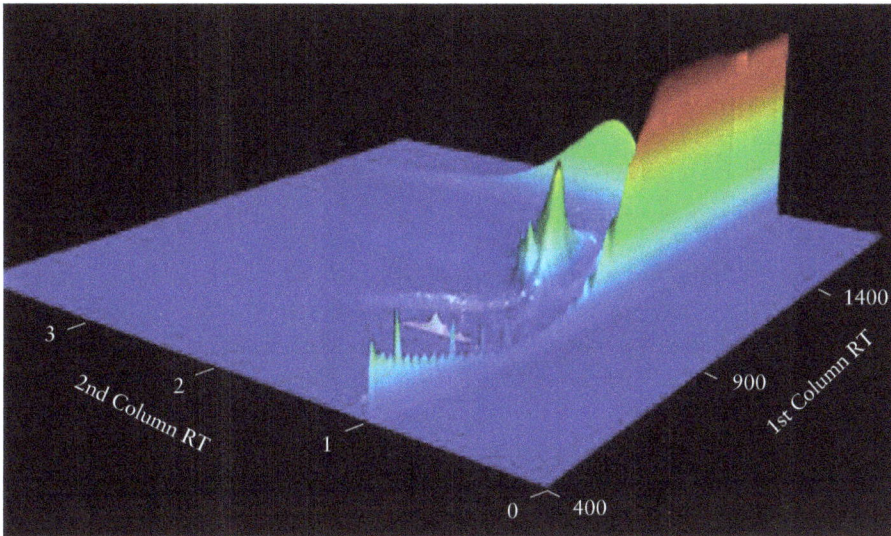

Figure 10.7: 3D GC×GC–ToF–MS chromatogram of the species present in an 'oily' bloom on the surface of a diene rubber sample (RT: retention time).

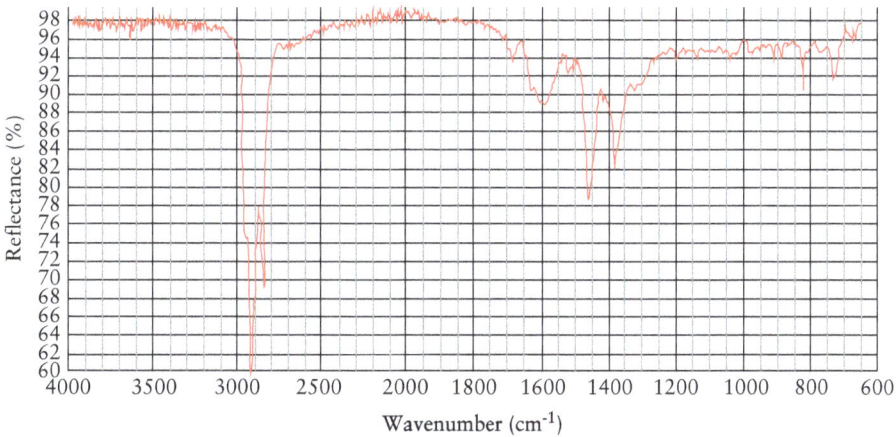

Figure 10.8: FTIR spectrum of an oily bloom present on the surface of a diene rubber sample.

surface of the rubber. The data in Figure 10.8 also suggested that the bloom was due to a naphthenic-type hydrocarbon oil. An oil of this type was one of the compounding ingredients used in the formulation of this rubber product.

Once a solution of the bloom constituents has been obtained, it is also possible to analyse it using other chromatographic techniques, for example, high-performance liquid chromatography (LC) or supercritical fluid chromatography. Also, on some occasions, it might be considered informative to use ion chromatography to investigate the possibility of salts (e.g., nitrates or sulfates) being present in a bloom.

Other analytical techniques capable of providing the structural and elemental data of blooms that are *in situ* on the surface of rubber compounds and products include:

- Secondary-ion mass spectrometry (SIMS)
- Laser-ionisation mass analysis (LIMA)

If larger quantities of bloom are available, it is possible to use techniques that can function with a couple of milligrams, such as differential scanning calorimetry (DSC). Use of a thermal technique like DSC can be beneficial if it is suspected that a bloom is composed of substances that have characteristic melting points, such as microcystalline waxes.

As discussed above, elemental sulfur can be responsible for blooms and its poor dispersion within a rubber compound can be a contributory factor. Jurkowski and co-workers [1] studied the dispersion of sulfur in rubber compounds using radioisotopes. They introduced the radioactive sulfur-35 isotope into compounds and measured the radioactivity by volumetric and surface analytical methods. This approach enabled the physical and chemical processes in the rubber to be investigated as well as the study of sulfur blooming. The research team also obtained experimental data on the effect that the mixing conditions and storage temperatures had on the generation of blooms.

10.2.3.3 Modelling of blooming

In addition to chemical analytical work, researchers have also used modelling to study the phenomena associated with blooming. An example of this work is that of Hoei [2], who developed a general kinetic model for the blooming of crystallisable additives in rubber and applied it to the analysis of the blooming of docosane in natural rubber that had been crosslinked using dicumyl peroxide. Use of this model enabled the cause of the blooming of docosame to be related to the melting of crystal particles as a result of a localised depression in their melting point. A reduction in elastic stress around the crystal particles was also believed to occur, and the data generated by the model enabled an initial diffusion coefficient of the docosane during blooming to be calculated.

10.3 Contaminants

10.3.1 Introduction

Contaminants can take many forms and can enter the product stream either in the ingredients used in the production, or during one of the stages of manufacture (e.g., mixing, extruding, moulding, storage or transportation).

Rubber products can fail, both aesthetically and physically, because of particulate contaminants and sometimes the physical failures can be catastrophic if the rubber part is essentially in maintaining a seal, or part of a mechanical support apparatus. This section will also cover contaminants that can become part of a rubber product in service, either due to physical attachment or, in the case of liquids, by being absorbed into the surface or throughout the bulk.

If the contaminants are in the form of solid, discreet entities (e.g., fibres), their detection in the product and removal are relatively easy. However, if they are in the form of absorbed liquids, their detection is sometimes dependent upon indirect methods, such as seeing if a drop in hardness is apparent, and their removal can be more involved (e.g., by volatilisation or extraction). The most commonly encountered contaminants and the techniques that have been found to be the most effective at identifying them are covered in Sections 10.3.2.1–10.3.2.3.

10.3.2 Identification of different types of contaminant

10.3.2.1 Particulates

Whether they are on the surface of a rubber product, or below it and have to accessed by sectioning, it is often possible to remove particulate contamination (e.g., using tweezers or the end of a scalpel blade) and transfer it to an analytical instrument.

A technique that is ideal for the analysis of particulate contamination is FTIR microscopy and its use, providing the particle is not black, will deliver a significant amount of structural and qualitative information on its composition. This is often enough to identify its source, but if the data are complex, indicating that a mixture of substances may be present (e.g., two or more inorganic compounding ingredients), it is useful to also apply EDX in order to provide complementary elemental information. It is then often possible using the combined data to identify the contamination and its source.

Other techniques can be used to compliment FTIR and EDX data to characterise the particulate matter more thoroughly:

– Optical microscopy with image capture
– Scanning electron microscopy (SEM)
– GC–MS

GC–MS can be very useful if a complex mixture of substances is present, but there are obvious practical constraints and it can be used only if the particle is composed of a soluble, organic substance (e.g., cure system or antidegradant system ingredient), within a certain MW range (e.g., 35 to 400 Da), and if a sufficient mass is present to enable detection.

If the particulate contamination is too small, or too well embedded in the product to allow removal, then it is usually possible to analyse it *in situ*. This can be carried out using a similar range of techniques to those shown above, although FTIR microscope analyses would have to be carried out in reflectance mode and the particle would have to have an area >10 um^2 in order for good-quality spectra to be obtained. It may also be possible to obtain structural information on the contaminant using SIMS and LIMA, although these are very sensitive, surface-specific techniques and care has to be taken on sample preparation and handling in order to avoid contamination.

If large particles are present, possibly in larger rubber products, it may be possible, if the particle can be removed and it has sufficient mass (e.g., >3 mg), to use thermogravimetric analysis (TGA) to obtain its bulk composition. Also, if a hyphenated TGA technique is available (e.g., TGA–IR or TGA–MS) then this will enable both quantitative and qualitative data on the particle to be obtained.

10.3.2.2 Fibres

FTIR microscopy is also an excellent tool for the analysis of fibrous contamination, particularly if it can be removed from the product, although the relatively fragile nature of some fibrous materials may make this difficult. Also, due to the large aspect ratio of fibres, the minimum area constraint referred to in Section 10.3.2.1 for FITR microscopy may be more problematic in the case of very fine fibres. Analysing the fibre *in situ* may be the best option in a number of cases and in addition to FTIR microscopy other microscopic techniques (e.g., optical and SEM) can make important contributions towards finding the origin of the contamination.

SEM and optical microscopy are particularly useful because they can differentiate synthetic fibres (e.g., polyester fibres) from natural fibres (e.g., cotton), and possibly the type of natural fibre, by their surface appearance: synthetic fibres are smooth and natural fibres have characteristic topographies. They can also show the exact positioning of the fibre in relation to the product surface (e.g., on the surface or partially embedded in the surface) and this may provide some clue as to which stage in its life the product encountered the fibre.

10.3.2.3 Liquids

A rubber product can absorb a solvent or other liquid (e.g., water) in service and this can cause a reduction in physical properties (e.g., hardness and tensile strength) and, possibly, failure. It is necessary in these cases to identify the fluid that has been absorbed so that appropriate measures can be taken (e.g., a change in the rubber's composition to make it more resistant). Identifying the fluid can also go a long way to solving a failure problem if some abuse of the rubber component is suspected in that it had been exposed to an environment it should not have come into contact with, or was not expected to come into contact with.

One important practical consideration of this kind of work is that, due to their inherent volatility, and the fact that their absorption is primarily a physical effect with only low intermolecular forces (e.g., van der Waals forces) at work, the majority of low-MW liquids (e.g., solvents) will be slowly lost from a rubber sample over time. Losses once the sample is in the laboratory can be minimised by refrigeration until the time of analyses.

There are a range of possible analytical options for identifying the liquid that the rubber product has absorbed and some of the principal ones, with their advantages and disadvantages, are presented below.

A relatively simple approach that can be used to identify an involatile liquid (e.g., an oil) in a rubber involves the extraction of the rubber with a solvent (e.g., acetone), removal of the solvent by drying (e.g., in an oven), and analysis of the dried extract by FTIR, either using a standard bench or a microscope. This approach has the major disadvantage that the solvent-extraction step will also remove many other substances from the rubber sample (e.g., oligomers, plasticiser, process aids, and substances relating to the cure system and antidegradant system) and these can mask the IR data from the absorbed liquid. Manipulation of the IR data using, for example, background subtraction of a control sample, may assist the identification process, but an approach that is more targeted may still be required, particularly if the level of the absorbed fluid is relatively low.

If problems are encountered with the solvent extraction–FTIR approach, another route that can be taken is to use a solvent to extract the rubber that is compatible with a chromatographic system and then, rather than drying down the extract and potentially losing the absorbed fluid, analyse it using that system. Chromatographic system-extraction solvent combinations that could be used are:

- GC–MS – acetone, methanol or dichloromethane
- LC–MS – methanol or acetonitrile

As discussed in Chapter 2, both of these chromatographic techniques have their strengths and weaknesses, and this will be a relatively costly and time-consuming route, but combining both techniques should provide a high chance of success.

The use of TGA to determine low-MW additives, such as plasticisers, has been covered in Section 3.4.1 and it can also be used to identify and quantify the presence of a fluid. One advantage of the TGA technique for this type of investigation is that it requires only a relatively small sample size (e.g., 5 to 10 mg) and so it is possible to use it, in a limited sense, to 'depth profile' the sample by quantifying the fluid at the surface of a sample and at different depths through its bulk. Due to its more sophisticated nature, TGA can also be more effective at identifying the presence of a low-MW fluid than the solvent extraction–FTIR approach.

The temperature at which an absorbed fluid will be lost from a rubber sample during a TGA experiment will depend upon its MW and its chemical properties, but the majority of solvents and water will show up in a TGA trace before, or at the start, of a weight loss event due to a plasticiser or oil. It has been shown that the use of a high-resolution programme or other approaches (e.g., use of a slower heating rate) can assist in improving the resolution between these events [3].

The use of sealed pans, which are punctured just prior to TGA analysis, is recommended to help prevent the loss of the absorbed liquid from the rubber sample, and use of a TGA–IR or TGA–MS combination can enable identification of the absorbed liquid and so may remove the need to use a more time-consuming and sophisticated technique, such as headspace GC–MS (see below) or extraction GC–MS and/or LC–MS (see above). If a hyphenated TGA option is not available to the analyst, or in more complex circumstances whereby a mixture of fluids has been absorbed, it is necessary to use a chromatographic approach.

As discussed in Chapters 4 and 5, the headspace GC–MS technique is a very powerful tool for identifying and, if standards are available, quantifying low-MW substances (e.g., plasticisers and cure system species) in rubber samples. It can, therefore, be very useful for identifying liquid contaminants if they are sufficiently volatile; the temperature limit (usually around 150 to 200 °C) of the transfer line from the headspace sampler to the GC–MS instrument is the limiting factor. The high sensitivity and excellent resolving power of the GC–MS technique mean that it is usually possible to identify the liquid in the presence of the other volatile substances (e.g., additives and their breakdown products) that will be in the headspace of the rubber sample.

References

1. B. Jurkowski, E. Koczorowska, W. Goraczko and J. Manuszak, *Journal of Applied Polymer Science*, 1996, **59**, 4, 639.
2. Y. Hoei, *Rubber Chemistry and Technology*, 2010, **83**, 1, 46.
3. M.J. Forrest in *Proceedings of the Polymer Testing Conference*, 7–11th April, Shawbury, UK, 1997, Paper No.5.

11 Analysis of rubber latices

11.1 Introduction

The majority of this book is concerned with the analysis of solid rubber samples or products (e.g., gum rubbers, compounded rubbers, foamed rubbers) but it was thought necessary, in order to fully reflect the diversity of rubbery materials, to include a section on the analysis of rubber latices. Latices can be readily split into two principal types:

- Natural rubber (NR) latices from *Hevea brasiliensis*.
- Synthetic rubber latices produced by the emulsion polymerisation of monomers, such as styrene, butadiene, chloroprene and acrylonitrile (ACN).

NR latex straight from *Hevea brasiliensis* usually has ≈30% solids content and one of the first processes that it undergoes is a concentration step (e.g., by centrifuging) to increase its solids content to ≈65%. Ammonia, at a 'high' or 'low' level is added to the latex to stabilise it and keep it free from micro-organisms. In the case of the low ammonia latices, at least one other preservative is added. The addition of these stabilisers results in NR latices being subdivided into the following classifications [1].

- High ammonia latex.
- Low ammonia with 0.2% sodium pentachlorophenate latex.
- Low ammonia with 0.25% boric acid and 0.05% sodium pentachlorophenate (LABA) latex.
- Low ammonia with up to 0.1% zinc diethyldithiocarbamate (ZDEC) latex.

With respect to synthetic latices, the polymerisation of organic monomers *via* the emulsion polymerisation method to create these latices is a huge technical subject in its own right and has been the subject of numerous books [2, 3]. This section will provide an overview of the analytical work that can be carried out on both natural and synthetic latices. However, the detailed technology and science associated with these products is outside the scope of this book, and references have been provided throughout to help the interested reader research the subject.

The rubber latex branch of the rubber products family is a very important one, including as it does many high-tonnage and instantly recognisable products, as well as those that have a lower profile, and so are less well known. Latex gloves and condoms fall into the former category and adhesives and elastic cord into the latter. There have been a number of textbooks describing in detail the manufacture of products from rubber latices, including those published by Blackley [4, 5]. There are many physical tests that have been developed to test the final latex products (e.g., burst pressure of condoms and permeation performance of safety

https://doi.org/10.1515/9783110640281-011

gloves) which are outside the remit of this book, but are described in the literature for those who are interested [4, 5].

This section is concerned with the analysis of rubber latices and within this subject there are two discreet areas:
- Analysis of the latex itself – including natural latex and synthetic latices.
- Analysis of products manufactured from both natural and synthetic latices.

It will concentrate on the chemical analytical tests used to characterise latex and to test its quality and composition. Some physical tests that are used for these purposes (e.g., total solids tests and viscosity tests) will also be included.

With regard to the chemical analysis of the final products produced from latices, because the water component of the latex is invariably removed (by coagulation and then by drying and vulcanisation) during the manufacturing process, what remains is a solid rubber product that can be subjected to the same analytical tests and testing protocols as other solid rubber products when it comes to the determination of, for example, the principal components and additives. The same techniques and approaches can also be applied for the purposes of reverse engineering, failure diagnosis and for demonstrating compliance with regulations.

One area of analysis and quality control which has been particularly active for NR latex in recent years is the determination of residual protein in products such as gloves (particularly medical gloves) and condoms due to the well publicised allergy and sensitisation problems. For example, the US Center for Disease Control and Prevention has estimated that up to 6% of the population has a sensitivity or allergy to the proteins in NR latex, and it is thought that there has been an increase in the sensitivity to the accelerators used in NR latex and synthetic rubber latex [6]. To address these problems and concerns, a number of tests that can be carried out on latex and latex-derived products have been developed. One of the tests that can be used on latex products is a latex allergy test and this particular test, as well as examples of the others that are available (e.g., protein test), have been included in Section 11.2.15.

11.2 Quality control tests

11.2.1 Sampling

The sampling of rubber latex is covered by the International Organization for Standardizations' ISO 123:2001, which describes the procedures to be adopted to obtain representative samples of latex contained in drums or in bulk tanks. In this standard, it is emphasised that the material of the container in which the latex is kept must be impermeable and chemically resistant to the latex. Also, space should be left in the container to allow for thermal expansion of the latex and the container should be kept closed as much as possible to prevent evaporation and interaction with the air. During handling operations, care should be taken to prevent the introduction of air into the latex because foam on the surface will rapidly form a skin.

11.2.2 Coagulum content

ISO 123:2001 specifies that if the coagulum of the latex exceeds 0.05%, it shall be filtered before any other tests are carried out on it. The coagulum content of rubber latex is defined as the material retained on stainless-steel wire cloth having an average aperture width of $180 \pm 15\,\mu m$ under the conditions of the test, which are provided in ISO 706:2004. The test method in ISO 706:2004 involves diluting 200 g of latex with soap solution and passing it through the filter, thoroughly washing the filter and any retained matter, and then drying to constant weight at 100 °C. The coagulum content is expressed as a percentage by mass of the latex.

11.2.3 Total solids content

The total solids content of rubber latex is described in ISO 124:2014 and is determined by drying ≈ 2.0 g of latex to constant mass at a temperature of 100 or 70 °C for NR latex. In the case of synthetic rubber latex, an additional option of 125 °C is available if a reduced pressure of 20 kPa is used.

11.2.4 Dry rubber content

The dry rubber content of NR latex is determined by coagulating ≈ 10 g of latex, after dilution with water, with dilute acetic acid. The coagulated rubber is digested, washed thoroughly with water, and pressed into a film of up to 2-mm thickness before drying to constant weight at 70 °C.

The dry rubber content of NR latex is a more accurate determination than the total solids content because a much larger mass of latex is tested and the dry rubber content is less hygroscopic that the total solids content. The difference between these two measurements can be regarded as the 'non-rubber solids' content of NR latex, and will include substances such as proteins.

11.2.5 Dry polymer content

The dry polymer content test method is applicable to high solids 'cold' polymerised synthetic rubber latices, particularly certain styrene-butadiene rubber (SBR) types.

The method, which is available in the British Standard, BS 3397:1976, measures the total styrene-butadiene content, including any other polymeric material present in the latex. To carry out the determination, ≈6 g of latex is coagulated by the addition of acetone, refluxed briefly and then washed and dried to constant weight at 100 °C.

11.2.6 Density

The method to determine the density of NR latex is provided in ISO 705:2015, and requires the use of a 50-cm^3 density bottle. It emphasises the importance of ensuring that no air is trapped and that the temperature is accurately recorded.

11.2.7 Determination of pH

The pH of rubber latices is measured using the test method provided by ISO 976:2013. This method uses a glass electrode and saturated calomel cell after standardising the pH meter with borax and potassium hydrogen phthalate solutions. A suitable combination electrode may be used in place of the single electrodes. The method should give a result that is accurate to 0.1 pH.

11.2.8 Alkalinity

The alkalinity of NR latex is defined as the percentage by mass of ammonia [or potassium hydroxide (KOH) if it has been used as a preservative] that it contains. The alkalinity of a latex is determined by titrating to pH 6.0 in the presence of a stabiliser. The method, described in ISO 125:2011, usually employs electrometric titration, but it is also possible to use methyl red as an indicator.

11.2.9 Potassium hydroxide number

This was the first test to be developed as an indicator of latex quality. The test is described in ISO 127:1984 and KOH number is defined as the number of grams of KOH equivalent to the acidic moieties combined with ammonia in latex containing 100 g of total solids. It is determined by potentiometric titration of NR latex with KOH solution, after adjustment of the alkalinity and dilution of the latex. The endpoint of the titration is the point of inflection of the titration curve of pH against the volume of KOH added.

The KOH number collectively determines volatile fatty acid (VFA), carbonate/bicarbonate [carbon dioxide (CO_2) number – Section 11.2.11], non-volatile acids

(NVA), higher fatty acid (HFA) and other acidic species. Given that HFA are stabilising, whereas the other acidic species are all destabilising, reduces the value of the KOH number as a measure of latex quality because it adds together both stabilising and destabilising components.

11.2.10 Volatile fatty acid number

This test is described in ISO 506:1992. The VFA number is determined after coagulating the latex with ammonium sulfate and acidifying the resulting serum by steam-distilling the serum in a Markham still and measuring the volatile acids in the distillate by titration with barium hydroxide solution. Prior to the titration, it is essential to purge the distillate of CO_2 with nitrogen. The most predominant VFA in natural latex is acetic acid.

11.2.11 Carbon dioxide number

The CO_2 number refers to the carbonate and bicarbonate content of NR latex. It can be determined by a gravimetric method [7] and other methods (e.g., the macrobaryta technique after acidifying stabilised latex).

Sundaram and Calvert [8] discussed the changes that can occur in the CO_2 number of field latex during a number of stages in its production, for example, before and after ammoniation, and the effects of centrifuging and storage of the concentrate prior to shipping, as well as changes that may occur during shipping. The CO_2 number is usually ≈0.12 greater than the VFA number, and so is an alternative indicator of the level of preservation of NR latex.

11.2.12 Non-volatile acids

NVA contribute more to the KOH number than any of the other acidic groups. NVA are determined by passing latex serum, obtained by coagulation with acetic acid, through a strong cation-exchange resin, followed by evaporation and acidimetric titration [9]. This method is time-consuming (≈4 h) and requires a measure of expertise.

11.2.13 Higher fatty acids

These compounds are determined, after extraction of latex total solids with acetone, by acidification with sulfuric acid and, after work-up, titration with KOH solution [9].

The VFA number, CO_2 number, HFA number and NVA number are all expressed in the same units as the KOH number.

11.2.14 Electrical conductivity

Electrical conductivity testing of latices has been of interest for a number of reasons:
-   As a measure of the quality of NR latex.
-   For the conductometric titration of synthetic latices for total soap content.
-   As an aid for the soap titration of synthetic latices for the determination of average particle size.

Its potential use in the first example arises because it has been established that as NR latex deteriorates its electrical conductivity tends to increase due to an increase in the ionic strength of the aqueous phase of the latex. However, the comparative insensitivity of electrical conductivity to the deterioration in quality means that determination of the VFA number (Section 11.2.10) is often regarded as more useful.

Measuring electrical conductivity has proved to be of practical use for both the determination of total soap, in which a diluted, acidified solution of latex is back-titrated with alkali, and in providing assistance to the soap titration method for average particle size determinations [4].

11.2.15 Latex allergy tests, protein tests and sensitivity tests

Although the Food and Drug Administration in the United States have stated that, at present, they are not aware of any tests that can show a product is completely free of NR latex proteins that can cause allergic reactions, tests are available to detect allergens [10]. These tests, which can be performed using bespoke testing kits, are an immunological test for measuring NR latex allergens from a variety of rubber products (e.g., gloves). Separate testing kits have been made available by manufacturers for measuring each of the major allergens: Hev b 1, Hex b 3 and Hex b 6.02. These tests function by using specific monoclonal antibodies that have been developed against the clinically relevant latex allergens present in NR latex products.

In addition to allergen tests, there are also standard test methods that have been developed by standardisation organisations [e.g., American Society for Testing and Materials (ASTM)] for the determination of proteins in NR latex products. Examples of these tests include:
-   ASTM D5712 – Modified Lowry/total protein – Standard for the quantification of total aqueous extractable protein associated with natural rubber, natural rubber latex, and elastomeric products.

- ASTM D6499 – Inhibition enzyme-linked immunosorbent assay (ELISA) is an immunological method which determines the amount of antigenic protein in natural rubber and its products by using rabbit antisera specific for natural rubber latex proteins.
- ASTM D7427 – Allergen ELISA is a test method that uses four ELISA kits to quantify four of the most common *Hevea* allergens (i.e., b 1, 3, 5 and 6.02) in natural rubber latex and its products.

Other tests that are available include:
- ASTM Guayule ELISA ASTM WG 25943 – Measures the levels of extractable Guayule proteins from commercial products. At the time of writing, this method is being validated by ASTM.
- ASTM D6214 – Powder content – Determines the average powder or filter-retained mass found on a sample of medical gloves.
- ASTM D7558 – Chemical sensitivity – Determines the amount of total extractable accelerators in natural rubber latex gloves and nitrile latex gloves.

11.3 Chemical stability tests

Chemical stability tests have been slow to achieve international acceptance because they have been found to be difficult to reproduce, have relatively poor repeatability, or of questionable relevance to industrial latex processes.

11.3.1 Natural rubber latices

Zinc oxide (ZnO) is added to NR latices in a number of industrial processes. Hence, the stability of the latex-to-zinc ammine ions (from the reaction of ZnO with the ammonia preservative) has been the focus of attention. Two particular tests which have gained prominence are the zinc stability time (ZST) test [11] and the zinc oxide viscosity test [12].

Other chemical stability tests for NR latex tend to be methods that are modifications of the ZST test or methods that use preformed zinc ammonium acetate.

11.3.2 Synthetic rubber latices

The chemical stability tests that have been proposed for synthetic latices are more varied and diverse. One widely regarded method involves the addition of a surfactant and ZnO dispersion to the latex, and measurement of the time taken to initiate coagulation. Other methods determine the amount of coagulum formed by the addition of an electrolyte (e.g., calcium chloride) or methanol.

11.4 Polymer composition of synthetic latices

11.4.1 Bound styrene content

The ISO 3136:1983 method for bound styrene is applicable only to straight SBR latices and not to modified versions such as carboxylated SBR latex. In comparison, the ISO 4655:1985 method applies to all SBR-type latices.

In the case of ISO 3136:1983, the solid polymer is first prepared by coagulating the latex with sodium chloride and sulfuric acid in the presence of methanol. The polymer is then dried, extracted with ethanol–toluene azeotrope, and then pressed into a thin sheet. The bound styrene value is obtained by refractive index measurements.

11.4.2 Total bound styrene content

ISO 4655:1985 describes two methods for the determination of total bound styrene. This standard can be applied to 'reinforced' SBR latices, whereas ISO 3136:1983 (Section 11.4.1) cannot. Both methods in ISO 4655:1985 start with the coagulation of the latex with isopropanol, followed by extensive washing and thorough drying of the coagulum. In the carbon/hydrogen method, the coagulum is burnt in a specified apparatus and the total bound styrene determined from the amount of CO_2 and water that are quantitatively absorbed. In the nitration method, the coagulum is nitrated and oxidised to convert its total bound styrene content to p-nitrodenzoic acid, which is then separated by multiple extractions and determined quantitatively by measuring its ultraviolet (UV) absorption. The two methods are capable of producing comparable results.

11.4.3 Bound acrylonitrile content

The ISO 3900:1995 method for bound ACN can be applied to nitrile rubber (NBR) latices, carboxylated versions of NBR latices, and nitrile–isoprene latices.

The preparative stage of this method involves extraction of an air-dried film of the nitrile latex with water to remove water-soluble nitrogen-containing material and drying to constant mass. The analytical procedure is then the same as that described in ISO 1656:1996 for the determination of the nitrogen content of NR.

11.4.4 Volatile monomers

Instrumental methods for the determination of residual monomers have largely replaced the older international standards that were essentially based on wet chemistry techniques, for example:

- ASTM D4026-06 for styrene in SBR latices.
- ISO 3899:2005 for ACN in NBR latices.
- ISO 3499:1976 for polyvinyl acetate latices.

One principal reason for this replacement is that gas chromatography (GC) methods are more sensitive and so give a lower limit of detection. The preferred technique is static headspace GC, in which the latex is heated in a sealed vial and the headspace analysed for the presence of the monomer. Calibration is usually achieved using the standard addition approach, where known amounts of the monomer are added to aliquots of the same latex to ensure that the headspace partitioning remains constant. For monomers such as ACN, a specific detector (a nitrogen phosphorus detector) can be used to achieve very low levels (i.e., ppb) of detection. This compares with the old ISO 3899:2005 method, which has a detection limit of only 100 ppm.

11.5 Particulate property tests

Test methods for particle properties cover determinations of viscosity, surface tension, soap content, soap deficiency and particle size.

11.5.1 Viscosity

There are three international standards for the determination of viscosity: ISO 1652:2011, ISO 2555:1989 and ISO 3219:1993.

ISO 1652:2011 is applicable for the determination of the viscosity of rubber latex. In the standard, the Brookfield L instrument is specified for low viscosities (≤200 mPa.s), the Brookfield R instrument for high viscosities (>2,000 mPa.s) and either instrument can be used for intermediate viscosities. The speed of rotation of the viscometer is restricted to 60 rpm for the Brookfield L instrument and 20 rpm for the Brookfield R instrument.

ISO 2555:1989 is for the determination of the viscosity of dispersions of resins and differs from ISO 1652:2011 in two respects: only the R version of the Brookfield instrument is specified and there is no restriction on the speed of rotation. However, speeds of 10 and 20 rpm are recommended.

ISO 3219:1993 is a general standard for the determination of the viscosity of dispersions of polymers. This standard uses a rotational viscometer working at a defined shear rate and seven shear rates from 1 to 250 s^{-1} are recommended.

The viscosity of latex increases with its total solids content so, as well as depending on the rate of shear used, both total solids content and shear rate should be appended to all viscosity results.

11.5.2 Surface tension (surface free energy)

ISO 1409:2006 describes the determination of the surface tension of rubber latex. A du Nouy tensiometer with a platinum ring of 40 or 60 mm in circumference is specified. The latex is required to be at a total solids content of ≤40%. The units of surface tension are mN/m. Surface free energy is synonymous and has the same numerical value as surface tension if expressed in units of mJ/m^2.

11.5.3 Soap content

The soap content of synthetic latices that have been produced using potassium oleate can be determined by potentiometric titration with sulfuric acid, after stabilisation with a non-ionic stabiliser and neutralised isopropanol, and adjustment of the pH to at least 11.0 with KOH. Two inflection points occur at around pH 9.4 and 5.0. The volume of acid in between these two inflection points is calculated for a latex sample and a blank, and the potassium oleate content is regarded as the difference between the two. Other methods exist for different fatty acid and rosin acid soap systems.

11.5.4 Soap deficiency

The soap deficiency of some latices can be determined by surface tension titration or conductometric titration with the same soap that is already present in the latex.

With surface tension titration, the surface tension is lowered linearly as soap is added. Once the surface of the polymer particles is saturated with adsorbed soap, a break in the line (i.e., an inflexion) occurs and further addition of soap has only a minor effect on surface tension. The amount of soap added to reach this inflexion point represents the soap deficiency of the latex.

With conductometric titration, the initial electrical conductance of the latex increases rapidly in a linear way with soap added. As with the former method, once saturation of the particle surface has occurred, the conductance increases more gradually. Again, the soap deficiency is the amount of soap added to reach the break in linearity.

11.5.5 Particle size

There are a number of experimental methods that can be used to determine particle size: electron microscopy, soap adsorption, light scattering, centrifugation, fractional creaming and counting methods.

To be amenable to electron microscopy, rubbery latex particles have to be hardened, for example, by bromination, to prevent distortion or coalescence.

There are a number of soap adsorption methods [13] and these require the latex to have a soap deficiency. The initial soap content of the latex and the molecular adsorption area of the soap need to be known.

Light-scattering methods measure the intensity of radiation that is scattered by the latex at various angles to the direction of incidence.

Centrifuge methods depend on the difference in density between the particles and the serum in which they are dispersed.

The Coulter Counter is a useful instrument for measuring the size distribution of latex particles in the diameter range 40 to 3,000 µm.

'Fractional creaming' was developed by Schmidt and Biddison [14]. The method uses the quantitative inverse relationship between the concentration of a creaming agent (sodium alginate) and the size of the creamed particles.

11.6 Miscellaneous chemical tests

11.6.1 Nitrogen and trace metals

The nitrogen content of NR latex is determined using the method described in ISO 1656:1996. This test applies only to the analysis of NR latices, and the nitrogen content that is obtained may be used to obtain an estimate of the protein content of a latex by applying a factor of 6.25.

There are international standards for the determination of copper (ISO 1654:1971), manganese (ISO 1655:1975), and iron (ISO 1657:1986). These tests can be applied to both NR and synthetic rubber latices.

More contemporary methods are also available. For example, a method for the determination of the nitrogen content of natural and synthetic rubber latex using a photometric method is described in ISO 8053:1995 and a sodium periodate photometric method for the determination of manganese is specified in ISO 7780:1998.

It is possible to determine other metals and elements in both types of rubber latex by using atomic absorption spectrometry (AAS) or one of the inductively coupled plasma-based methods. For example, the determination of metals in rubber latex samples by AAS is described in the six-part standard ISO 6101.

11.6.2 Zinc diethyldithiocarbamate

It is possible to use ZDEC as an auxiliary preservative in NR latex and to estimate the amount of ZDEC in a latex sample by quantitatively decomposing it to carbon disulfide, converting this compound into a water-soluble dithiocarbamate (e.g., diethylammonium diethyldithiocarbamate), and then forming the corresponding copper derivative which is estimated absorptiometrically.

11.6.3 Boric acid

ISO 1802:1985 describes the determination of the boric acid content of the LABA type of NR latex. The test involves complexing the boric acid with mannitol and titrating the liberated hydrogen ions with alkali. Importantly, LABA types of NR latex have a stabiliser system that employs boric acid and sodium pentachlorphenate as secondary stabilisers, with ammonia as the primary stabiliser.

11.6.4 Determination of additives and compounding ingredients

Because of its colloidal properties, rubber latex is compounded using a different approach to that used for solid rubber. For example, water-soluble materials can be added to the latex as solutions, whereas insoluble liquids are added as emulsions, and insoluble solid materials after they have gone through a grinding stage to reduce their particle size to ≤5 μm. If these constituents are added, their concentrations must be adjusted to control viscosity, reduce ionic shock and their pH has to be matched to that of the rubber latex. In general, the types of additives that are used to stabilise and cure rubber latex are the same as those used in solid rubbers, although there will be specific classes of additive that are only used in latex (e.g., preservatives, surfactants and stabilisers).

The types of antioxidants and curatives that are commonly used in rubber latex formulations are summarised in Table 11.1.

Once the latex has been destabilised, dried and cured and a solid product produced, the identification of the antioxidants and curatives can be undertaken using the same, or similar, analytical techniques and approaches to those described in Chapters 2 and 4. For example, Faridah [15] reviewed the high-performance liquid chromatography (HPLC) and thin-layer chromatography procedures that can be used to detect and identify the amount of residual accelerators and antioxidants in rubber products and rubber latex dipped products. The article concentrated on the different extraction procedures that can be used to isolate the analytes from the rubber products, and provided the types of accelerators and antioxidants, and their residues, that can be detected by HPLC.

Table 11.1: Examples of antioxidants and curatives used in rubber latices [15].

Antioxidants used in rubber latex	Curatives used in rubber latex
Hindered phenols	Sulfur
Zinc 2-mercaptotoluimidazole	ZnO
Styrenated phenols	Guanadines
p-Phenylenediamines	Thioureas
Paraffinic waxes	Thiazoles
Dithiocarbamates (residual quantities from cure)	Dithiocarbamates
Network-bound antioxidants	Thiurams
	Xanthates

Reproduced with permission from D. Hill in *Proceedings of Latex 2006*, 24–25th January 2006, Frankfurt, Germany, Rapra Technology Ltd, Shawbury, UK, 2006, Paper No.8. ©2006, Rapra Technology Ltd [15]

Another very important class of ingredient that is used in the compounding of rubber latex is stabilisers. Some examples of the stabilisers that can be used in rubber latex have been provided by Hill [16], and are listed below:
- Alkali (e.g., KOH and ammonium hydroxide)
- Fatty acid soaps (e.g., potassium laurate)
- Fatty alcohol ethoxylates
- Alkyl aryl sulfonates
- Alkyl sulfates
- Rosin acid soaps
- Proteins (e.g., casein)

Specific information on the stabilisers that have been used in a rubber latex can be obtained using some of the analytical techniques described in Chapter 2. Guidance on the methods and approaches that can be employed are included in general polymer analysis books such as those published by Crompton [17] and publications specifically addressing latices, for example, those published by Blackley [4], Daniels and Anderson [18], Warson and Finch [19] and Poehlein and co-workers [20].

Some bulk test methods to determine stabilisers and their effect on the bulk properties of latices are described in Sections 11.2 and 11.5. Finally, the identification and characterisation of the polymer within the rubber latex is possible using the relevant parts of Chapter 2, Chapter 3 and Section 11.4.

11.7 Characterisation studies undertaken on rubber latices

The quality control tests and analytical methods described in Sections 11.2–11.6 have provided an overview of the work that can be carried out to characterise and evaluate specific properties of rubber latices and the additives and ingredients that are used in their formulations. This section will provide some examples of the work that has been published over the last 15 years to characterise rubber latices and their constituents. Often such studies are an essential part of research programmes aimed at developing new products with attractive properties. On other occasions they have been carried out in order to extend and enhance the knowledge that exists in certain areas (e.g., chemical properties or chemical structure). It, therefore, performs a similar function to Chapter 5, which provides examples of published reverse engineering programmes that have been carried out on solid rubber products, in that it demonstrates the multi-technique approach that analysts use in order to obtain as much data as possible on test samples.

Yan and co-workers [21] synthesised zinc ion self-crosslinkable polyacrylate (PA) latices that can be cured at room temperature. They used a seeded semi-continuous emulsion polymerisation method to produce the latices and ZnO as the crosslinking agent. In order to control the amount of crosslinking that took place within the latex, the quantity of ZnO that was employed was related to the quantity of methacrylic acid (MAA) monomer. Once the latices were produced, they were characterised using a range of analytical techniques that included Fourier–Transform infrared (FTIR), differential scanning calorimetry (DSC), transmission electron microscopy (TEM) and thermogravimetric analysis (TGA). The results obtained showed that, within a certain concentration range of MAA, the average particle size decreased with increasing amounts of MAA, and the stability of the latex increased. The optimum values of the variables were found to be:

- Proportion of MAA within the monomer feed – 12%.
- Amount of ZnO – 25% mole fraction of MAA.
- Temperature at which ZnO is introduced – 60 °C.

The TEM results showed that the latex particles were coarse spherical particles that were surface-enriched with carboxyl groups, and the zinc ions were dissociated as a zinc ammine complex in the aqueous phase. The FTIR data confirmed that the chelate crosslinking occurred between zinc ions and carboxylic acid during the film-forming process, with the resulting films exhibiting excellent hardness and the ability to be sanded. The DSC thermograms revealed that the glass transition temperature (T_g) of the latices increased as a function of the formation of a coordinate structure, and TGA analyses confirmed that the use of ZnO enhanced the thermal stability of crosslinkable PA latices.

A range of self-crosslinking latices has also been produced by researchers from the Ministry of Eduction in China and Nankai University [22]. The research team used seed emulsion polymerisation with various amounts of vinyl triethoxysilane in conjunction with methyl methacylate, n-butyl acrylate and styrene. The properties of the latices were characterised by employing a combination of techniques (e.g., FTIR, DSC and TGA) and films from the latices were prepared and their mechanical properties and water repellency assessed. Together, the results provided the workers with information on the structure, degree of crosslinking and thermal properties of the latices.

Microemulsion photopolymerisation has been used to prepare silicone-modified styrene-butyl acrylate copolymer latices by a group of Chinese researchers [23]. The group used a combination of FTIR, UV, TEM, TGA, water swelling tests and acid/alkali resistance tests to study the effect that modifying the monomer-to-surfactant mass ratio and the silicone content had on the properties of the latices and coatings that were produced from them using the casting technique. The data obtained showed that the latex particles were almost spherical and had a uniform particle size. Also, as the monomer-to-surfactant mass ratio or silicone content increased, the acid, alkali and water resistance also increased.

A team from Wuhan University of Technology [24] synthesised the silicone monomer methacryloxypropyltris(trimethylsiloxy)silane and used it as a monomer in a co- polymerisation with methyl methacrylate. The co-polymerisation reaction was used to form emulsifier-free latices of silicone acrylate copolymer using sodium 3-allyloxy-2-hydroxypropanesulfonate, potassium persulfate and ultrasonic emulsification. The emulsifier-free latices that resulted were analysed using a wide range of techniques (e.g., spectroscopic and microscopic) and were found to be better dispersed and have a lower surface free energy than latices produced using conventional emulsion polymerisation.

Yuan and co-workers [25] prepared, by emulsion polymerisation, a silicone-modified styrene-butyl acrylate copolymer. This was achieved by grafting silicone oil to a styrene-butyl acrylate copolymer through a reaction with a functional monomer (gamma-propylmethacrylate trimethoxysilane). The surface properties, hydrophobicity, water repellency and T_g of the latices were determined by FTIR, contact angle measurements, water absorption measurements and DSC. As well as yielding primary characterisation information, the results also enabled the team to evaluate the effects that the grafted silicone oil had on the properties of the latices.

McMahan and co-workers [26] prepared guayule and NR latices and characterised them to determine their solids content, pH, particle size, molecular characteristics, nitrogen content and sulfur content. The team also cast films from the two types of latices and determined their curing, rheological and dynamic mechanical properties using the Advanced Polymer Analyzer 2000. Within the programme of work, a comparison was also made of the curing properties and mechanical properties of films that had been subjected to heat treatment.

Stephen and co-workers [27] studied the thermal degradation and ageing behaviour of microcomposites of NR, carboxylated styrene-butadiene rubber (XSBR) latices and their blends. During their work they used TGA, SEM, X-ray photoelectron spectroscopy and mechanical property tests, before and after ageing, to examine the effect of clay and silica microfillers on the thermal stability of the latices in the blends. The results obtained showed that the fillers improved the thermal stability and ageing resistance of the XSBR latices and their blends. During the study, the surfaces of samples were exposed to ion beam irradiation and the surface analysis results showed some redistribution of elements, but no changes in the binding energies.

The chemical structures and the thermal and thermooxidative stabilities of the gel and sol fractions from chlorinated natural rubber latex (CNR) have been analysed by a Chinese group of researchers [28] using the following techniques:

- Chemical analysis
- High-resolution pyrolysis GC–mass spectrometry
- Differential thermal analysis
- TGA

From the data that resulted from the test programme, the researchers concluded that it was the presence of the carbonyl group and tertiary C–Cl group that resulted in the CNR from latex having worse stability than CNR produced from solution. Also, the thermal and themooxidative stabilities of the sol fraction were better than those of the gel fraction.

Nawamawat and co-workers [29] studied the structure of branch-points in NR after washing fresh and aged NR latices by centrifugation with sodium dodecyl sulfate. The team found that the nitrogen content of the NR isolated from both the fresh and aged latex decreased to 0.2% after centrifuging thrice, a value similar to that of enzymatically deproteinated NR. The gel content of the NR from the two types of latex was found to decrease in a stepwise manner, to ≈6% for the aged latex and ≈3% for the fresh latex. In addition, the decrease in the molecular weight and the Huggins constant brought about by washing was considered to indicate a decrease in the branch points in the NR latex and to suggest that the proteins participated in gel formation. Analysis of the proteins that had been removed by the washing process by gel electrophoresis showed that the majority had not been degraded.

Rubber products that are used as protective clothing (e.g., gloves) are often manufactured from latex and one of the most important properties is resistance to permeation by liquids (e.g., solvents). A research team from New Zealand [30] evaluated an attenuated total reflectance (ATR)–FTIR method for measuring the permeation of a number of chemicals (e.g., solvents, a commercial pesticide mix and a volatile solid) through gloves that were made of different rubbers (e.g., polychloroprene, NBR and NR). Prior to conducting the measurements, the team ensured that good contact existed between the sample and the ATR crystal by using a low-pressure gas. The results showed that it was possible to measure the permeation of the chemicals through the samples by collecting a series of FTIR spectra, but that the diffusion coefficients could only be estimated. Their study showed that the ATR method had both advantages and disadvantages over the traditional two-compartment cell method that is used for permeation work.

A Malaysian team [31] carried out work on NR latex products to investigate the effect that residual chemicals have on their biocompatibility and physical properties. The approach involved the preparation of several in-house compounds of differing chemical composition and testing these using liquid chromatography and tensile tests to identify an optimum formulation, which contained a low level of residual chemicals and had acceptable physical properties. The biocompatibility of the specimens produced from this optimum formulation was also evaluated using an *in vitro* cytotoxicity method. The effect of pre-washing the products with acetone on the residual chemical content, physical properties and biocompatibility was also investigated, and comparative work on commercial, powder-free gloves was included in the test programme.

Potter [32] reported a pilot study to investigate the use of various chemical, physical and spectroscopic methods to assess the chemical changes occurring in NR latex condoms aged under different conditions. The tests employed included ATR–FTIR spectroscopy, determination of residual accelerators, and the assessment of crosslink type and crosslink density using solvent swelling and chemical probes. During the ageing process, the condoms were packed in air-impermeable aluminium foil laminate packaging. The comparative results obtained on un-aged and aged condoms are summarised below:

- ATR–FTIR did not detect any significant differences in chemical structure.
- Residual accelerator data did not provide information on the chemical changes that might be occurring during ageing, but was useful background information.
- Statistically significant changes in crosslink density and crosslink type were observed depending upon the ageing conditions.

The conclusion reached from the study was that chemical probe studies (e.g., using propane-2-thiol) offered significant promise as a method for investigating the chemical changes that occur as NR latex condoms age. The study also provided further evidence to that already in the public domain that high-temperature ageing (i.e., at 70 °C) resulted in changes that were not typical of those that were seen at intermediate and ambient temperatures.

Bluemich and co-workers [33] described how a bar-magnet nuclear magnetic resonance–mobile universal surface explorer was designed with a new coil which made it particularly useful for investigating samples with a thickness <1 mm and for examining the surfaces of polymer samples. The team applied the technique to the study of a number of different samples and reported on the findings. The examples included in their study are shown below:

- Swelling and drying of a latex membrane exposed to cyclohexane vapour
- Surface damage on rubber samples
- Drying of a thin, sprayed adhesive layer

References

1. K.O. Calvert in *Polymer Latices and their Application*, Ed., K.O. Calvert, Applied Science Publishers Ltd., Barking, UK, 1982, Chapter 2.
2. D.C. Blackley in *Emulsion Polymerisation: Theory and Practice*, Applied Science Publishers Ltd., Barking, UK, 1975.
3. *Chemistry and Technology of Emulsion Polymerisation*, 2nd Edition, Ed., A.M. van Herk, John Wiley & Sons Limited, Chichester, UK, 2013.
4. D.C. Blackley in *Polymer Latices Science and Technology*, Volumes 1–3, 2nd Edition, 1997, Chapman and Hall, London, UK, 1997.
5. *Polymer Latices and their Application*, Ed., K.O. Calvert, Applied Science Publishers Ltd., Barking, UK, 1982.
6. *Latex Testing*, Bureau Veritas Group, Paris, France. http://www.bureauveritas.com/services +sheet/latex-testing.
7. K.O. Calvert and R.K. Smith, *Rubber India*, 1974, **February**, 31.
8. P. Sundaram and K.O. Calvert in *Proceedings of International Rubber Conference RRI Malaysia*, Kuala Lumpur, Malaysia, 1975, **5**, 313.
9. K.O. Calvert, *Plastics and Rubber: Materials and Applications*, 1977, **2**, 2, 59.
10. *Don't be Misled by 'Latex Free' Claims*, US Food and Drug Administration, Silver Spring, MD, USD, 2015. https://www.fda.gov/ForConsumers/ConsumerUpdates/ucm342641.htm.
11. J.L.M. Newnham, K.O. Calvert and D.J. Simcox in *Proceedings of Natural Rubber Research Conference*, Kuala Lumpar, Malaysia, 1961, p.668.
12. H.G. Dawson, *Rubber World*, 1956, **135**, 239.
13. S.H. Maron, M.E. Elder, I.N. Ulevitch and C. Moore, *Journal of Colloid Science*, 1954, **9**, pages 89, 104, 263, 347, 353 and 382.
14. E. Schmidt and P.H. Biddison, *Rubber Age*, 1960, **88**, 484.
15. H.A.H. Faridah, *Malaysian Rubber Technology Developments*, 2005, **5**, 1, 8.
16. D. Hill in *Proceedings of Latex 2006*, 24–25th January 2006, Frankfurt, Germany, Rapra Technology Ltd, Shawbury, UK, 2006, Paper No.8.
17. T.R. Crompton in *Determination of Additives in Polymers and Rubbers*, Smithers Rapra Technology Limited, Shawbury, UK, 2007.
18. E.S. Daniels and C.D. Anderson in *Emulsion Polymerisation and Latex Applications*, Rapra Review Report No.160, Rapra Technology Limited, Shawbury, UK, 2003.
19. H. Warson and C.A. Finch in *Applications of Synthetic Resin Latices*, Volumes 1–3, John Wiley & Sons, Chichester, UK, 2001.
20. *Proceedings of the NATO Advanced Study Institute on Polymer Colloids*, 28th June–8th July 1982, University of Bristol, Bristol, UK, Eds., G. W. Poehlein, R.H. Ottewill and J.W. Goodwin, Springer Science + Business Media, Dordrecht, The Netherlands, 1983.
21. W. Yan, X. Zhang, Y. Zhu and H. Chen, *Iranian Polymer Journal*, 2012, **21**, 9, 631.
22. T.Y. Guo, C. Xi, G.J. Hao and M.D. Son, *Advances in Polymer Technology*, 2005, **24**, 4, 288.
23. T. Wan, C. Wu, X.L. Ma, L. Wang and J. Yao, *Polymer Bulletin*, 2009, **62**, 6, 801.
24. C. Zhang, L. Hu and Y. Hu, *ACS Polymer Preprints*, 2009, **50**, 2, 599.
25. J. Yuan, G. Gu, S Zhou and L. Wu, *High Performance Polymers*, 2004, **16**, 1, 69.
26. C.M. McMahan, K. Cornish, H. Pawlowski and J. Williams in *Proceedings of the Spring 167th ACS Rubber Division Meeting*, 16–18th May, San Antonio, TX, USA, American Chemical Society Rubber Division, Akron, OH, USA, 2005, Paper No.62.
27. R. Stephen, A.M. Siddique, F. Singh, L. Kailas and S. Jose, *Journal of Applied Polymer Science*, 2007, **105**, 2, 341.
28. Y. Dan, L. Sidong, Z. Jieping and J. Demin, *China Synthetic Rubber Industry*, 2003, **26**, 1, 47.

29. K. Nawamawat, J.T. Sakdapipanich, D. Mekkriengkrai and Y. Tanaka, *Kautschuk, Gummi, Kunststoffe*, 2008, **61**, 10, 518.
30. K. O'Callaghan, P.M. Fredericks and D. Bromwich, *Applied Spectroscopy*, 2001, **55**, 5, 555.
31. M.S.N. Qamarina, K.L. Mok, A.Y. Tajul and R.N. Fadilah, *Journal of Rubber Research*, 2010, **13**, 4, 240.
32. B. Potter in *Proceedings of the 6th Latex and Synthetic Polymer Dispersions*, 23–24th March, Amsterdam, The Netherlands, Smithers Rapra Technology Ltd, Shawbury, 2010, Paper No.7.
33. B. Bluemich, V. Anferov, S. Anferova, M. Klein and F. Fechete, *Macomolecular Materials and Engineering*, 2003, **288**, 4, 312.

12 Conclusions

Since publication of the *Rapra Review Report* version of this book, there has been a significant number of developments in the field of analytical chemistry. These have resulted in considerable improvement in the results that can now be achieved by the rubber analyst; this extensive up-to-date review has reflected these developments and provided many practical examples of their application to rubber samples.

This review has also provided an in-depth description of every important area of rubber analysis. It has dealt with the various compositional objectives, both specific (i.e., single substance) and collective (i.e., bulk composition), which need to be undertaken to ensure consistently high quality in manufacturing and to understand and characterise rubber samples and products. It has also covered the important themes (e.g., H&S and regulatory studies) which have had a major influence on the nature of the work that is being carried out today in a large number of R&D and test house laboratories and which, in many cases, has led to the setting up of specialist teams. With the increasing proliferation and importance of regulations, and as pressures continue to grow and every effort is made to reduce risk to society, it is to be expected that such activities will increase in the coming years, with analysts having a crucial function within surveillance and enforcement work.

Also covered in this book are the analytical applications, such as reverse engineering, cure state studies, failure diagnosis work, and the analysis of contaminants, which are so important for industry because they seek to develop and improve the products that they put into the marketplace and to understand why, on some occasions, whether due to abuse, poor manufacturing or incorrect choice, rubber products can fail to perform satisfactorily in service.

It is hoped that the strong emphasis on the practical application of the various analytical techniques, procedures and methods to solve problems and answer questions will enhance this book's use to practicing analysts and, in some ways, it can be looked upon as taking the form of a 'how to' guide to rubber analysis. The many examples of the information that has been obtained by carrying out certain tests in particular ways should also have the benefit of illustrating the capability of these approaches if applied to real samples to achieve stated goals and objectives.

The additional reading references provided in each section should greatly assist the analyst who requires more information, both specific and general, on any particular topic and who would like, or requires, more theoretical background on how particular analytical instruments and experimental methods function and achieve results.

Although an attempt has been made to include as much information as possible, the subject of rubber analysis is so large that it is inevitable that certain subject areas, analytical techniques and methods will have been excluded, or only given a cursory mention; again, the references provided will assist here by filling in any gaps.

https://doi.org/10.1515/9783110640281-012

For the future, it is to be expected that R&D activities by the instrument manufacturers, aimed towards more ever more sophisticated, faster, sensitive and higher resolution equipment, will continue and that these advances will continue to be of benefit to rubber analysts and further enhance the already significant contribution they make to industry and academia.

Due to the time and money constraints and pressures that are endemic in the modern world, one area which is expected to continue to attract attention is the development of techniques (e.g., those based on mass spectrometry) that can be used for the direct analysis of rubber samples, hence by-passing the often time-consuming and expensive sample preparation steps (e.g., solvent extraction or ashing) that have traditionally been employed.

With regard to rubber technology itself, in some industrial sectors there have been developments aimed at simplifying the composition of rubber products, an example being the development of relatively 'clean' halobutyl rubber compounds for pharmaceutical applications to reduce the level and range of extractable and leachable compounds. In general, there have also been a number of changes to eliminate particular classes of additive (e.g., high-polyaromatic hydrocarbon content extender oils), or reduce potentially harmful breakdown products (e.g., nitrosamines from certain accelerators used in sulfur vulcanisation systems).

However, despite these changes, rubber technology is a very complex science and it is clear that, even with the development of new and more advanced analytical techniques, the analysis of rubber compounds and products will continue to present the analyst with complex and difficult challenges. It is in the meeting and overcoming of these challenges that makes the role of the rubber analyst one of the most satisfying within the analytical chemistry profession.

Appendix 1: Standard nomenclature system for rubbers

The nomenclature system used in the rubber industry is based on International Organization for Standardisation, ISO 1629. The last letter of the identification code defines the basic group to which the polymer belongs, whereas the first letters provide more specific information relating to the polymer's structure.

Appendix 1.1 – M group: Rubbers having a saturated carbon–carbon main chain

CFM   Fluorochlorocarbon rubbers (e.g., copolymers of vinylidene fluoride and chlorotrifluoroethylene).

CM   Chlorinated polyethylene – two grades are available; one containing 36% chlorine by weight and the other 42% chlorine by weight.

CSM   Chlorosulfonated polyethylene. The polymers contain varying amounts of chlorine (20–45%) and sulfur (0.5–2.5%). The optimum values are Cl (30%): S (1.5%).

EPDM   Terpolymer of ethylene, propylene and a small amount of a third diene monomer allowing the use of a sulfur-based cure system.

EPM   Copolymer of ethylene and propylene having monomer ratios of 70–30 and 30–70 *wt/wt*.

FPM   Fluorocarbon rubbers having fluoro/fluoroalkyl groups on the saturated carbon backbone (e.g., copolymers of hexafluoropropylene and vinylidene fluoride and copolymers of vinylidene fluoride and 1-hydropentafluoropropylene).

IM   Polyisobutylene.

Appendix 1.2 – O group: Rubbers having carbon and oxygen in the main chain

CO   Polyepichlorohydrin.
ECO   Copolymer of epichlorohydrin and ethylene oxide.
GPO   Copolymer of propylene oxide and allyl glycidyl ether.

Appendix 1.3 – Q group: Silicone rubbers

MQ   Polydimethylsiloxane – can be an oil, wax or rubber depending on molecular weight.
MFQ   Fluorinated MQ – degrees of fluorination can vary.
MPQ   MQ with the presence of some phenylmethyl siloxane units in the backbone.
MPVQ   MPQ with the presence of some vinylmethylsiloxane units in the backbone.

https://doi.org/10.1515/9783110640281-013

Appendix 1.4 – R group: Rubbers having an unsaturated carbon backbone

ABR   Copolymers of butadiene and methyl methacrylate. Also, terpolymer of butadiene, methyl methacrylate and acrylonitrile and tetrapolymer of butadiene, methyl methacrylate, acrylonitrile and styrene.

BIIR   Brominated isobutylene-isoprene rubber 2–3% *w/w*.

BR   Polybutadiene – can vary greatly in *trans*-1,4, *cis*-1,4 and vinyl content. General purpose rubbers are usually 90% *cis*-1,4 or approximately 45% *cis*-1,4, 45% *trans*-1,4 and 10% vinyl.

CIIR   Chlorinated isobutylene-isoprene rubber 2–3% *w/w*.

CR   Polychlorobutadiene – there are two principal types: G types, which are amber copolymers with sulfur and have a molecular weight of ≈100,000 and W types, which are white homopolymers and have a molecular weight of ≈200,000.

IIR   Copolymer of isobutylene and a small amount of isoprene (i.e., butyl rubber) to provide cure sites.

IR   Synthetic *cis*-1,4-polyisoprene. *Cis*-1,4 level 90–99% with the remainder *trans*-1,4 and vinyl.

NBR   Copolymer of acrylonitrile and butadiene – acrylonitrile content of the random copolymer can vary between 15 to 50% to change the degree of oil resistance. Two important modified versions of nitrile rubber are hydrogenated NBR (in which the double bonds in the butadiene segments are removed by hydrogenation) and carboxylated NBR which, in addition to acrylonitrile and butadiene units, contains some carboxyl groups.

NR   Natural rubber – NR consists of 96% polymer which is virtually 100% *cis*-1,4 with *trans*-1,4 and vinyl content at <0.1%. The other 4% is made of natural products such as proteins. There is a wide variety of NR grades [e.g., Rib Smoked Sheet (RSS) and Standard Malaysian Rubber (SMR)]. These denominations are accompanied by numbers denoting purity and quality (e.g., RSS 1 and SMR 10). There are also modified natural rubbers available, such as epoxidised natural rubber (ENR).

SBR   Copolymer of styrene and butadiene – the styrene level in the random copolymer can vary between 10 and 80%, but the general purpose grade contains ≈23%. Many grades are available and these are identified by a specific coding system.

Appendix 1.5 – T group: Rubbers having carbon, oxygen and sulfur in the main chain

EOT   OT monomers polymerised with ethylene dichloride (e.g., polyethylene disulfide and polybutyl ether disulfide).

OT   Copolymer of *bis*-chloroalkyether (or formal) with sulfur (e.g., *bis*-2-chloroethylformal, sulfur and 1,2,3-trichloropropane as a cure site monomer).

Appendix 1.6 – U group: Rubbers containing carbon, oxygen and nitrogen in the main chain

AU   Polyester-type polyurethanes.
EU   Polyether-type polyurethanes.

A wide range of polyurethanes can be produced by making use of a number of different isocyanates, polyols and chain extenders.

Appendix 1.7 – Others rubbers – General designations and descriptions

ACM   Acrylic rubbers.Polynorbornene. Polyfluorophophazene rubbers. Ethylene acrylic rubber.
EVA   Ethylene vinyl acetate (elastomeric in the range 40–60% vinyl acetate).

Appendix 1.8 – Other polymer-related acronyms which are commonly encountered in the rubber industry include

HNBR   Hydrogenated nitrile rubber.
TPR   Thermoplastic rubber (e.g., MG Rubber, NR that has had methyl methacrylate grafted onto it to produce a material containing 30 or 49% methyl methacrylate).
XNBR   Carboxylated nitrile rubber.

Appendix 1.9 – Other other naturally occurring polyisoprenes

Balata   100% *trans*-1,4 polyisoprene
Chicle   Mixture of 25% *cis*-1,4 and 75% *trans*-1,4 polyisoprene
Guayule   100 *cis*-1,4 polyisoprene
GuttaPercha   100% *trans*-1,4 polyisoprene

Appendix 2: International rubber analysis standards (ISO)

Appendix 2.1 General ISO standards

Standard No.	Title
565	Test sieves – Metal wire cloth, perforated metal plate and electroformed sheet – Nominal sizes of openings
1382	Rubber - Vocabulary
1407	Rubber – Determination of solvent extract
1408	Rubber – Determination of carbon black content – Pyrolytic and chemical degradation methods
1629	Rubber and latices – Nomenclature
2194	Industrial screens – Woven wire cloth, perforated plate and electroformed sheet – Designation and nominal sizes of openings
2781	Rubber, vulcanised – Determination of density
3104	Petroleum products – Transparent and opaque liquids – Determination of kinematic viscosity and calculation of dynamic viscosity
3105	Glass capillary kinematic viscometers – Specifications and operating instructions
3865	Rubber, vulcanised or thermoplastic – Methods of test for staining in contact with organic material
4645	Rubber and rubber products – Guide to the identification of antidegradants – Thin-layer chromatographic methods
4648	Rubber, vulcanised or thermoplastic – Determination of dimensions of test pieces and products for test purposes
4650	Rubber – Identification – Infrared spectroscopic methods
4661-1	Rubber, vulcanised or thermoplastic – Preparation of samples and test pieces – Part 1: Physical tests
4661-2	Rubber, vulcanised - Preparation of samples and test pieces – Part 2: Chemical tests
5945	Rubber – Determination of polyisoprene content
6528-1	Rubber – Determination of total sulfur content – Part 1: Oxygen combustion flask method
6528-2	Rubber – Determination of total sulfur content – Part 2: Sodium peroxide fusion method

(continued)

(continued)

Standard No.	Title
6528-3	Rubber – Determination of total sulfur content – Part 3: Furnace combustion method
7269	Rubber – Determination of free sulfur
7725	Rubber and rubber products – Determination of bromine and chlorine content – Oxygen flask combustion technique
8054	Rubber, compounded or vulcanised – Determination of sulfide sulfur content - Iodometric method
9924-1	Rubber and rubber products – Determination of the composition of vulcanisates and uncured compounds by thermogravimetry – Part 1: Butadiene, ethylene-propylene copolymer and terpolymer, isobutene-isoprene, isoprene and styrene-butadiene rubbers
9924-2	Rubber and rubber products – Determination of the composition of vulcanisates and uncured compounds by thermogravimetry – Part 2: Acrylonitrile-butadiene and halobutyl rubbers
9924-3	Rubber and rubber products – Determination of the composition of vulcanisates and uncured compounds by thermogravimetry – Part 3: Hydrocarbon rubbers, halogenated rubbers and polysiloxane rubbers after extraction
10398	Rubber – Identification of accelerators in cured and uncured compounds
10638	Rubber – identification of antidegradants by gas chromatography/ mass spectroscopy
11089	Rubber, raw synthetic – Determination of anti-degradants by high-performance liquid chromatography
15671	Rubber and rubber additives – Determination of total sulfur content using an automatic analyser
15672	Rubber and rubber additives – Determination of total nitrogen content using an automatic analyser

Appendix 2.2 Latex ISO standards

Standard No.	Title
123	Rubber latex – Sampling
124	Latex, rubber – Determination of total solids content
125	Natural rubber latex concentrate – Determination of alkalinity

(continued)

(continued)

Standard No.	Title
126	Latex, rubber, natural concentrate – Determination of dry rubber content
127	Rubber, natural latex concentrate – Determination of KOH number
498	Natural rubber latex concentrate – Preparation of dry films
506	Rubber latex, natural, concentrate – Determination of volatile fatty acid number
705	Rubber latex – Determination of density between 5 and 40 °C
706	Rubber latex – Determination of coagulum content (sieve residue)
976	Rubber and plastics – Polymer dispersions and rubber latices – Determination of pH
1147	Plastics/rubber – Polymer dispersions and synthetic rubber lattices – freeze–thaw cycle stability test
1652	Rubber latex – Determination of apparent viscosity by the Brookfield test method
1657	Rubber, raw and rubber latex – Determination of iron content – 1,10-Phenanthroline photometric method
1802	Natural rubber latex concentrate – Determination of boric acid content
2004	Natural rubber latex concentrate – Centrifuged or creamed, ammonia-preserved types – Specification
2005	Rubber latex, natural, concentrate – Determination of sludge content
2008	Rubber latex, styrene-butadiene – Determination of volatile unsaturates
2028	Synthetic rubber latex – Preparation of dry polymer
2438	Rubber latex, synthetic – Codification
3136	Rubber latex – Styrene-butadiene – Determination of bound styrene content
6101	Rubber – Determination of metal content by atomic absorption spectrometry (5 parts)
7780	Rubbers and rubber latices – Determination of manganese content – Sodium periodate photometric methods
8053	Rubber and latex – Determination of copper content – Photometric method
11852	Rubber – Determination of magnesium content of field natural rubber latex by titration
12000	Plastics/rubber – Polymer dispersions and rubber lattices (natural and synthetic) – Definitions and review of test methods
12243	Medical gloves made from natural rubber latex – Determination of water-extractable protein using the modified Lowry method

(continued)

(continued)

Standard No.	Title
13741-1	Plastics/rubber – Polymer dispersions and rubber latices (natural and synthetic) – Determination of residual monomers and other organic components by capillary-column gas chromatography – Part 1 Direct liquid injection method
13741-2	Plastics/rubber – Polymer dispersions and rubber latices (natural and synthetic) – Determination of residual monomers and other organic components by capillary-column gas chromatography – Part 2 Headspace method
13773	Rubber – Polychloroprene latex – Determination of alkalinity

Appendix 2.3 Carbon black ISO standards

Standard No.	Title
1124	Rubber compounding ingredients – Carbon black shipment sampling procedures
1125	Rubber compounding ingredients – Carbon black – Determination of ash
1126	Rubber compounding ingredients – Carbon black – Determination of loss on heating
1138	Rubber compounding ingredients – Carbon black – Determination of sulphur content
1304	Rubber compounding ingredients – Carbon black – Determination of iodine adsorption number – Titrimetric method
1437	Rubber compounding ingredients – Carbon black – Determination of sieve residue
1867	Carbon black for use in the rubber industry – Specification for sieve residue
1868	Rubber compounding ingredients – Carbon black – Specification limits for loss on heating
3858-1	Carbon black for use in the rubber industry – Determination of light transmittance of toluene extract – Part 1: Rapid method
3858-2	Carbon black for use in the rubber industry – Determination of light transmittance of toluene extract – Part 2: Method for product evaluation
4652-1	Rubber compounding ingredients – Carbon black – Determination of specific surface area by nitrogen adsorption methods – Part 1: Single-point procedures

(continued)

(continued)

Standard No.	Title
4656-1	Rubber compounding ingredients – Carbon black – Determination of dibutyl phthalate absorption number – Part 1: Method using absorptometer
5435	Rubber compounding ingredients – Carbon black – Determination of tinting strength
6209	Rubber compounding ingredients – Carbon black – Determination of solvent extractable material
6810	Rubber compounding ingredients – Carbon black – Determination of surface area – CTAB adsorption methods
6894	Rubber compounding ingredients – Carbon black – Preparation of samples for determination of dibutyl phthalate absorption number (compressed sample)
12245	Carbon blacks used in rubber products – Classification system

Appendix 2.4 Standard ISO test formulations

Standard No.	Title
1658	Natural rubber (NR) – Evaluation procedure
2302	Isobutene-isoprene rubber (IIR) – Evaluation procedures
2322	Styrene-butadiene rubber (SBR) – Emulsion and solution-polymerised types – Evaluation procedures
2475	Chloroprene rubber (CR) – General-purpose types – Evaluation procedure
2476	Rubber, butadiene (BR) – Solution polymerised types – Evaluation procedures
3257	Rubber compounding ingredients – Carbon black – Method of evaluation in styrene butadiene rubbers
4097	Rubber, ethylene-propylene-diene (EPDM) – Evaluation procedure
4659	Rubber, styrene-butadiene (carbon black or carbon black and oil masterbatches) – Evaluation procedure

Appendix 2.5 Statistical standards

Standard No.	Title
2602	Statistical interpretation of test results – Estimation of the mean – Confidence interval
2854	Statistical interpretation of data – Techniques of estimation and tests relating to means and variances
2859	Sampling procedures for inspection by attributes (5 parts)
3207	Statistical interpretation of data – Determination of a statistical tolerance interval
3301	Statistical interpretation of data – Comparison of two means in the case of paired observations
3494	Statistical interpretation of data – Power of tests relating to means and variances
3534	Statistics – Vocabulary and symbols (3 parts)
5725	Accuracy (trueness and precision) of measurement methods and results (6 parts)

Appendix 2.6 Raw and compounded rubber standards

Standard No.	Title
247	Rubber – Determination of ash
248	Rubbers, raw – Determination of volatile matter content
249	Rubber, raw natural – Determination of dirt content
1407	Rubber – Determination of solvent extract
1434	Natural rubber in bales – Amount of bale coating – Determination
1656	Rubber, raw natural, and rubber latex, natural – Determination of nitrogen content
1657	Rubber, raw and rubber latex – Determination of iron content – 1,10-Phenanthroline photometric method
1795	Rubber, raw natural and raw synthetic – Sampling and further preparative procedures
2000	Rubber, raw natural – Specification
2453	Rubber, raw styrene-butadiene, emulsion polymerised – Determination of bound styrene content – Refractive index method
2454	Rubber products – Determination of zinc content – EDTA titrimetric method

(continued)

(continued)

Standard No.	Title
3899	Rubber – Nitrile latex – Determination of residual acrylonitrile content
3900	Rubber – Nitrile latex – Determination of bound acrylonitrile content
4655	Rubber – Reinforced styrene-butadiene latex – Determination of total bound styrene content
4660	Rubber, raw natural – Colour index test
5945	Rubber – Determination of polyisoprene content
6101	Rubber – Determination of metal content by atomic absorption spectrometry (5 parts)
6235	Rubber, raw – Determination of block polystyrene content – Ozonolysis method
7270	Rubber – Identification of polymers (single polymers and blends) – Pyrolytic gas chromatographic method
7780	Rubbers and rubber latices – Determination of manganese content – Sodium periodate photometric methods
7781	Rubber, raw styrene-butadiene – Determination of soap and organic-acid content
8053	Rubber and latex – Determination of copper content – Photometric method
11089	Rubber, raw synthetic – Determination of anti-degradants by high-performance liquid chromatography
11344	Rubber, raw synthetic – Determination of the molecular-mass distribution of solution polymers by gel permeation chromatography
12492	Rubber, raw – Determination of water content by Karl Fischer method
19050	Rubber, raw, vulcanised – Determination of metal content by ICP– OES
22768	Rubber, raw – Determination of the glass transition temperature by differential scanning calorimetry

Appendix 3: Specific gravities of rubbers and compounding ingredients

Appendix 3.1 Specific gravities of polymers

Polymer	Specific gravity (g/cm³)
Acrylic rubbers	1.1
Bromobutyl rubber	0.91
Butyl rubber	0.91
Chlorobutyl rubber	0.91
Chlorosulfonated polyethylene	1.11–1.28*
Epichlorohydrin copolymer	1.27
Epichlorohydrin homopolymer	1.38
Ethylene-propylene diene monomer rubber	0.86
Ethylene-propylene rubber	0.85
Ethylene-vinyl acetate copolymer	0.94
Fluorocarbon rubbers	1.41–1.86*
High styrene-butadiene rubber	1.05
Natural rubber	0.92–0.93
Nitrile rubbers	0.95–1.00*
Nitrile/polyvinyl chloride blends	1.0–1.11*
Polybutadiene rubbers	0.91–0.93
Polychloroprene rubbers	1.23–1.25*
Polyisoprene rubber	0.92–0.93
Polynorbornene	0.96
Polysulfide rubbers	1.27–1.60*
Styrene-butadiene rubbers	0.94

* According to grade

Appendix 3.2 Specific gravities of compounding ingredients

Ingredient	Specific gravity (g/cm^3)
Aluminium silicate	2.10
Antimony trioxide	5.40
Barium sulfate	4.30
Bitumen	1.04
Calcined china clay	2.50
Calcium carbonate	2.65
Calcium oxide (lime)	2.19
Calcium silicate	2.10
Carbon black	1.80
Carnauba wax	0.99
China clay (kaolin)	2.60
Cotton fibre	1.05
Diethylene glycol	1.12
Factice	1.05
Graphite	2.04
Indene-coumarone resin	1.09
Iron oxide	4.5–5.1
Lanolin	1.08
Litharge	9.30
Magnesium carbonate	2.21
Magnesium oxide	3.60
Mica	2.80
Mineral oil	0.91
Oleic acid	0.90
Paraffin oil	0.80
Paraffin wax	0.90
Pine tar	1.08
Rosin	1.07
Silica	1.95
Stearic acid	0.85
Sulfur	2.05

(continued)

(continued)

Ingredient	Specific gravity (g/cm^3)
Talc (magnesium silicate)	2.80
Titanium dioxide (anatase)	3.90
Titanium dioxide (rutile)	4.20
Zinc oxide	5.55
Zinc stearate	1.10

Abbreviations

3D	Three-dimensional
6PPD	N-(1,3-dimethyl butyl)-N'-phenyl-*p*-phenylenediamine
AAS	Atomic absorption spectrometry
ACN	Acrylonitrile
AES	Atomic emission spectroscopy
AET	Analytical evaluation threshold
Am-Ep	Amine-epoxy
AP	Atmospheric pressure
APCI	Atmospheric pressure chemical ionisation
API	Atmospheric pressure ionisation
ASTM	American Society for Testing and Materials
ATD	Automated thermal desorption
ATR	Attenuated total reflectance
BET	Brunauer–Emmett–Teller
BHT	Butylated hydroxyltoluene
BPOG	BioPhorum Operations Group
BR	Polybutadiene
BS	British Standard
BT	Benzothiazole
CB	Carbon black(s)
CF	Carbon fibre
CBS	N-cyclohexyl-2-benzothiazole sulfenamide
CCS	Collisional cross-section
CFR	Code of Federal Regulations
CGNR	Chemically grafted with cardonol
CIIR	Chlorobutyl rubber
CNR	Chlorinated natural rubber latex
COC	Cyclic olefin copolymers
CoE	Council of Europe
CR	Polychloroprene
CSM	Chlorosulfonated polyethylene rubber
CTAB	Cetyltrimethylammonium bromide
CTNR	Carboxyl-terminated nitrile rubber
DAE	Distillate aromatic
DBP	Dibutyl phthalate
DCBS	N,N-dicyclohexyl-2-benzothiazyl sulfenamide
DCM	Dichloromethane
DDA	Dynamic dielectric analysis
DEA	Dielectric analysis
DIN	Deutsches Institut für Normung e.V.
DMA	Dynamic mechanical analysis
DOTG	N,N''-di-*ortho*-tolyl guanidine
DPG	Diphenyl guanidine
DSC	Differential scanning calorimetry
DTDM	4,4'-Dithiodimorpholine
E&L	Extractables and leachables

https://doi.org/10.1515/9783110640281-014

EC	European Commission
EDQM	European Directorate for the Quality of Medicines & Healthcare
EDX	Energy-dispersive X-ray spectroscopy
EI	Electron ionisation
ELISA	Enzyme-linked immunosorbent assay
ELS	Evaporative light scattering
EMA	Ethylene methyl acrylate copolymer
EPDM	Ethylene propylene diene monomer
EPM	Ethylene propylene monomer
EU	European Union
EV	Efficient vulcanisation
EVA	Ethylene-vinyl acetate
FDA	Food and Drug Administration
FEA	Finite element analysis
FEF	Fast extrusion furnace
FID	Flame ionisation detector
FR	Fluorocarbon rubber
FSA	Food Standards Agency
FT	Fourier–Transform
FTIR	Fourier–Transform infrared
GC	Gas chromatography
GPC	Gel permeation chromatography
GRG	General rubber goods
GTM	Gas transfer mould
HAF	High-abrasion furnace
HFA	Higher fatty acid
HPLC	High-performance liquid chromatography
HRAM	High-resolution accurate mass
HSE	Health and Safety Executive
IC	Ion chromatography
ICBA	International Carbon Black Association
ICH	International Council for Harmonisation of Technical Requirements for Pharmaceuticals for Human Use
ICP	Inductively coupled plasma
IGC	Inverse gas chromatography
IMS	Ion-mobility spectrometry
IPPD	Isopropyl p-phenylenediamine
IR	Infrared
ISO	International Organization for Standardization
LABA	Low ammonia with 0.25% boric acid and 0.05% sodium pentachlorophenate
LC	Liquid chromatography
LDPE	Low-density polyethylene
LIMA	Laser-ionisation mass analysis
MAA	Methacrylic acid
MALDI	Matrix-assisted laser desorption/ionisation
MALLS	Multi-angle laser light scattering
MBS	2-(4-Morpholinothio)benzothiazole
MBT	2-Mercaptobenzothiazole
MBTS	2,2′-Dithiobis(benzothiazole)

MDHS	Method for Determination of Hazardous Substance
MDR	Moving die rheometer
MMT	Montmorillonite
Mn	Number average molecular weight
MPTD	Dimethyl diphenyl thiuram disulfide
MS	Mass spectrometry
MS2	Tandem mass spectrometry
MW	Molecular weight(s)
Mw	Weight average molecular weight
MWD	Molecular weight distribution
Mz	Z average molecular weight
NBR	Nitrile rubber(s)
NIR	Near-infrared
NMR	Nuclear magnetic resonance
NOX	Nitrosating agents
NP	p-nonylphenol
NR	Natural rubber
NVA	Non-volatile acids
NVR	Non-volatile residue
ODR	Oscillating disc rheometer
OES	Optical emission spectrometry
OML	Overall migration limits
PA	Polyacrylate
PAH	Polyaromatic hydrocarbons
PAN	N-phenyl-1-naphthylamine
PBT	Polybutylene terephthalate
PC	Polycarbonate
PDI	Polydispersity index
PDMS	Polydimethylsiloxane
PE	Polyethylene
PI	Polyisoprene
PP	Polypropylene
PQRI	Product Quality Research Institute
PS	Polystyrene
PTBP	p-*tert*-butylphenol
PTOP	p-*tert*-octylphenol
PU	Polyurethane
PVC	Polyvinyl chloride
PVI	Prevulcanisation inhibitor
Py	Pyrolysis
QC	Quality control
R&D	Research and development
REACH	Registration, Evaluation, Authorisation and Restriction of Chemicals
RI	Refractive index
RIC	Reconstructed ion current
RPA	Rubber process analyser
RSD	Relative standard deviation
RT	Retention time
SANS	Small-angle neutron scattering

SBR	Styrene-butadiene rubber(s)
SBS	Styrene-butadiene-styrene
SEC	Size exclusion chromatography
SEM	Scanning electron microscopy
SFC	Supercritical fluid chromatography
SFM	Scanning force microscopy
SIMS	Secondary-ion mass spectrometry
SML	Specific migration limits
SPM	Scanning probe microscopy
SUS	Single-use systems
TAC	Triallyl cyanurate
TBBS	N-*tert*-butyl-2-benzothiazole sulfenamide
TBSI	N-*tert*-butyl-2-benzothiazole sulfenimide
TBTD	Tetrabutylthiuram disulfide
TC	Technical Committee
TCT	Thermal desorption cold trap injector
TD	Thermal desorption
TDEC	Tellurium diethyldithiocarbamate
TEA	Thermal energy analyser
TEM	Transmission electron microscopy
TETD	Tetraethylthiuram disulfide
T_g	Glass transition temperature
TGA	Thermogravimetric analysis
THF	Tetrahydrofuran
THM	Thermally-assisted hydrolysis and methylation
TIC	Total ion current
TLC	Thin-layer chromatography
T_m	Melting temperature
TMA	Thermal mechanical analysis
TMQ	Poly-2,2,4-trimethyl-1,2-dihydroquinoline
TMTD	Tetramethylthiuram disulfide
TMTM	Tetramethylthiuram monosulfide
ToF	Time-of-flight
TPE	Thermoplastic elastomer(s)
TS	Tensile strength
UPLC	Ultra-performance liquid chromatography
US	United States
USP	United States Pharmacopeia
UV	Ultraviolet
UV-Vis	Ultraviolet visible
VFA	Volatile fatty acid
VTA	Variable temperature analysis
XRD	X-ray diffraction
XRF	X-ray fluorescence spectroscopy
XSBR	Carboxylated styrene-butadiene rubber
ZDEC	Zinc diethyldithiocarbamate
ZDMC	Zinc dimethyldithiocarbamate ZnO Zinc oxide
ZST	Zinc stability time

Index

https://doi.org/10.1515/9783110640281-015

www.ingramcontent.com/pod-product-compliance
Lightning Source LLC
Chambersburg PA
CBHW080906220326
41598CB00034B/5487